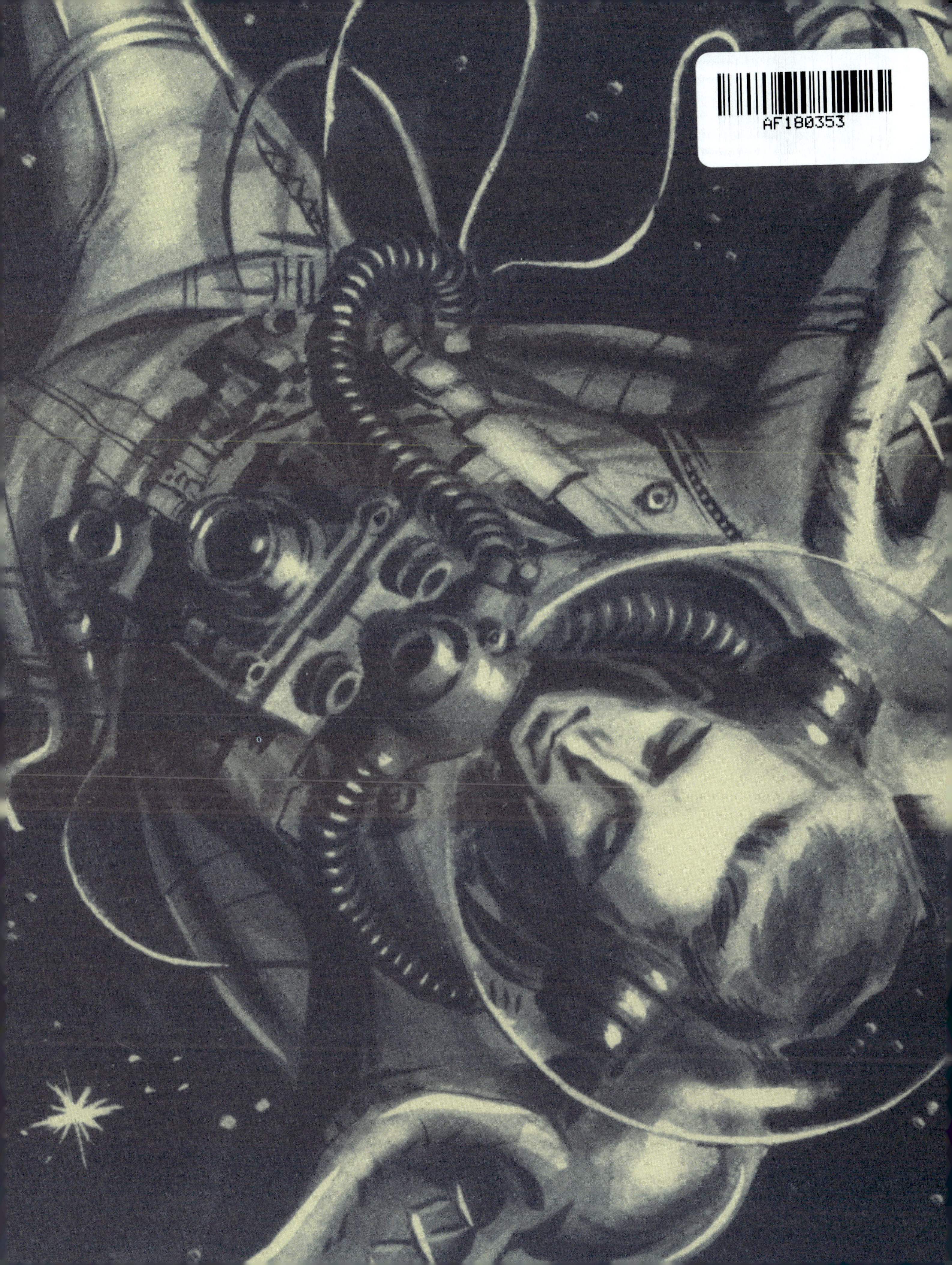

WALLY WOOD
GALAXY ART AND BEYOND

BY ROGER HILL

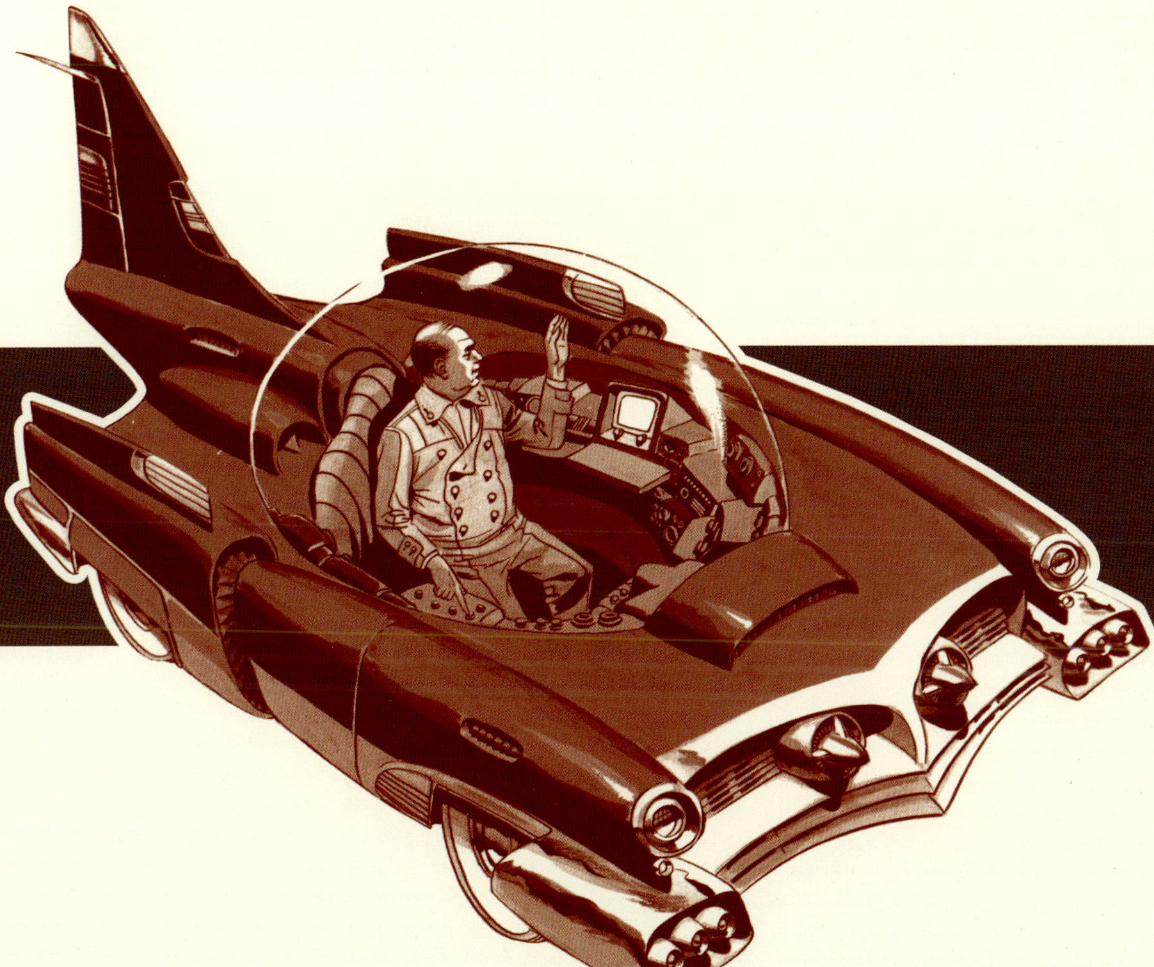

IDW

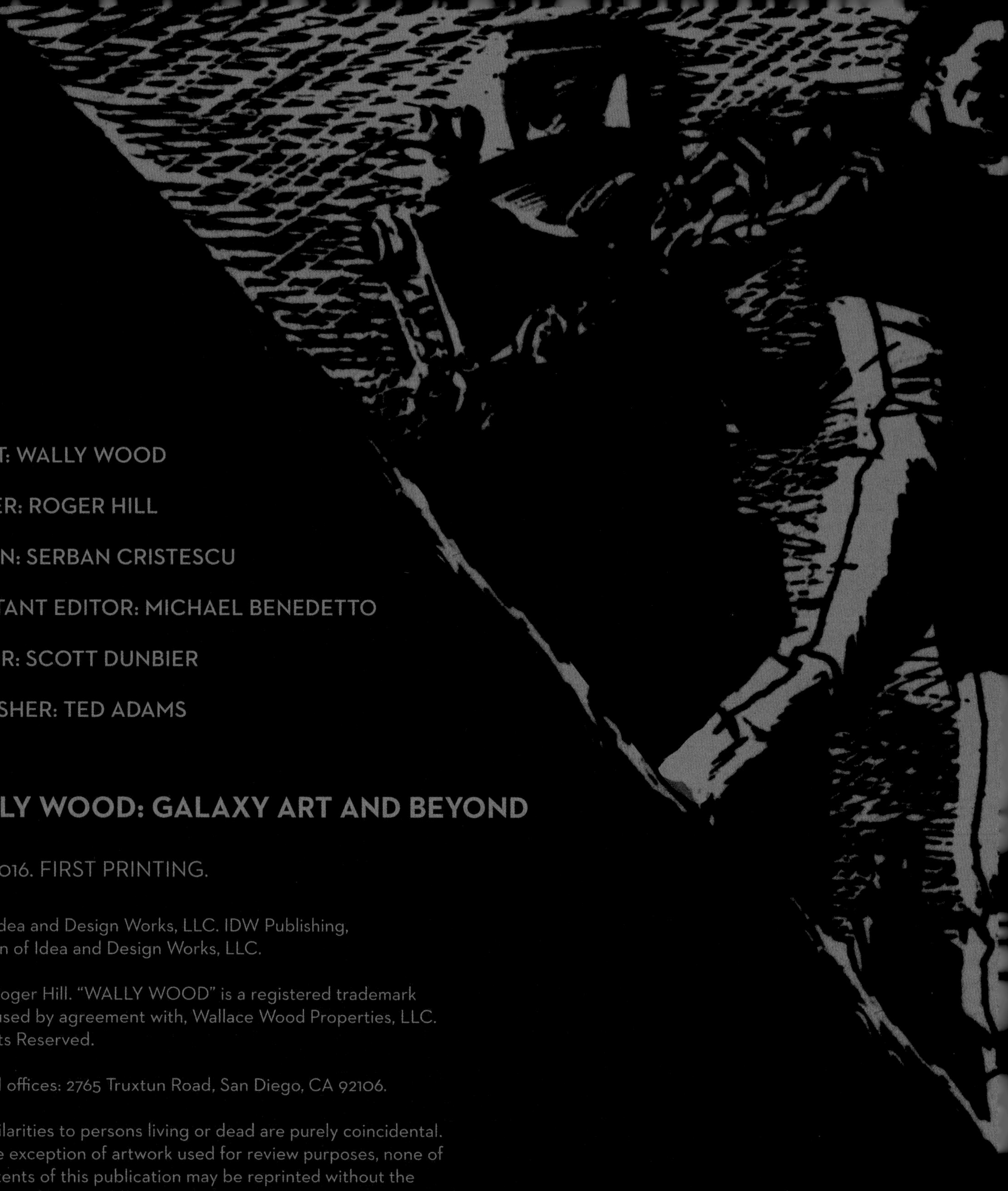

ARTIST: WALLY WOOD

WRITER: ROGER HILL

DESIGN: SERBAN CRISTESCU

ASSISTANT EDITOR: MICHAEL BENEDETTO

EDITOR: SCOTT DUNBIER

PUBLISHER: TED ADAMS

WALLY WOOD: GALAXY ART AND BEYOND

MAY 2016. FIRST PRINTING.

Editorial offices: 2765 Truxtun Road, San Diego, CA 92106.

IDW

ISBN: 978-1-61377-184-6

19 18 17 16 1 2 3 4

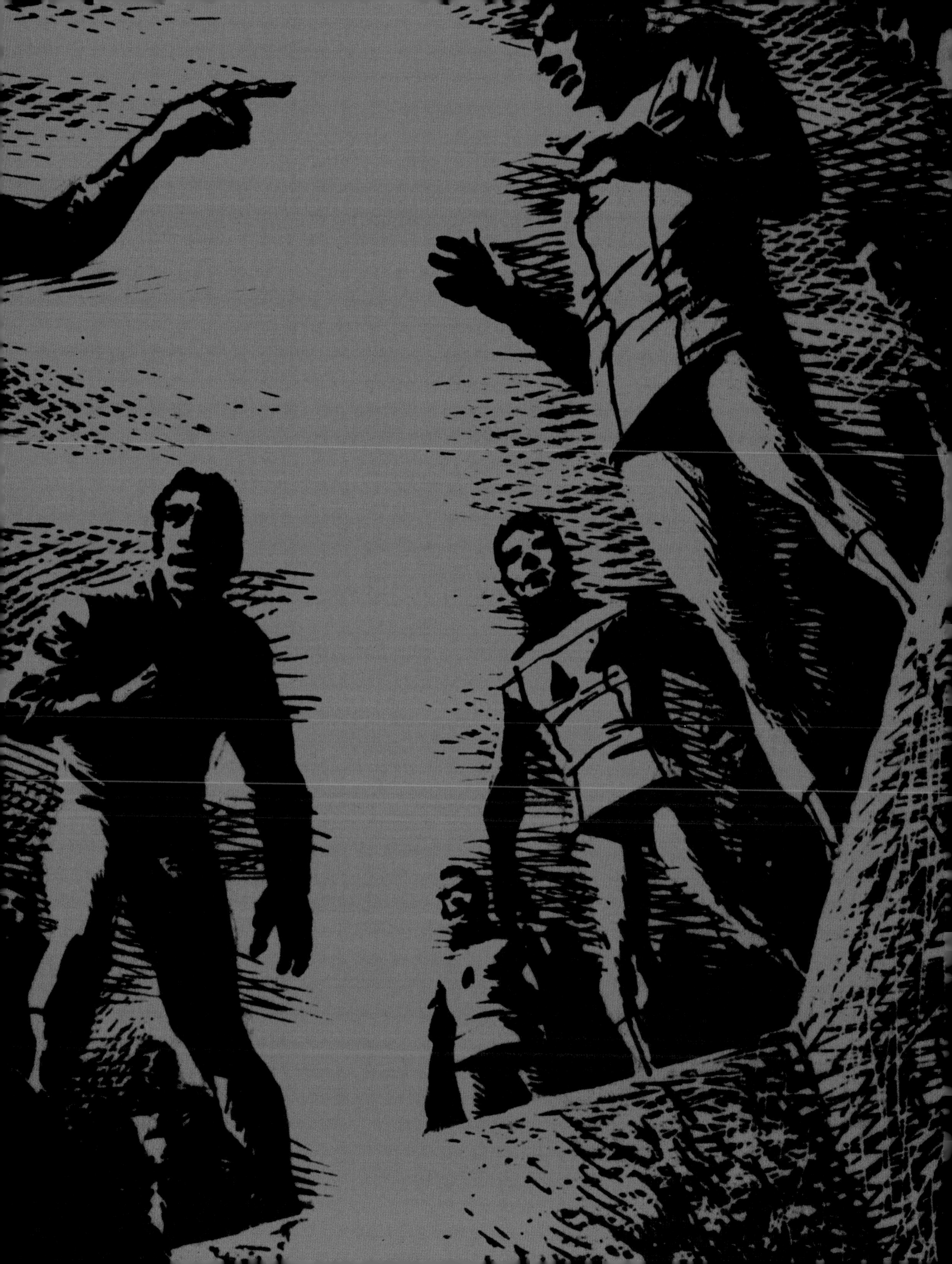

ACKNOWLEDGEMENTS

This book has been many years in the making and will hopefully fulfill the expectations of Wally Wood aficionados everywhere who have waited so long for a book like this to happen. That is, a book that encompasses all of Wood's science-fiction illustration art created for the pulps and digest magazines that he worked on from 1953 to 1968. Other books have come along during that time, collecting an assortment of these images, but never ALL of them together in one place. This is what we have tried to do. It was not an easy task to find all of these images, especially since my original intent was to photograph or scan them all from the original art in order to maintain the highest degree of reproduction. Obviously that idea went out the window when I realized that these originals were scattered all across the globe, locked into collections. Luckily, Robert M. Guinn, publisher of *Galaxy* magazine, had saved most of the art his company published rather than throwing it out. But even his holdings of art were scattered over thirty years ago when he cleaned out most of his *Galaxy* inventory through art dealers or collectors.

Over the past thirty years of collecting these originals myself, and searching and talking with other collectors, I was able to locate nearly half of Wood's original interior illustrations and all but three of his cover paintings. That's why the reproduction in this book will be noticeably better than most other books issued up to this point showcasing this genre of Wood's work. This task was not accomplished without the help of many people, most of whom are longtime collectors, including Woody himself. Yes, even the artist had managed to retrieve a number of his originals from Guinn early on before the publisher liquidated his archives. In fact, at least four of Woody's *Galaxy* cover paintings were returned to the artist, and even though he later gave them away to friends and relatives over the years, they are all reproduced from the original art in this book. Luckily Wood was also in the habit of making good continuous-tone stats of some of his originals before turning in the art to the publishers. We have had access to these stats thanks to Bill Pearson of the Wally Wood Estate.

Major contributors to this massive project include Alex and Phyllis Eisenstein, Glynn and Suzanne Crain, Doug Ellis and Deb Fulton, Jerry Weist and Dana Hawkes, and many others who unselfishly contributed one or two pieces from their collections to make this the finest book it could be of its kind. These folks went through a lot of trouble to photograph or scan some of the beautiful art in their collections. A special tip of the hat goes to Jeff Killian, who did much of the scanning work on images for this book.

Lastly I'd like to thank my editor Scott Dunbier and assistant editor Michael Benedetto for their untiring determination to see that the production quality of this book was the best it could be for our worldwide audience of Wally Wood fans.

Enjoy!

Roger Hill

March 2016

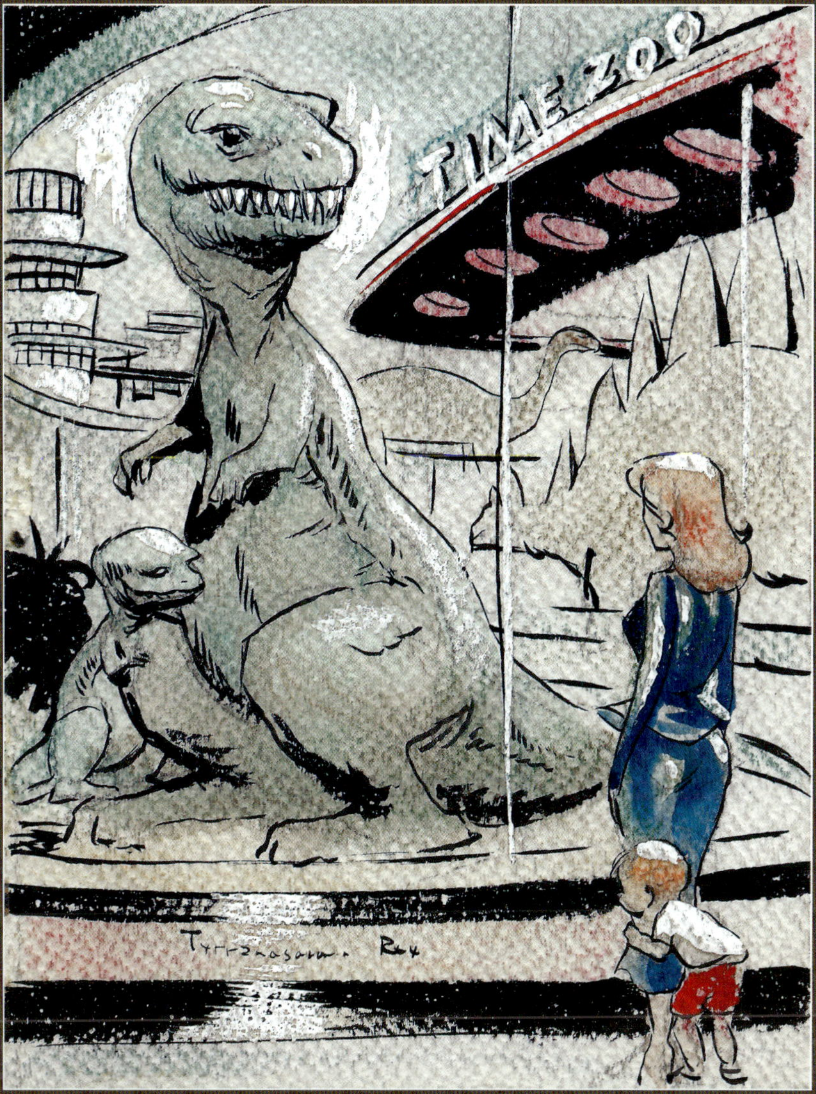

This book is dedicated to
Dan Adkins, Bill Pearson, and Glenn Wood,
three of the best friends Wally Wood ever had
during his lifetime.

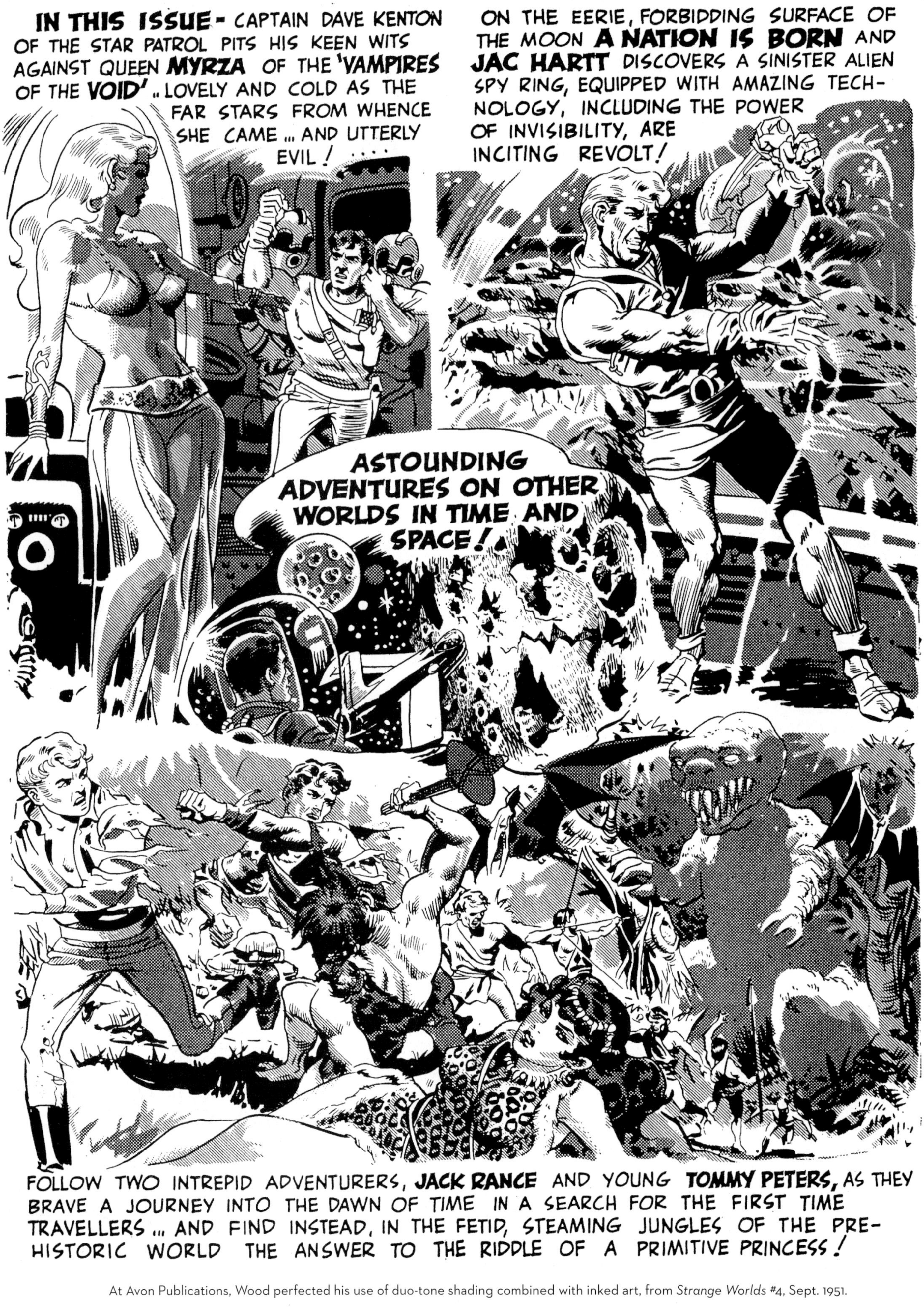

At Avon Publications, Wood perfected his use of duo-tone shading combined with inked art, from *Strange Worlds* #4, Sept. 1951.

WALLY WOOD: GALAXY ART AND BEYOND

Introduction by Roger Hill

Wallace Allen Wood (1927 – 1981) continues to be recognized today as one of the most creative and prolific comic artists who ever worked in the field. In 2010, the artist was awarded The City of Palma Award for Comic Art, followed by the largest single exhibition of American comic art and related illustrative art ever mounted for a comic-book artist at the Casal Solleric in Palma, Spain. Much has been written about Wood's involvement with the comics field.

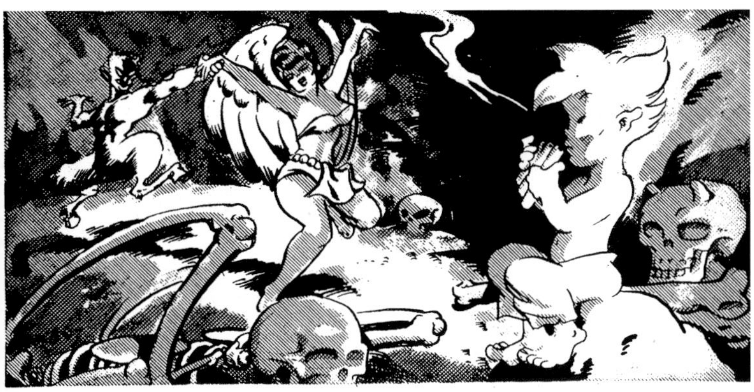

Early childhood duotone illustration by Wood, circa early 1940s.

In 1957, when Wood began contributing interior illustration work for *Galaxy Science Fiction* magazine, many readers were not aware that he had previously ventured into the low-paying field of pulp magazine art, but as a comic-book artist rather than an illustrator. Although this work does not relate to the theme of this book, it is mentioned here as a historical footnote of Wood's "pulp magazine" work. This happened in 1950 for Trojan Magazine's *Incorporated*, which was operated by Frank Armer, who had been involved in the field of pulp magazines since around 1925. During the 1930s, Armer was in partnership with Harry Donenfeld, who eventually went on to become one of the largest comic-book publishing moguls in the history of the genre (publishing such titles as *Superman, Batman,* and *Wonder Woman*). Donenfeld and Armer produced a line of "Spicy" pulp magazines, titled *Spicy Detective Stories, Spicy Mystery Stories, Spicy Adventure Stories,* and *Spicy Western Stories*. Because of the racy, violent content of Armer's magazines, and under increasing pressure from local civic groups, the publisher eventually cleaned up the stories and titles he published.

In 1934, the company had introduced comic-strip characters into some of their magazines. When Wally Wood's partnership with Harry Harrison dissolved in early 1950, Wood left EC Comics to freelance on his own. Harrison kept the EC account but within two months lost the job. Realizing that Wood was the more creative talent behind their earlier efforts, EC publisher Bill Gaines hired Wood back. During this brief absence from EC, Wood produced several eight-page western stories for *Trojan*. Characters such as "Six-Gun Smith," "Lariat Lucy," and "Fighting Bob Dale" all appeared in *Six-Gun Western* and *Fighting Western* magazines. These stories, printed in black and white and dated May and June 1950, represent Wood's first pulp work. A year later, these same stories wound up being reprinted in color in Trojan's new *Western Crime Busters* comic book.

Wood's first official entry into the field of science-fiction illustration came in 1953 when he provided his first illustration work to Fiction House Publications for their pulp magazines *Jungle Stories* and *Planet Stories*. Unless you were already an established "pulp" artist, most newcomers to this field were usually paid between

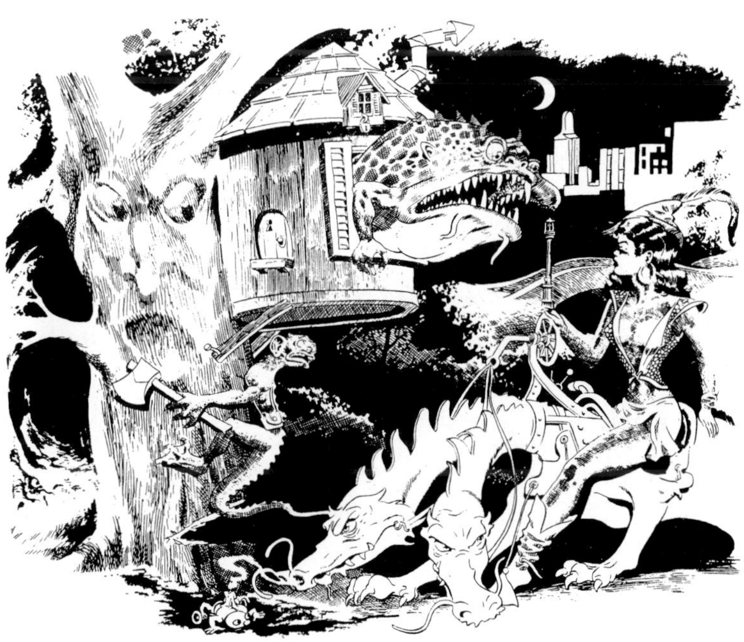

Early Wood fantasy drawing, circa 1949-1950, used as a portfolio sample piece.

ten to fifteen dollars per illustration. Established artists like Virgil Finlay and Hannes Bok, both recognized as some of the best in their genres, were paid more, but not by much. No one was going to earn substantial incomes working for these magazines, other than the cover artists themselves who were paid as much as ten to fifteen times more for cover art. For Wood, who was already an established veteran of comic books and earning $32 per page working on the popular EC titles, this early pulp work almost certainly had to be a cut in pay. The artist had never worked for Fiction House before, and one has to wonder how Wood made the connection and why he decided to work for such a low rate.

Fiction House had been around since 1921 when entrepreneurs Jack Kelly and Jack Glenister formed the company and put out their first pulp magazine titled *Action Stories*. Over the next thirty-four years, the company would grow to become a major competitor in the field of pulp fiction magazines, helping many unknown writers become successful authors. Their line of action-oriented titles included *Air Stories, Baseball Stories, Detective Book Magazine, Fight Stories, Jungle Stories, Planet Stories, Wings,* and several others. In 1938, under the direction of Thurman Scott, who had succeeded his father-in-law Jack Glenister to the publishing throne of Fiction House, the company ventured into the arena of comic books, introducing some of the very same pulp-magazine titles to the four-color comic-book format with the word "*Comics*" now in place of "*Stories*." The 1940s proved to be their best years of productivity and sales, but by 1953, the financially troubled company was reprinting most of its comic titles and two years later was out of business. We may never know the reason Wood took on such low-paying work, but if nothing else, it may have been a way to break up the monotony of doing comics, at least for a little while. Jack O'Sullivan was serving as magazine editor at the time Wood worked there.

Regardless of the low pay, the four illustrations Wood produced for Fiction House, three for *Planet Stories* and one for *Jungle Stories*, are some of the more finely detailed drawings the artist did during the 1950s. In comparison, these full-page illustrations easily hold their own against most EC science-fiction covers created

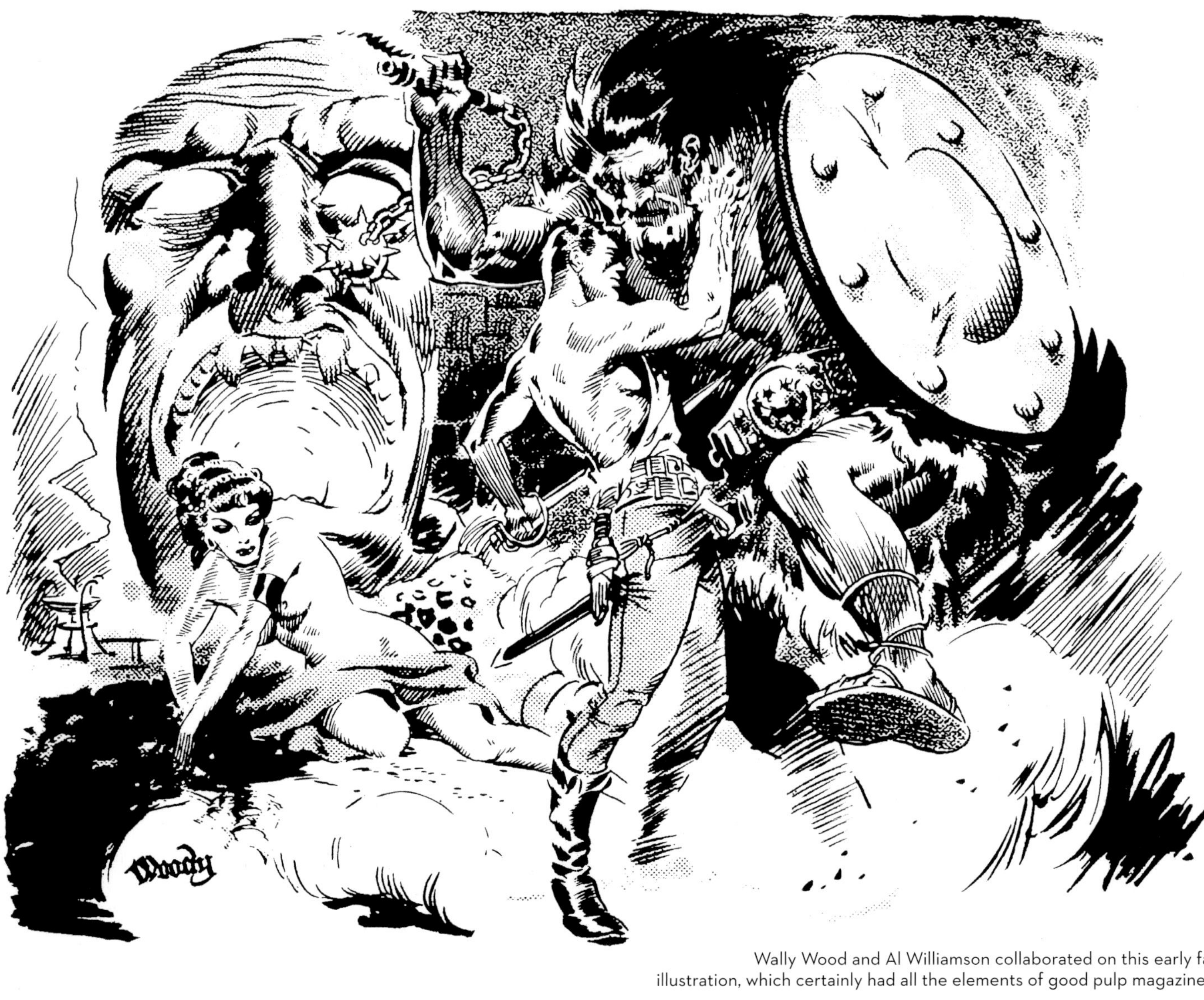

Wally Wood and Al Williamson collaborated on this early fantasy illustration, which certainly had all the elements of good pulp magazine work.

by Wood. In some cases, one could even say they are superior since each drawing was uninhibited by comic-book title or logo interference. Just pure art, that's all. Wood always loved drawing women, and for his first contribution to the January 1953 issue of *Planet Stories*, he provided a full-page illustration for "War Drums of Mercury Lost," written by John W. Jakes, showing a plethora of beautiful women sitting at the feet of a marbled throne while two armored men in the foreground engage in a fistfight. Two issues later, comments from the readers began to show up in "The Vizigraph," which served as the letters column for *Planet*. One reader exclaimed: "This WOOD that did the illo for WAR DRUMS OF MERCURY LOST is very, very good. Hang on to him." For *Jungle Stories*, spring 1953, Wood spared no expense on detail and worked up a powerful illustration for the story "Gorilla! Gorilla!" written by Anson Slaughter, depicting a gorilla rescuing another gorilla from a cage with big-game hunters shooting at them in the background.

His next illustration, for "Un-Reconstructed Woman," written by Hayden Howard, published in the September 1953 issue of *Planet*, showing a woman running through the woods with a space-suited man hot on her trail, titillated readers even more with the suggestion of full nudity, something Wood could never get away with in the comics. This illustration got the attention of several male readers who wrote into the magazine with favorable comments published throughout 1954: "You can send me the illo by Wood in the September issue. You should buy more of his work. His illos prove that he puts out a heck of a lot of work on each one. Nice." Another letter, written together by a group of fans wrote: "This is a bit too late to talk of Wood's illo in the September ish, but having been somewhat followers of him (thru the E.C. line), and regarding him as THE leading cartoonist, we were pleased pink to see that he was finally called upon to draw for a prozine. The response from the readers proves that others like his work too. More Wood—maybe even (oh joy) a cover!" Another fan opined: "Why not have Wood do a cover? Though he is known chiefly as a cartoonist, I think he could rank with Finlay if given an opportunity."

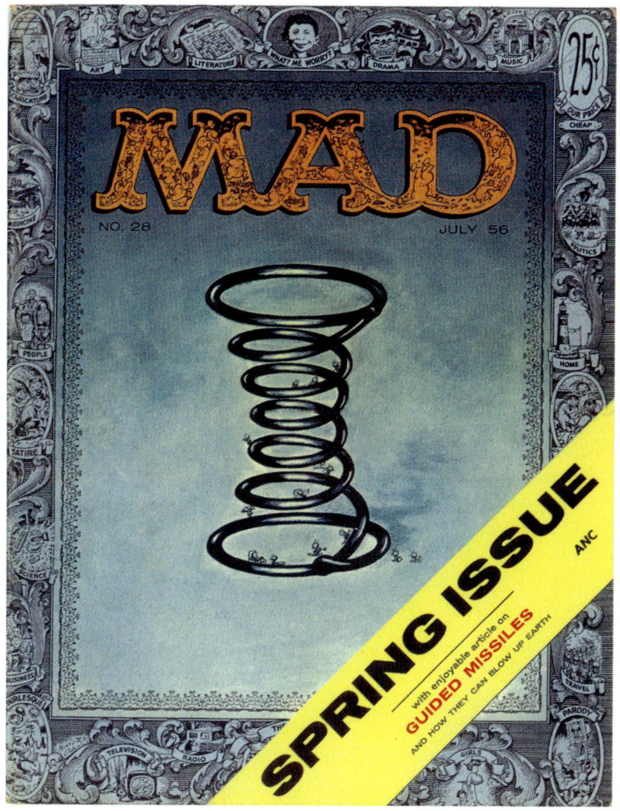

Wood's first published color painting from the cover of *Mad* #28, July 1956.

Like most pulp magazine covers of the time, Fiction House pulp covers were painted in oils or acrylics by established illustrators. There is no evidence that Wood ever submitted a cover painting to *Planet Stories*. In fact, the only painting Wood seems to have done early on in his career was a fantasy oil painting created for and gifted to EC publisher Bill Gaines during the early 1950s. This painting, depicting a nude woman and two warriors fighting a giant alien octopus, hung in Gaines' office for several years. While this painting may not have been quite up to pulp standards of the day, it certainly showed imagination and promise. It would be a few more years before Wood picked up his brush and tried his hand at another cover painting, this time for *Mad* magazine's "spring issue," published in July 1956, depicting...a giant spring! At *Mad*, the artist was a natural at tonal artwork, having worked with different gray-tone shading techniques since childhood when he drew his own comics following the styles of Hal Foster, Roy Crane, Will Eisner, and other favorite newspaper strip artists. Woody loved utilizing different mechanical methods of shading, which allowed the artist to add

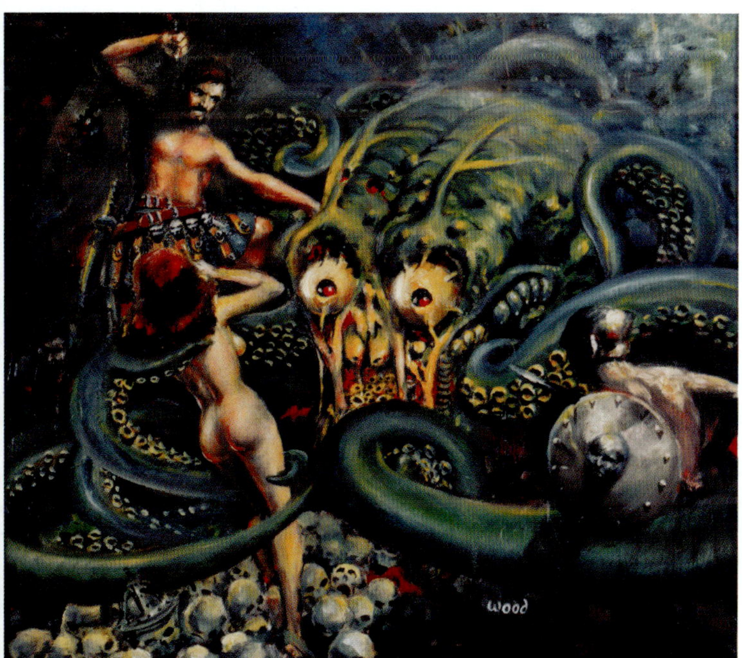

Wood's earliest known painting of any kind was this colorful fantasy piece created for EC publisher Bill Gaines during the mid-1950s.

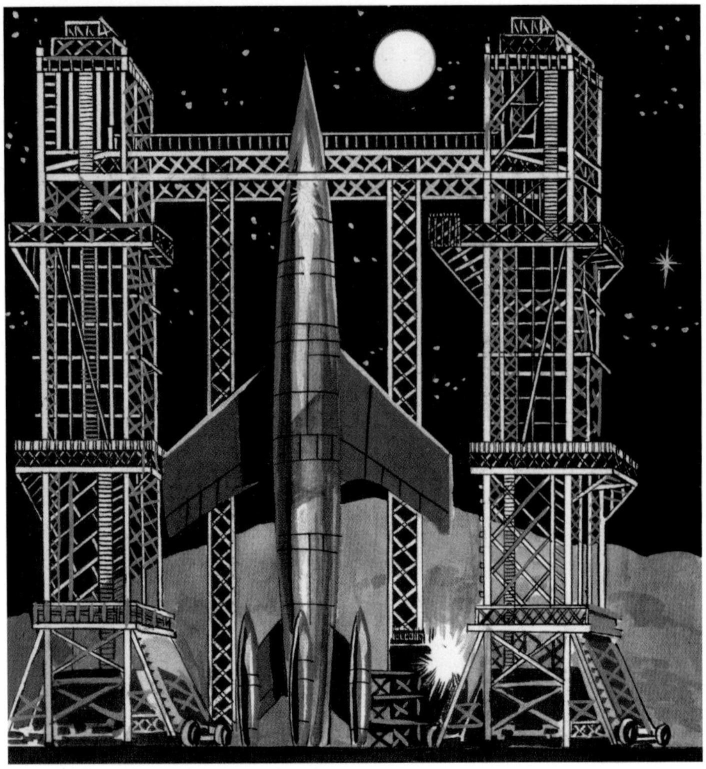

Wood's first published ink-wash illustration appeared in Will Eisner's *Spirit* Sunday newspaper section, dated June 29, 1952.
Opposite: Galaxy's first editor Horace L. Gold. Portrait by Bill Nelson.

a greater dimensional depth to his work. This tonal experience came in most handy when working for *Galaxy* later on.

Wood's earliest known published ink-wash, or gray-tone painting, can be traced back to Will Eisner's syndicated Sunday newspaper strip called *The Spirit,* which the artist contributed to during the summer of 1952. Three drawings, using ink-wash shading, were created by Wood to look like half-toned photographs in a mocked-up series of newspaper articles announcing that a trip to the moon was now feasible and about to happen. The first of these three ink-washed illustrations showed up in *The Spirit* Sunday section dated June 29, 1952, depicting a spectacular spaceship, named *Orville Wright #1*, being built on a launch pad. The caption below asks the question, "Will this be the first rocket ship to reach the moon?" Sunday installments for July 13 and July 20 each contained one additional ink-wash photo drawing by Wood, and on August 3, 1952, Wood introduced the first seven-page "Outer Space" story as Denny Colt, alias of "The Spirit," blasted off from Earth on man's first journey to the Moon. Wood helped on eight Sunday sections for this series of outer-space stories before bowing out to concentrate more on his EC work.

Unfortunately, by the end of 1954, as public condemnation of comic books snowballed into the formation of a Comics Code Association to oversee and self-regulate comic-book contents, comic publishers were hit hard and began reducing their output. Within a year, the new standards of decency effectively put an end to most of EC's comic titles that Wood was heavily involved with. In late 1955, as the era of EC Comics came to an end, twenty-eight-year-old Wally Wood must have wondered what the future held in store for him. One can only imagine his thoughts as his last comic-book cover, drawn for EC's last science-fiction comic, hit the comic racks. *Incredible Science Fiction* #33, dated Jan. – Feb. 1956, was Bill Gaines' last comic book and marked the end of a period of job security Wood had enjoyed with Gaines since 1950. The comic-book industry was crumbling around him, and luckily for Wood, *Mad*, now in magazine format, was there to pick up some of the slack. The only difference was that Wood now had to produce tonal artwork in shades of gray using a diluted ink-wash or other methods. He also indulged in the use of Zip-A-Tone and Craftint, commonly referred to as "Duo-Tone" art paper. Zip-A-Tone came in hundreds of different screens and shading patterns, and Wood mostly preferred a variety of lines or dots. These patterns, printed on a mylar sheet of plastic, could be laid down and adhered over the artwork then trimmed with an Exacto blade by the artist to leave only the areas he wanted shaded. Craftint or Duo-Tone art paper was specially prepared art paper with two different shading patterns embedded in the surface. To bring out a light or dark shading pattern effect, the artist would apply one or two different chemicals in selected areas with a brush. Wood enjoyed using all these methods, which allowed him to literally paint in shades of gray on the art board. Most of EC's past comic artists produced work for *Mad* magazine and then eventually moved on to other companies or jobs. Wood stayed at *Mad* longer than any of the original artists and turned out some genuine masterpieces that are heralded today as bona fide classics. During the next six years, Wood practically ignored the four-color comics industry.

Horace Leonard Gold (1914 – 1996) deserves much more attention here than the pulp publishing houses previously discussed simply because Wood did more illustrative work for *Galaxy* magazine than all the others put together. This body of work is considered some of Wood's finest and extends over into *If Science Fiction* as well. Both magazines were created under Gold's editorship. Gold and Wood probably had very little in common on the social front, but their commercial backgrounds may have crossed paths several times during their careers. When Wood arrived in New York City in July of 1948, he had but one thing on his mind as far as employment went; he was there to break into the comics field. And once he found his first job, that's what he did for the next nine years. By that time,

Horace Gold, thirteen years older than Wood, had been around for quite awhile, chalking up a track record in the field of pulp-magazine and comic-book writing and editing. Gold sold his first science-fiction story in 1934, titled "Inflexure," which appeared in Street and Smith's *Astounding Science Fiction Stories* under the name of Clyde Crane Campbell. Other stories followed under the Campbell name, and at the same time, he tried to break into the slicks using the name of Julian Gray without much success.

From 1935 to 1938, what with writing sales being few and far between, Gold ventured from one job to another, trying his hand at selling shoes, upholstering furniture, floor scraping, door-to-door sales, and other job opportunities that came his way. In 1938, he started selling stories under his own name to *Astounding* again, and when *Unknown* debuted in 1939, also under the editorship of John W. Campbell, he sold several more short stories. Between 1939 and 1941, he took an assistant editor's job with the Standard chain of magazines. He also became a freelance writer, providing stories for DC's *Superman* and *Batman* comics and for Fawcett Publications' *the Adventures of Captain Marvel*. From about 1940 to 1949, Gold provided plot ideas to Superman co-creator Jerry Siegel for *Superman* stories, then in 1943 formed a partnership with fellow writer Ken Crossen, whom he worked with on *Superman* and *Batman* scripts. As a duo, they wrote stories for various pulp titles and also provided radio drama scripts for *Mystery Theatre* and *The Kate Smith Hour*.

In 1944, shortly after Gold and Crossen formed their own comics publishing company called Spark Publications, Gold was drafted by the Army. By the time he returned home in late 1946, Spark Publications had filed for bankruptcy and was out of business. Gold returned to writing for the comics and doing occasional radio shows, pulps, and slicks. That is, until 1949, when he was approached by a past Spark Publications employee named Vera Cerutti with a proposition. Cerutti was then working for a foreign company named World Editions Publishing and needed help in putting out a magazine for the American market. When the company showed an interest in publishing a new title, Gold suggested a science-fiction magazine and wound up getting the editor's job and the okay to pay the highest price ever paid for science-fiction writing: three to four cents per word, as opposed to the two-cents-a-word rate then prevalent in the market. After contracts were signed with the American News Company for distribution, Gold hired Washington Irving Van der Poel, Jr. as Art Director. Gold also came up with two possible titles for the magazine: *Galaxy Science Fiction* and *If Science Fiction*.

Gold had Frank Conley, the ex-husband of Gold's wife, letter the logos for each title and, with Van der Poel working up mockups for the cover layouts, invited a number of editors, artists, writers, and fans over to Harry and Evelyn Harrison's apartment for a secret ballot voting on which title, layout, and lettering they liked the best. At this time, Harry Harrison was a comic book artist who was sharing a studio in Columbus Circle with Wally Wood and other assistants producing mostly romance stories for Fox Publications. Wood may have actually been included with the visitors to Harrison's apartment that fateful day to see the beginnings of this new magazine. A year later, Wood and Harrison went their separate ways, citing artistic differences as the leading cause of the breakup. In later years, Harrison would prove himself a much better writer than he ever was an artist.

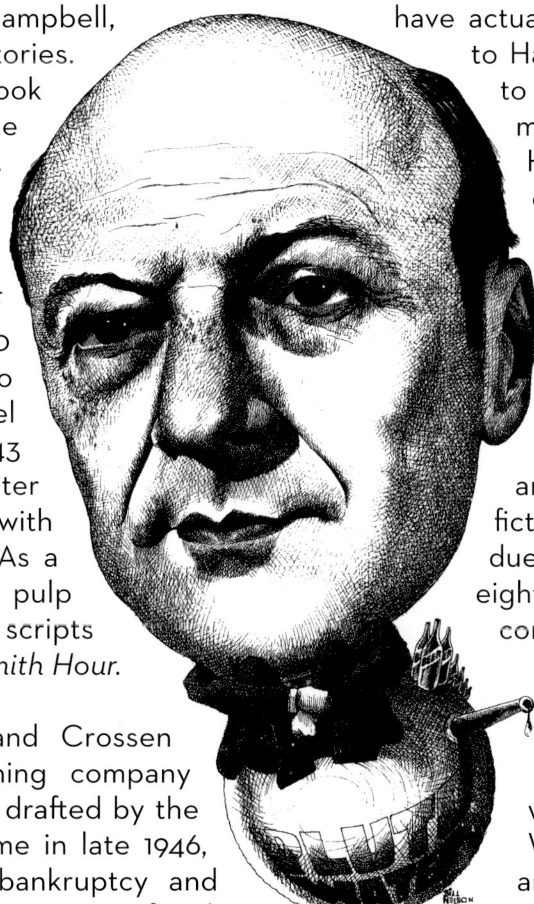

In the May 1, 1950, issue of *Fantasy Times*, Horace Gold made the announcement that the new science-fiction magazine he was editing was due out on the newsstands in about eight weeks. It would be a monthly, containing 160 pages and selling for twenty-five cents a copy. He also announced that he wanted good artists, whether they had illustrated science fiction before or not, and was seeking the addresses of Hannes Wesso and Elliot Dold, two veteran artists who had proven their ability in pulp magazines of the 1930s and 1940s. Apparently Gold never found them, for when the first issue of *Galaxy Science Fiction* finally appeared five months later, cover dated October 1950, other artists had been employed to illustrate the stories. Artists such as David Stone, Paul Calle, and Paul Pierre were providing the interior illustrations for the new magazine with top writers such as Clifford D. Simak, Theodore Sturgeon, Richard Matheson, Fritz Leiber, Groff Conklin, Isaac Asimov, and others providing the stories. The *Galaxy* cover design was also a major break from traditional science-fiction magazines in that the cover image was shifted to the right and down, leaving an L-shaped blank area of white on the left side with room at the top for the logo, date, and price.

Another press release to fandom, before *Galaxy*'s debut, came in June where Gold discussed the magazine's new cover design:

"It is completely unlike any S-F magazine—dignified, appealing, something no reader will be ashamed to be seen with. The format may be a handicap to us at first, but it will be spotted instantly once established. The inside illustrations are also a 100-degree departure from S-F art ruts. Our cover stock is really something to see: the most beautiful, highly enameled on the market: it takes color like nothing else anywhere."

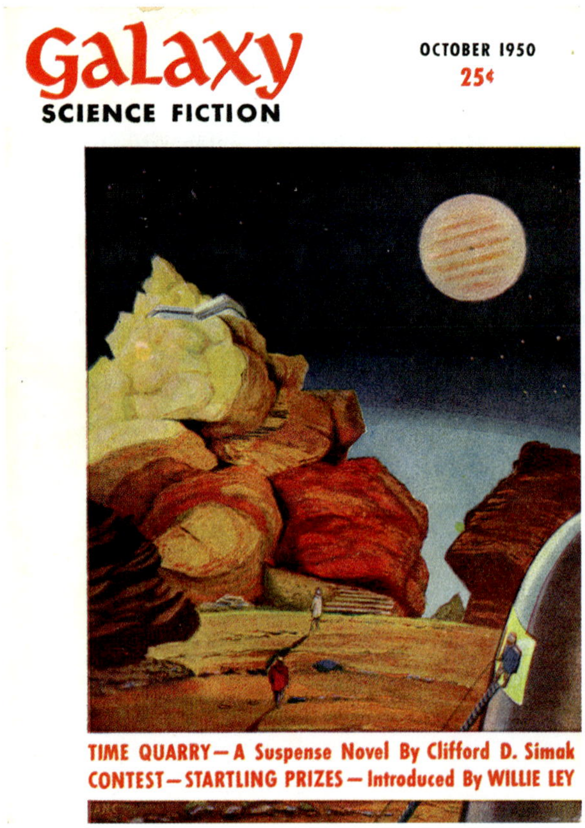

Galaxy was also the first magazine to use a new, revolutionary continuous-tone engraving process developed by one of the country's foremost experimenters in color reproduction. Like photography, the process used a granular surface technique rather than mechanical dots and allowed an imperceptible blending of the moiré pattern, which eliminated blurring through more accurate registration. In his first-issue editorial, titled "For Adults Only," Gold stated, "*Galaxy* will make a compound break with both the lurid and the stodgy traditions of S-F magazine publishing. From cover design to advertising selections, *Galaxy Science Fiction* intends to be a mature magazine for mature readers... mature in reading; age alone is no assurance of maturity. *Galaxy* will not be slanted at the comic addicts and lovers of juvenile S-F. Neither will it be random selections from science textbooks and journals. My opinion is that there is a huge segment of our S-F population which is seeking stories that do not offend or drown the imagination. We had planned to bring out a companion bi-monthly, but have since changed to a potentially more satisfactory arrangement. *Galaxy* will have a companion reprint book, which will issue, every other month, a complete, unabridged, uncut novel, exactly as it appeared in hard covers. This book will be digest size, like *Galaxy*, and similar in format so they can easily be identified. Price is 25 cents."

The new series was labeled as the *Galaxy Science Fiction Novels*. The first novel in this series was a reprint of "Sinister Barrier," a story written in 1943 by Eric Frank Russell. Within five issues, *Galaxy* was in the black and on its way to becoming one of the most popular science-fiction magazines being published. John W. Campbell, editor of *Astounding Science Fiction*, was forced to raise his pay rates per word to match Gold's. When the sixth issue appeared in March of 1951, *Galaxy* introduced ink-wash illustrations for the first time to their reading audience, as drawn by artist John Bunch. Reproduction on these illustrations was not quite as clear as works in black and white, although wash-work by Vincent Napoli and Don Sibley in the following issue showed much improvement. Covers up to this point had been painted by David Stone, Don Sibley, Don Hunter, John Bunch, and Chesley Bonestell, and for the June 1951 issue, Ed Emshwiller debuted with his first professional sale, a beautiful painting depicting three aliens studying the relics of an extinct race. Emshwiller, or Emsh as he signed his work, went on to become one of the more popular and prolific cover artists for *Galaxy* and other science-fiction magazines during the 1950s. Unfortunately, readers were unhappy about most of the artwork appearing inside *Galaxy* and wrote letters of complaint to editor Gold. It appeared fans enjoyed the more finished, detailed ink-wash drawings but rejected the drawings that looked like sketches. This was Gold's and Art Director Van der Pohl's idea to bring a fresh, new look to the face of science fiction. Obviously, it wasn't working, and by the June 1951 issue, Gold was ready to address this and other concerns in his editorial about some of the headaches plaguing him:

"Other headaches are distribution, newsstand display, rocketing costs, ruckuses over ads, sweating good stories into better stories, and improving art, which has been the biggest single gripe of readers. We could have avoided the battle over art by either eliminating it, a bad practice for circulation, or hunching down to the general science fiction level, a cowardly evasion since most of it is so damned unsightly."

Most, if not all of *Galaxy*'s early artistic decisions on covers, interior illustrations, and artists were guided by Art Director Van der Poel who, before coming to *Galaxy*, had never worked in the science-fiction field and had been a director for industry trade papers like *Iron Age*. On top of these problems, during the summer of 1951, *Galaxy's* printer defaulted on their paper supply, which forced Gold to go looking for a new supplier. After calling every printer he could find in the Manhattan yellow pages, he finally stumbled across the only man who would agree to supply him with the paper he needed.

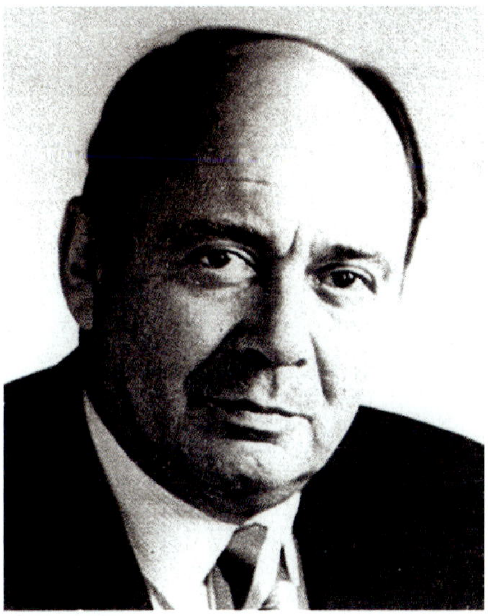

Galaxy publisher Robert M. Guinn, circa 1960s.

Robert M. Guinn (1911 – 1989), a printing broker who was familiar with *Galaxy*'s quick success toward becoming one of the top-selling science-fiction magazines, agreed to step up to the plate. Guinn had started out as an account executive in 1931 and by 1932 had become the advertising manager for the Acme Offset Company. During the 1930s, he worked his way up the ladder of success in several companies and in 1942 opened his own printer brokerage firm. Unfortunately, the only paper that Guinn could supply Gold with for a while was more like blotter paper than newsprint. In September, when disagreements within World Editions led to a disruption in *Galaxy*'s distribution, showing a decline in sales, *Galaxy* was put up for sale for $3,000. Gold, who had become good friends with Guinn, talked him into submitting a bid, which was accepted. Shortly thereafter, when World Editions realized they'd made a tremendous mistake, they tried to buy the company back. But Guinn wasn't selling, and over the coming years, he and Gold became even closer as partners. A good working relationship developed as Guinn left policies and other decisions up to Gold but included him in distribution and advertising matters. Working together, the two men kept *Galaxy* on the path to success, right up there with fandom's favorite magazine *Astounding Science Fiction*, edited by John W. Campbell Jr. Some historians of science fiction conclude that by 1954, because of the success of *Galaxy*, twenty-two more science-fiction magazines had appeared on the newsstands. In September of 1952, Gold announced, "With the November issue, *Galaxy* will be printed with the new printing method 'photo-offset.' A better brand of paper will be used also." These are some of the important changes that would eventually showcase some of Wally Wood's finest illustrative artwork five years later.

While *Mad* magazine continued to provide steady work for Wood, it wasn't enough to keep the artist as busy as he wanted to be. In late 1956, after turning in several features for the first issue of Harvey Kurtzman's ill-fated *Trump* magazine, Wood decided to branch out and began looking for work in the field of commercial magazine illustration. He quickly connected with Dugent Publishing Company, where over the next year or two, he provided ink-wash and full-color gag cartoon illustrations for their *Dude, Gent,* and *Nugget* men's magazines. He also landed a job at Fawcett Publications, a company he had drawn romance comic-book stories for in 1950. For their January 1957 issue of *Mechanix Illustrated*, Wood provided the lead illustration for an article titled "You'll Own 'Slaves' by 1965." This prediction of the human race's future dependence on robots was

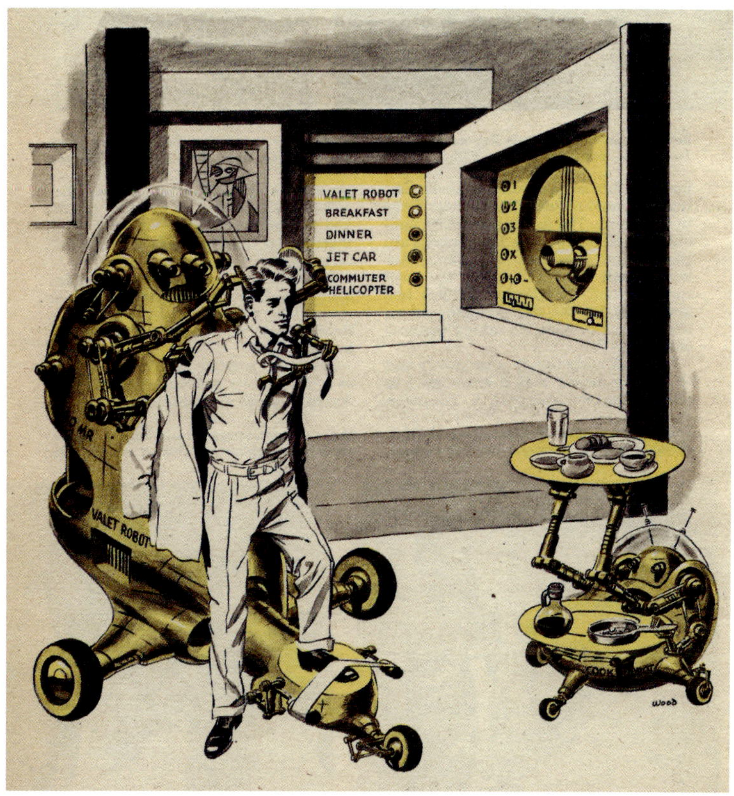

"You'll Own 'Slaves' by 1965" was an ink-wash job Wood created for the January 1957 issue of *Mechanix Illustrated*.

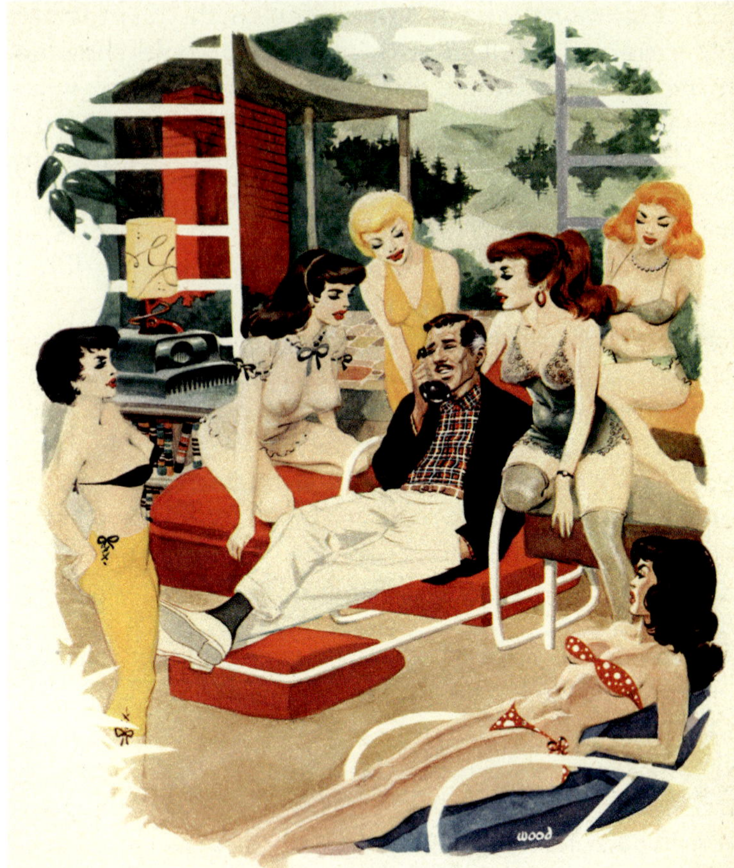

One of Wood's full-color men's magazine gag-cartoons, for *Gent*, August, 1957.

written by Otto Binder, one of the more prolific comic book and pulp-magazine writers of the 1940s. Binder had also written for EC Comics and other companies during the early 1950s, knew Wood's artistic ability, and may have arranged for Wood's involvement with this job. Wood's depiction of two valet robots waiting hand and foot on a man of the future is accompanied by the text stating, "Robots will dress you, comb your hair, and serve meals in a jiffy." Wood's knack for drawing realistic machinery and gadgetry achieved a convincing look into the future of mankind. The drawing was executed in beautiful ink gray tones of wash.

Next stop for Wood was Gnome Press, a small-press publishing company formed in 1948 by two local New York science-fiction fans named Martin Greenberg and David Kyle. The New York company became known mostly for publishing science-fiction classics in hardcover between the years of 1948 and 1962. In 1950, the company started reprinting the *Conan the Barbarian* stories by Robert E. Howard that had previously been serialized in *Weird Tales* magazine during the 1930s. By 1955, Gnome had exhausted their supply of Howard's *Conan* stories, and two years later, a Swedish fan by the name of Björn Nyberg became inspired enough to write a new tale about the sword and sorcery hero. L. Sprague de Camp, who had helped rewrite four of Howard's adventure novels for Gnome's *Tales of Conan* book two years earlier, edited the final manuscript before it went to press. Wood got the job to provide the dust-jacket illustration for this book, titled *The Return of Conan*, which was issued during the early months of 1957. One of the artist's early sketch preliminaries for this cover shows a figure of Conan standing with sword lowered getting ready for battle. Luckily, this rather static and tame version was replaced by Wood's revised cover depiction, executed in ink-wash tones, depicting the barbarian with sword flailing in full battle against soldiers with bodies lying in the background. Although critics argue that Wood's figure of Conan appears to be somewhat stiff, which it is, the final cover stands today as Wood's only interpretation of Howard's most famous action hero. The final dust jacket was printed with tones of red overlaid to give it a little bit of color. Being a very small publishing company and without much working capital, Gnome Press utilized monochromatic production applications for dust jackets on their books in order to keep costs at a minimum. This process allowed for the use of one color or shades of one color, adding a little more punch to a book that retailed for $3.00.

Wood turned out two more monochromatic paintings for other Gnome Press dust jackets during 1957. This included *Colonial Survey* by Murray Leinster, issued in June, and *The Shrouded Planet* by Robert Randall—a pseudonym of Robert Silveberg and Randall Garrett. For *Colonial Survey*, Wood took a different approach to painting, first rendering the image in pencil on illustration board then finishing the inks on an overlaid sheet of acetate. Using the acetate as a guide, he then applied different tones of orange gouache in various areas, covering his pencils on the board. The final monochromatic mixed-media painting worked well and created a strong composition, depicting a downward shot of a spaceman standing on the surface of a planet, looking upward, surrounded by several objects that play into the story.

Wood decided to venture further into the world of science-fiction illustration during the summer of 1957. By this time, most science fiction and other genre pulp-sized magazines had been replaced with the more popular digest-sized magazines. *Astounding Science Fiction, Fantasy and Science Fiction,* and *Galaxy* were the top-selling science-fiction magazines out of approximately thirty or so titles left. Most of the important science-fiction stories introduced during the 1950s debuted in these magazines. Wood could have worked for all of them, but for reasons unknown, he chose to seek work at the offices of *Galaxy*, located at 421 Hudson Street in Manhattan. Here, he met with *Galaxy*'s Art Director Van der Poel and received his first assignments: three interior ink-wash illustrations for a Theodore Sturgeon story titled

"The Pod in the Barrier," published in the September 1957 issue and a full-color painting that would show up on the cover of the October issue. These first ink-wash illustrations turned out to be some of Wood's finest efforts, showing the kind of lush detail he was known for putting into his previous EC science-fiction work. No doubt the artist wanted to make a good impression with these first efforts, and it showed. This issue also contained three ink-wash illustrations by cartoonist Don Martin, who had started working for *Galaxy* seven months earlier and was already becoming a favorite over at *Mad* magazine for his kooky-looking people and special brand of humor. It should also be noted that other illustrations in Wood's debut issue were provided by Leo and Diane Dillon, a husband-and-wife team of artists whose ink-wash illustrations had first shown up in the previous March issue and who would become good friends with Wood in the years that followed.

Wood's cover assignment for that October issue didn't appear to be near as difficult or complicated as the interior illustrations. The scene: a mountain-climbing man wearing cold-weather gear that includes a typical Wood space helmet, breathing hose, and oxygen tanks strapped to his back, standing solitary near the top of Mount Everest, watching other climbers inch their way up toward a pyramid-shaped alien. The story, written by Frederik Pohl and C. M. Kornbluth, titled "Wolfbane," was serialized in two parts for the October and November issues. Wood also provided three interior illustrations for each part, beautifully rendered in ink-wash tones. At the time Wood entered the digest magazine field, established artists such as Ed Emsh, Virgil Finlay, Jack Gaughan, and a few others were already supplying most of the visuals for readers of *Galaxy*. Wood's contributions to *Galaxy* must have been very popular with the readers and editor, as his work shows up in all the remaining issues published in 1957, along with a cover painting for the November-released *Galaxy Novel* #30, titled "Double Jeopardy," written by Fletcher Pratt.

As 1958 rolled around, Wood found time to contribute a two-page horizontal illustration for the story "Everything's Different up There" by Genevieve Haugen, published by Ziff-Davis in the February 1958 issue of *Amazing Stories*. The artist had last worked for Ziff-Davis in 1950 when he drew a nine-page story for their science-fiction comic *Amazing Adventures*. Ziff-Davis Publishing had surfaced in Chicago in 1927 then later moved to New York City where they became one of the leading publishers of hobbyist magazines. *Amazing Stories*, created by Hugo Gernsback in 1926, had been sold to Ziff-Davis in 1938 and was under the editorship of Paul W. Fairman at the time Wood worked for them. *Amazing Stories* ran for nearly eighty years and changed hands several times before ceasing publication in 2005. After another change of ownership in 2011, two final issues were published online in 2012 in support of various licensing agreements to secure trademark registration requirements, all in preparation for the resumption of regular publication in 2013.

In March 1958, Wood kicked into high gear, beginning with the cover art for *Galaxy Novel* #31, titled "Shambleau," written by C. L. Moore. This was the first installment in a series of "Northwest Smith" stories written for *Weird Tales* and first published during the 1930s. Wood's cover, which received a positive reception from readers, was beautifully composed and painted, demonstrating that the artist was improving. Most of these paintings were executed with acrylic paint on illustration board usually in a size of about nine by twelve inches, although on occasion, he would work a little larger. Wood's next cover for the August 1958 issue of *Galaxy*, showing firemen of 2025 A.D. fighting small

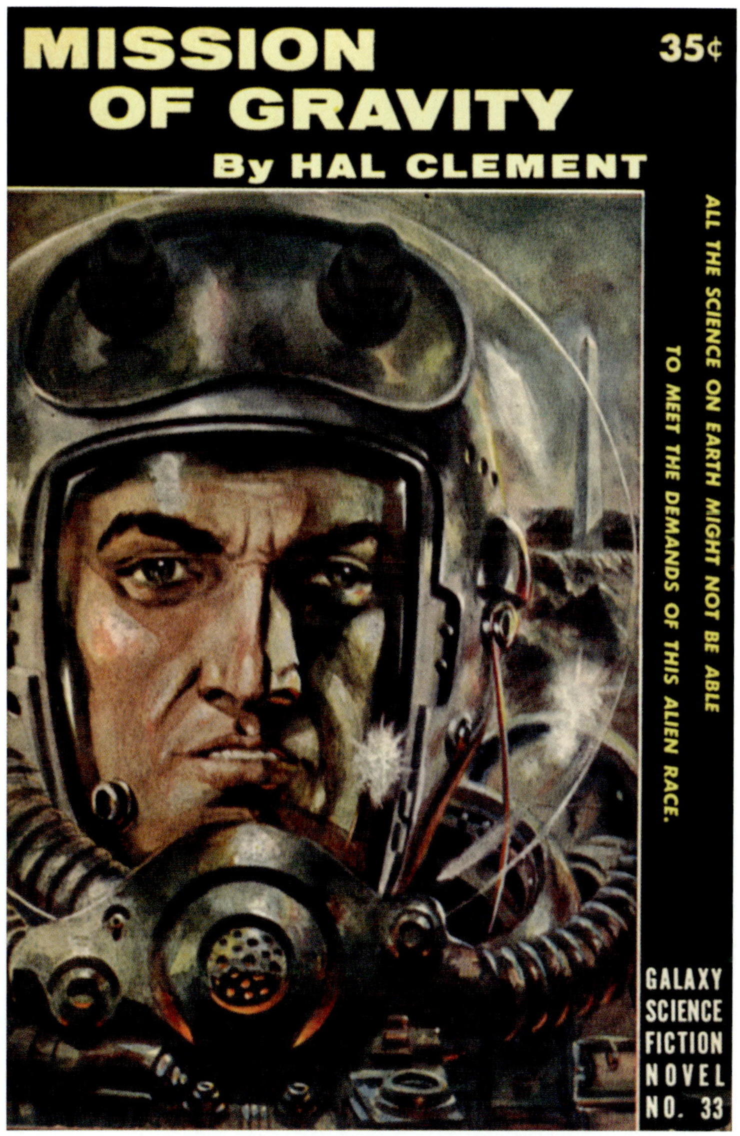

Cover of *Mission of Gravity* by Hal Clement, *Galaxy Novel* #33

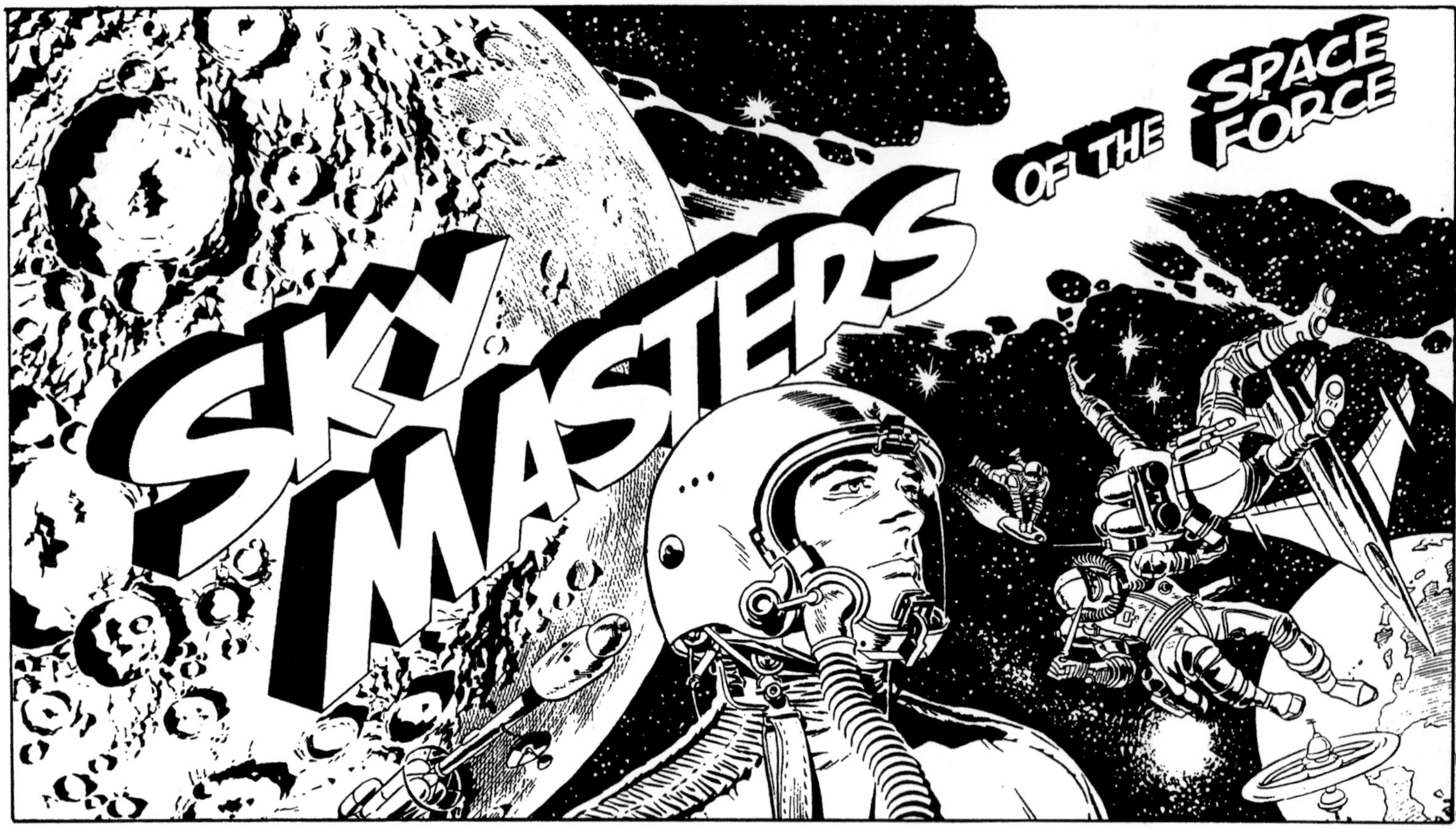

industrial blasts from a distance of two hundred yards, looked like a scene out of an atomic war with several of Wood's classic mushroom clouds rising up out of the destruction. That same year, the artist cranked out the cover art for *Galaxy Novels* #32, 33, 34, and 35, which beginning with #32 made a conversion from digest size to paperback format. These covers, more than any of his past work, resonated Wood's vision of technology and gadgetry, of science fiction and the future. *Mission of Gravity* by Hal Clement, *Galaxy Novel* #33, shows us a large portrait of a bubble-helmeted spaceman, with all the little nuts, bolts, hoses, and other protuberances we'd expect to see in suits made for space travel. He also managed to work in a little bit of sex with his cover for *Galaxy Novel* #35, featuring *The Forever Machine*, written by Mark Clifton and Frank Riley, depicting a nude woman—in the shadows—standing amidst two men and surrounded by classic Wood machinery on all sides.

The June 1, 1958, issue of *Science Fiction Times* reported the results of their first annual poll taken among the fans on various aspects of the science-fiction magazines then being published. In the category for "Best Interior Illustrations," Wally Wood made the list of favorites. Kelly Freas took first place, Ed Emsh came in second, and Wood took third. They also reported that out of all the many different artists working in the field at that time, only these three artists were even mentioned. Wood's reputation was growing among science-fiction fandom, and even editor Horace Gold himself commented on Wood's abilities in a press release to *Science Fiction Times* in August where he stated:

"'Time Killer' by Robert Sheckley, begins in the October 1958 issue of *Galaxy Science Fiction*. 'Time Killer' is illustrated by Wallace Wood, whom some of our readers have been kind enough to compare favorably with the great science-fiction artists of the past, despite his unfortunate handicap of being able to draw figures. Good taste—and recollections of the manner in which you have previously treated similar announcements—forbid my saying more."

Wood turned out twelve drawings for "Time Killer," during its serialization from October 1958 to February 1959, and even provided more interior illustrations for other stories in these issues. Where a schedule like this might overload other artists or illustrators, it didn't come close to matching the workload that Wood had endured during his EC days. On top of the *Galaxy* and Gnome Press work, Wood teamed up with Jack Kirby during the summer of 1958 to begin working ahead on a new science-fiction daily and Sunday newspaper strip called *Sky Masters of the Space Force*. With Kirby doing the penciling and Wood providing the inks, the strip debuted through the George Matthew Adams Syndicate on September 8, 1958. Wood also returned to comic books at this time, inking Kirby's "Challengers of the Unknown" stories for National Publications (DC).

When artist Dan Adkins came to New York City during the summer of 1958, he had already provided hundreds of drawings to science-fiction fanzines all across the country. Adkins had grown up reading EC comics and admiring the work of Wally Wood and other top artists employed there, and during the 1960s, he would become an assistant to Wood on some of his science-fiction illustrative work and for the comics. Now twenty-one years of age and seeking a career as a commercial artist, Adkins wrote about his sojourn to visit various New York publishers in the second issue of the fanzine titled Flafan. This gives us a pretty clear picture of what Wood himself was dealing with during those times:

"Been in New York for a couple weeks now. The pro-editors turned out to be very un-impressible. They pay peanuts for artwork. *Astounding* pays the highest rate—twenty-five bucks for an illo. Big deal. Of course I had been told before that science-fiction artists didn't make much. *Galaxy* pays a lousy ten bucks. Ray Palmer paid that good. And Palmer never gave me all the instructions Van Der Poel does. He seems to think the stuff in *Galaxy* is art—wants me to let go—get a lot of free line-wash and in plain words—be messy. He doesn't think Freas is any good cause he uses style and he don't even know what stf mags Lowndes puts out. To sum it up, I may appear in *Galaxy, Amazing, Fantastic, Infinity,* and perhaps *Venture*. The others haven't room for a new artist. Met Wally Wood—nice guy. Got his phone number, seen several of his originals, etc. But I used to want to work for the pros pretty bad—now that I know it's no big deal—I don't much care."

Amidst all the work Wood juggled for *Galaxy*, newspaper strips, and comic books in 1958, he also managed to make time for two more creations for Gnome Press, providing dust-jacket designs for *The Survivors* by Tom Godwin, released in late February, and *Undersea City* by Frederik Pohl and Jack Williamson, following in June. *Undersea City* was an extremely successful monochromatic effort on Wood's part with various shades of blue tones applied. Unfortunately, the publisher destroyed the depth and effectiveness of this piece by running it in a flat, straight tone of bluish-gray. For *The Survivors*, Wood turned in a large black-and-white inked drawing, which Gnome also turned into a one-color tone of blue during the final production process. The artist never worked for them again, and by 1962, Gnome was out of business. With the September 1958 issue, Gold changed the title of *Galaxy Science Fiction* to *Galaxy Magazine*, mentioning that the term "science fiction" scared many people away from buying.

Other changes were also coming to *Galaxy*. On January 23, 1959, Bob Guinn made the announcement in *Science Fiction Times* that his company, Digest Productions Corporation, had just bought *If Science Fiction* from The Quinn Company. If was a digest-sized magazine launched in March of 1952 by James L. Quinn, who hired Paul W. Fairman to be his editor. Three issues later, with discouraging circulation figures coming in, Quinn fired Fairman and took over the editorship himself for the next five years, with Larry T. Shaw assisting. After Guinn purchased the magazine, it was turned over to Horace Gold to edit. The magazine continued as a bimonthly, alternating with *Galaxy*, which went to the bimonthly schedule as well. The publisher then decided to cut costs by raising the cover price to fifty cents while increasing the page count. Further cost-cutting lowered authors' pay rates from three cents a word down to one-and-a-half cents a word, and this caused circulation to fall from 90,000 to 80,000 within a couple of months. This maneuver saved *Galaxy* over $12,000 a year. None of these changes affected Wood who, by the end of 1958,

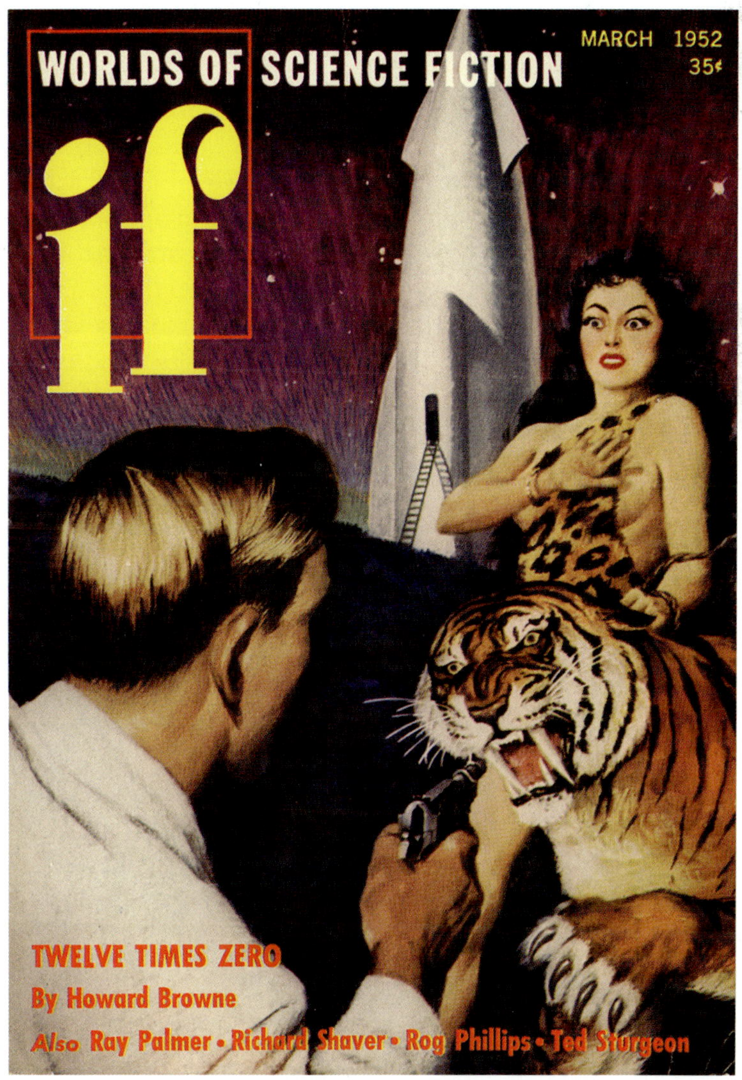

Cover for first issue of *If Science Fiction* magazine for March 1952.
Opposite: Wood accepted the job of inking Jack Kirby's syndicated newspaper strip *Sky Masters of the Space Force* in 1958. It lasted for eighteen months.

had contributed to nearly every single issue that year, providing a total of sixty interior illustrations and six full-color cover paintings.

The year 1959 looked bright for Wood as his *Galaxy* cover painting for the February issue gave him the chance to tackle a subject matter he excelled at better than most...robots! This cover accompanied the novella inside the issue written by Clifford D. Simak, titled "Installment Plan." The image depicts robots building robots with a man running up from behind. Normally *Galaxy* cover art was not obscured by anything, but this particular issue had a banner running across the top right corner of the image announcing "NEW! 196 pages." The issue is notable since Wood provided three illustrations each for four separate stories inside, making his contributions a total of twelve drawings. It is the heaviest Wood-illustrated issue he ever did working for the magazine. If readers thought this cover was good, they had a real surprise in store for them with the April issue. Wood knocked himself out with this one, which is recognized today as a genuine Wally Wood classic among Wood aficionados and labeled as the "poker playing" cover. Here the artist captured a scene right out of Star Wars, nearly twenty years before the movie's debut, by depicting a round table surrounded by poker-playing aliens, robots, and an Earthman. Several characters are wearing bubble helmets with breathing apparatuses in place, and smoking no less. Chips are stacked here and there with the robot's winnings stacked the highest. One alien, who is obviously upset and losing, is laying his phaser weapon on the table while others look at the hands they've just been dealt. The tiniest alien at the lower left of the table is holding a full house. The only things missing here are drinks and bottles of booze, which perhaps Wood felt would have been pushing the envelope a bit too much. Inside this issue, Editor Gold described the painting as, "Cover by Wood showing a full house beats togetherness." The idea for this painting may have actually come from Gold himself who, shortly after *Galaxy* started, began holding regular Friday night poker games at his apartment with writers, other editors, and musical composers. A. J. Budrys and Jerry Bixby were regulars, along with *Argosy* editor Robert Stein, and others like Ted Sturgeon, Fred Pohl, and Lester del Rey. Eventually, too many people were showing up, which frustrated Gold since only seven people can really play poker at one time. Coincidentally, Wood's painting only had seven figures around the poker table.

Wood strayed briefly over to Columbia Publications for a one-time job of doing two illustrations for editor Robert W. Lowndes' *Science Fiction Stories* digest magazine. The May 1959 issue contains two of the finest, most detailed pen-and-ink line drawings Wood did for the digest magazines, both featuring gorgeous women and handsome spacemen. These drawings accompanied a story written by Robert Silverberg titled "There's No Place Like Space." After that, it was back to painting a cover for the August 1959 issue of *Galaxy*. This turned out to be another masterpiece that, like the previous cover, didn't accompany any story inside. A skin diver, with spear gun in hand, is seen coming face to face with a bubble-helmeted alien emerging from a flying saucer sitting on the ocean floor. The alien seems to be more surprised than the diver, and on the contents page inside, Gold titled the cover "1 Down, 1 Up." Wood's last cover painting for *Galaxy* depicted a scene from Christopher Grimm's story "Someone to Watch Over Me," published in the October 1959 issue. Once again, Wood brought a nude female form into the picture, walking toward a bubble-helmeted spaceman standing next to his ship. Utilizing an avant-garde approach, Wood employed a form of colorful cubism to depict the background by using a multitude of viewpoints to represent the subject in greater context. Most of these background surfaces intersect at different angles, which removes a coherent sense of depth. Oddly, no other Wood art was found inside this issue. When 1959 came to a close, Wood's contributions were nearly half of what he had contributed in 1958, with thirty-seven interior illustrations and four cover paintings. *If Science Fiction*, owned now by Bob Guinn, took up some of the slack on the interior drawings, with covers usually being handled by Virgil Finlay, Ed Emsh, or other artists. Finlay and Emsh were seasoned veterans of science-fiction illustration and highly respected by editors, publishers, and fandom. When the names of nominations came out for the coveted annual Hugo Awards in 1959 for "Best Professional Artist," Wood's name was on the list, along with four other artists. The Hugo, named after Hugo Gernsback,

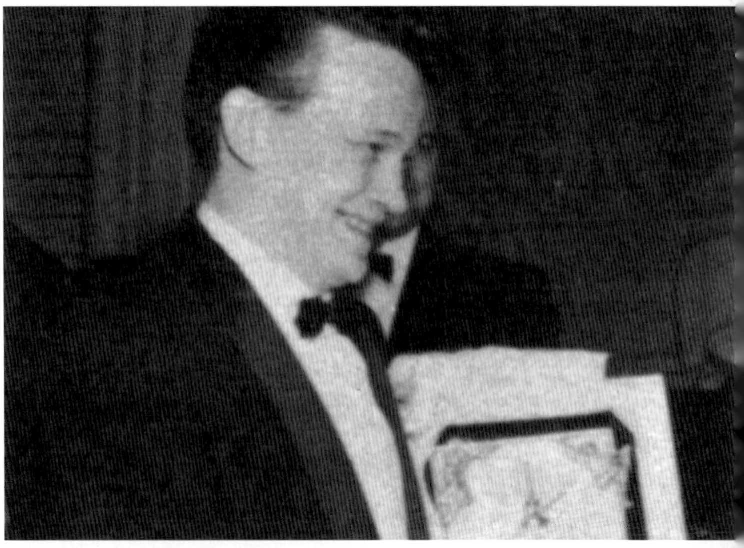

Wally Wood receiving the National Cartoonists Society Award for Best Comic Artist in 1957.

commonly referred to as "The father of Science Fiction," is given each year for the best writers, artists, etc., of works related to science fiction or fantasy released in the previous calendar year. In 1953, *Galaxy* had shared the prestigious award for "Best Magazine" with John W. Campbell's *Astounding*. Wood's Hugo nomination was for the work he did in 1958, but he was beat out by the popular Frank Kelly Freas.

On February 1, 1960, J. Harry Vincent, writer of the "Cosmic Reporter" column for *Science Fiction Times*, reported the following:

"We were most pleased to hear that Van der Poel has left *Galaxy* and *If*. His ideas of what science-fiction art should be left quite a bit to be desired. With the exception of Wood, Emsh, and one or two others, any artist who worked for them had to produce such monstrous artwork that it actually turned readers away. Sci-fi art should not be 'Mo Dern,' or 'Off-Trail.' It should be straight illustration."

Sam Ruvidich, a man who pretty much gave the artists free rein, replaced Van der Poel on both magazines. Wood plowed through 1960, turning out twenty-two interior drawings for *Galaxy*, twelve illustrations for *If*, and no covers for either magazine. Gone were the lush ink-wash drawings that Wood had excelled with early on, most being replaced now with straight pen-and-ink line drawings. When the Hugo Award nominations for "Best Professional Artist" were released for the previous year of 1959, the names of Ed Emsh, Virgil Finlay, Frank Kelly Freas, Mel Hunter, and Wally Wood were on the list. This time, Ed Emsh took home the top honors. Although Wood never received any of the prestigious Hugo Awards during his tenure working for the science-fiction magazines, he did take home the National Cartoonists Society Award for Best Comic Artist in 1957 and 1959 for his work on *Mad* magazine, which he provided on a steady basis. Besides his work for Guinn in 1960, he also slipped in two illustrations for the September 1960 issue of *Fury* magazine, published by the J. B Publishing Corporation of New York. These lead-in illustrations were used for the stories titled "Hang Our Kameraden and We Wipe Out Paris!" and "Pastor Parkhurst's Joy-Girl Paradise." Both drawings were executed with ink-wash and depicted several prime examples of ladies displaying exceptional female pulchritude.

Eventually, *Galaxy* was being distributed all over the world, reaching other countries like Australia, Belgium, Canada, Finland, France, Germany, and Great Britain, and in December of 1960, Guinn announced that the first Braille edition of *Galaxy* was being distributed to all libraries for the blind. The following year of 1961 showed a drastic drop in Wood illustrations, with only four appearing in *Galaxy* and two in *If*. One Italian artist by the name of Salvataggio was apparently so inspired by one of Wood's interior ink-wash illustrations from the October 1960 American edition of *Galaxy* that he painted a new full-color cover interpretation based on that drawing for the June 1961 Italian edition. The year 1962 came and went with even fewer Wood illustrations, and then Wood's output spiked upward to fourteen drawings in 1963. After that, his contributions tapered off significantly. In the spring of 1964, Wood had a dispute with *Mad* magazine's Al Feldstein and Bill Gaines over the quality of a job he turned in. The argument ended with Wood quitting and walking out, never to return. This feature, titled "Comics for Publications That Don't Have Comics," scripted by Frank Jacobs, showed up in *Mad* #88, July 1964, subsequently illustrated by Bob Clarke. Shortly after Wood's departure from *Mad*, he started working again for comic books, first tackling an entire book of war stories for Charlton Publications located in Derby, Connecticut. He then signed on with Stan Lee at Marvel to produce the *Daredevil* comic book for several issues, eventually bowing out to begin work on his popular *Thunder Agents* series for Tower Publications in the fall of 1965.

It should be noted that during the years of 1965 to 1968, Wood continued to secure digest illustration work from time to time, though most of these illustrations were created with the help of assistants. The two most notable assistants were Dan Adkins and Ralph Reese. Eventually, these artists went on to provide their own science-fiction illustrations for *Galaxy* and other digest magazines and produced work for comics with Wood and on their own. In 1966, Wood and Adkins collaborated on a final cover painting for *Worlds of If*, depicting a scene from "Earthblood," written by Keith Laumer and Rosel G. Brown. The story was serialized from April through July 1966, with the Adkins-Wood team also providing thirteen interior illustrations. When the April issue came out, the cover was incorrectly credited to artist Gray Morrow with the interior illustrations mistakenly credited to Norman Nodel. A personal anecdote from Dan Adkins in a 1970 issue of the fanzine *Yandro* specifically speaks about the creation of this cover painting:

"Let me tell you the story behind the cover for *Worlds of If Science Fiction*, April 1966 issue. It all began, as most covers do, with a telephone call. The phone was ringing at a building on 73rd Street in Manhattan. Up on the sixth floor of this building, several people were sitting around doing their thing, which was drawing.

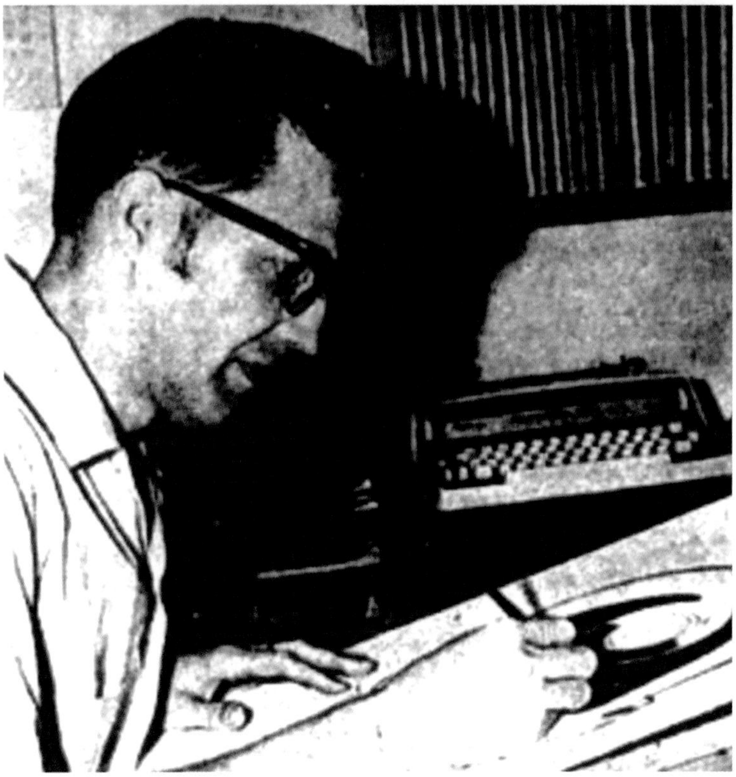

Wood's closest assistant during the *Galaxy* years - Dan Adkins.

Wally Wood heard the ringing as he was laying out the next cover for *Dynamo*, a superhero comic book. The phone had been ringing for twenty years as far as Wally was concerned. It never seemed to stop.

Ralph Reese heard the ringing as he ruled the borders for the next *Thunder Agents* story. Tony Coleman heard the phone ringing while putting some of the background details in the current *Thunder Agents* story with his brush. He wished it was a girl calling for him. If only he was back in jolly old England. Wally's wife heard the phone as she colored the proof for a *Thunder Agents* cover.

I heard it too, and put down the page I had been penciling to pick up the phone, which was beside my chair. It was the art editor at *Galaxy*. I gave the phone to Wally and he talked to her for maybe fifteen minutes. We were going to paint a cover and do some black and white illustrations.

Three days later, we got the manuscript in the mail and I took it home with me, read it and did sketches. Wally found the ones I had done for the inside illustrations to be okay, and I finished penciling them. He then inked most of the penciling, and I finished up the backgrounds. But he wasn't too happy with what I had come up with as far as the cover sketches.

Wally read the script and did some sketches. Tony Coleman helped on the coloring and Ralph took the works down to *Galaxy*. They okayed one of the colored sketches and Wally painted the cover with some suggestions from his wife on coloring. We all liked the end result. I found myself wishing Wally would do more painting. He was even better at it than I had thought, and it seemed so easy for him to do.

This was all done while we worked on the comic pages, which was the main source of money for us all. Ralph took the finished cover down to *Galaxy* and came back with it. Came back with it? What kind of deal was this? The art editor, who was Jo Ann something or other, had dashed off a note on the border of the painting. The background is too confusing... it looks like mud. Looks like mud! The girl is full of shit, I thought. That may have been kind words to what Wally thought. Anyway, he wasn't touching it. The hell with this two-bit account. Or why spoil a good painting because some gal thinks she knows what art is all about. That was true. Jo Ann was very pretty, indeed, but she didn't know anything about art except what she liked.

Wally gave her a call and learned she wanted brighter colors, like a poster or such. Anyway, after we calmed down and thought about it, we decided to fix it up. Or Wally decided I was to fix it up if we were to make any money on the deal at all.

I took it home and painted in some planets in the sky and added a lot of bold color to the monster. I jazzed up the background weeds and bushes a bit. It sort of looked nice and cheap when I got done.

I took it into Wally. It looks cheap, he agreed. *Galaxy* bought it. They thought it was pretty groovy. And of course when it was finally printed, Gray Morrow got credit for it. Which made us all happy. And they ran a blurb over most of the background, anyhow."

Between the years 1967 and 1969, probably due to a trade-off arrangement with publisher Guinn, a number of quarter-page ads for Wood's new *Witzend* fanzine began showing up in *Galaxy* and *If*. Horace Gold kept *Galaxy* profitable for many years and through a period of hard times for science-fiction pulp and digest magazines. When Wood started working for *Galaxy* in 1957, there were twenty-nine science-fiction magazines on the stands. By 1962, only seven were still being published. When Gold was injured in an automobile accident in 1960 and could no longer continue his editorial regimen, Fred Pohl (1919 – 2013) volunteered to help out until Gold recovered. Unfortunately, Gold was in a lot of pain and gradually lost weight until he had to be hospitalized in late 1960. His name appeared one last time on the editorial masthead of the April 1961 issue of *Galaxy* and he was replaced by Fred Pohl as Managing Editor by the June issue. In May of 1962, Guinn announced in *Science Fiction Times* that Sol Cohen (1910 – 1988) would take over as publisher of *Galaxy* and *If* on June 11. He also stated that the company would now expand both in science fiction and other fields that he had an interest in and that he hoped to go monthly with both his science-fiction publications.

Sol Cohen had been involved in the field of science fiction since 1949 when he was hired by Avon Publications specifically to work on developing their comic-book line. Wood had worked for Cohen during the early 1950s when Avon issued such popular comic titles as *An Earth Man on Venus, The Mask of Fu Manchu, Flying Saucers, Space Detective,* and others. Cohen also edited Avon's first science-fiction digest magazine called *Avon Science Fiction and Fantasy Reader* and worked on their science fiction/fantasy paperback line. After leaving Avon in 1959 and after a short stay at Hillman Books during the early 1960s, Cohen was hired as an assistant to Donald Wollheim at Ace Publications.

Under Cohen and Pohl's guidance, a new bimonthly called *Worlds of Tomorrow* was created in April of 1963. Guinn now had the distinction of holding more science-fiction titles than any other publisher in the business. All three magazines rose to the top of their field, with *If* winning the Hugo Award for "Best Professional Magazine" for three years running, 1966 to 1968. Wood provided illustrations for all three magazines during Pohl's editorship, with his last contributions showing up in *Galaxy* in August 1967 and in *If* in December of 1968. Jack Gaughan, Gray Morrow, and newcomers such as Vaughn Bode, Jeff Jones, and Roger Brand took up most of the slack with Woody's absence. Even the grand old master of detail, Virgil Finlay, returned briefly to work for *Galaxy*. Dan Adkins and Ralph Reese, both of the Wood Studio, also continued to provide cover paintings and interior drawings from time to time. *Worlds of Tomorrow* was discontinued in May of 1967, and Pohl edited his last issue of *Galaxy* in May of 1969 after returning from a trip to Brazil and being informed that Bob Guinn had sold all his magazines to publisher Arnold Abramson. Under the new publisher, both *Galaxy* and *If* struggled along until the new owners merged the two magazines together at the end of 1974. With sales gradually on the decline, *Galaxy* ceased publication during the late 1970s. In 1994, the title reemerged briefly as a magazine-sized publication under the editorship of Horace Gold's son, E. J. Gold. He produced eight issues between January 1994 and March 1995 before it too ended. Horace Gold passed away on February 21, 1996, and will always be remembered as an innovative editor who, for a while at least, paid the highest rates and brought new themes of sociology, psychology, and satire to science fiction during the 1950s.

Between 1957 and 1968, Wally Wood turned out over two hundred interior illustrations for the science-fiction digest magazines and a dozen cover paintings for *Galaxy* magazine and Gnome Press books. While some critics may say the science and technology of Wood's work is outdated by today's standards, the truth is, it must be studied, analyzed, and appreciated for the era it was created in. During the 1950s and 1960s, Wood was a visionary of the future, more so than many of his fellow artists. His knack for understanding and drawing complicated gadgetry, space ships, space travel and the universe, alien or human life forms, and the structures of other worlds came from an imagination that was so inspired, he spent a lifetime at the drawing table trying to put it all down on paper. It would appear that Wood was driven by an inner turmoil and strength that made him a prisoner of his own environment. Eventually, it broke his health and his will to live. Hopefully, many of these images from Wood's pen and brush, reproduced in this book and preserved for posterity for many years to come, will capture the hearts and minds of many science-fiction and fantasy aficionados who have not yet discovered his unique talents. He passed away in Los Angeles on November 2, 1981, leaving behind a significant body of work and many followers who passionately collect his art and published works.

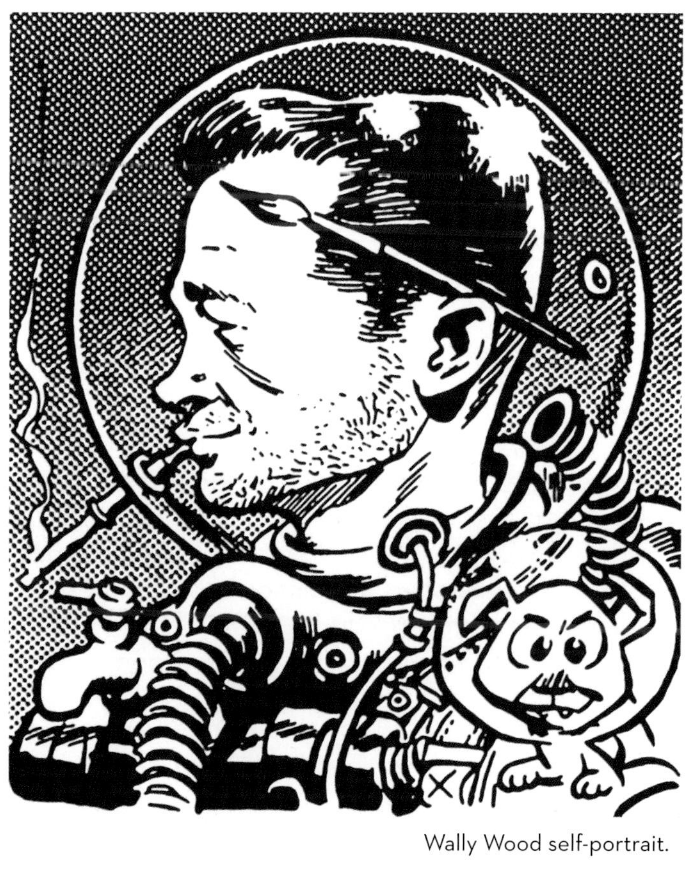

Wally Wood self-portrait.

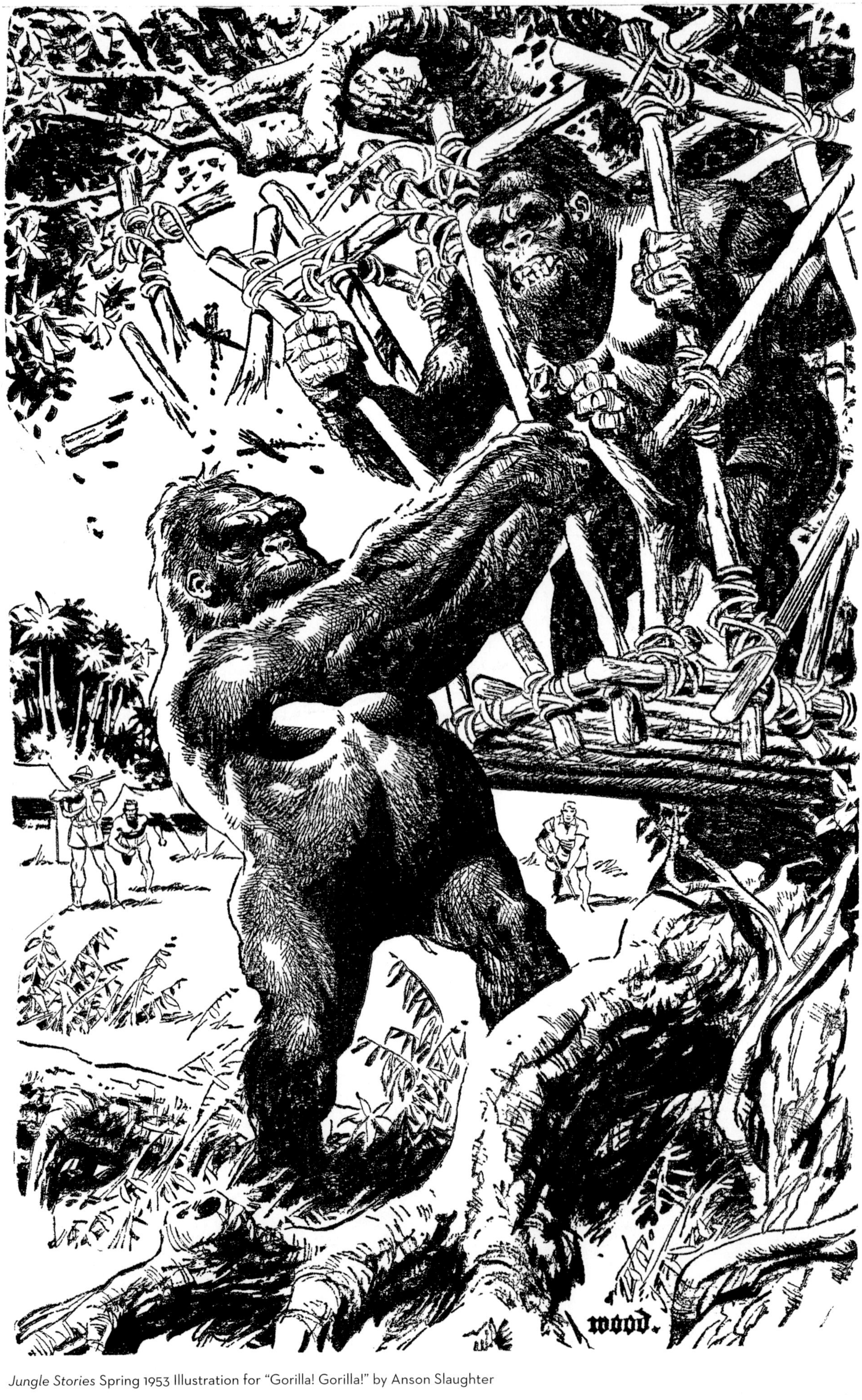

Jungle Stories Spring 1953 Illustration for "Gorilla! Gorilla!" by Anson Slaughter

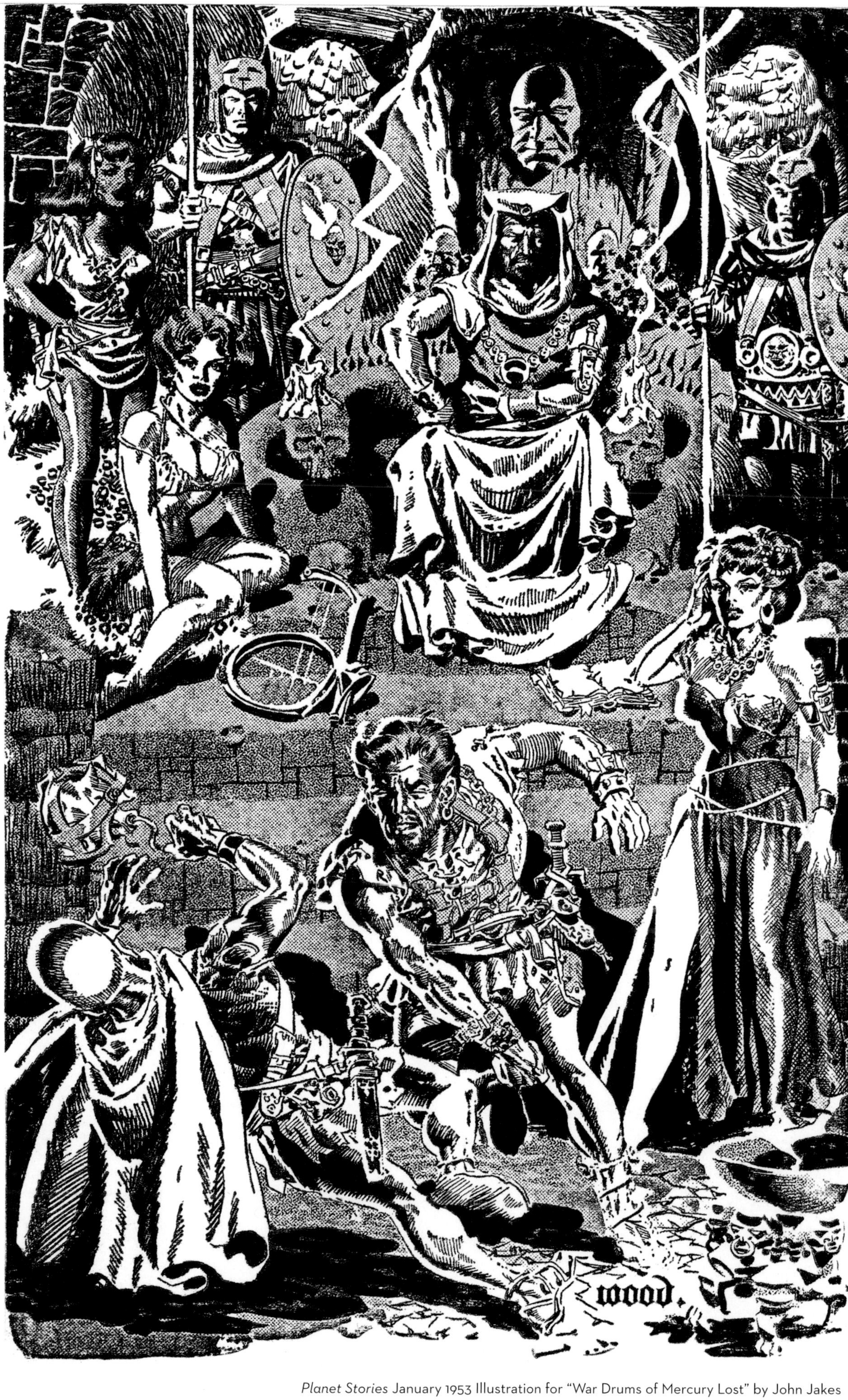

Planet Stories January 1953 Illustration for "War Drums of Mercury Lost" by John Jakes

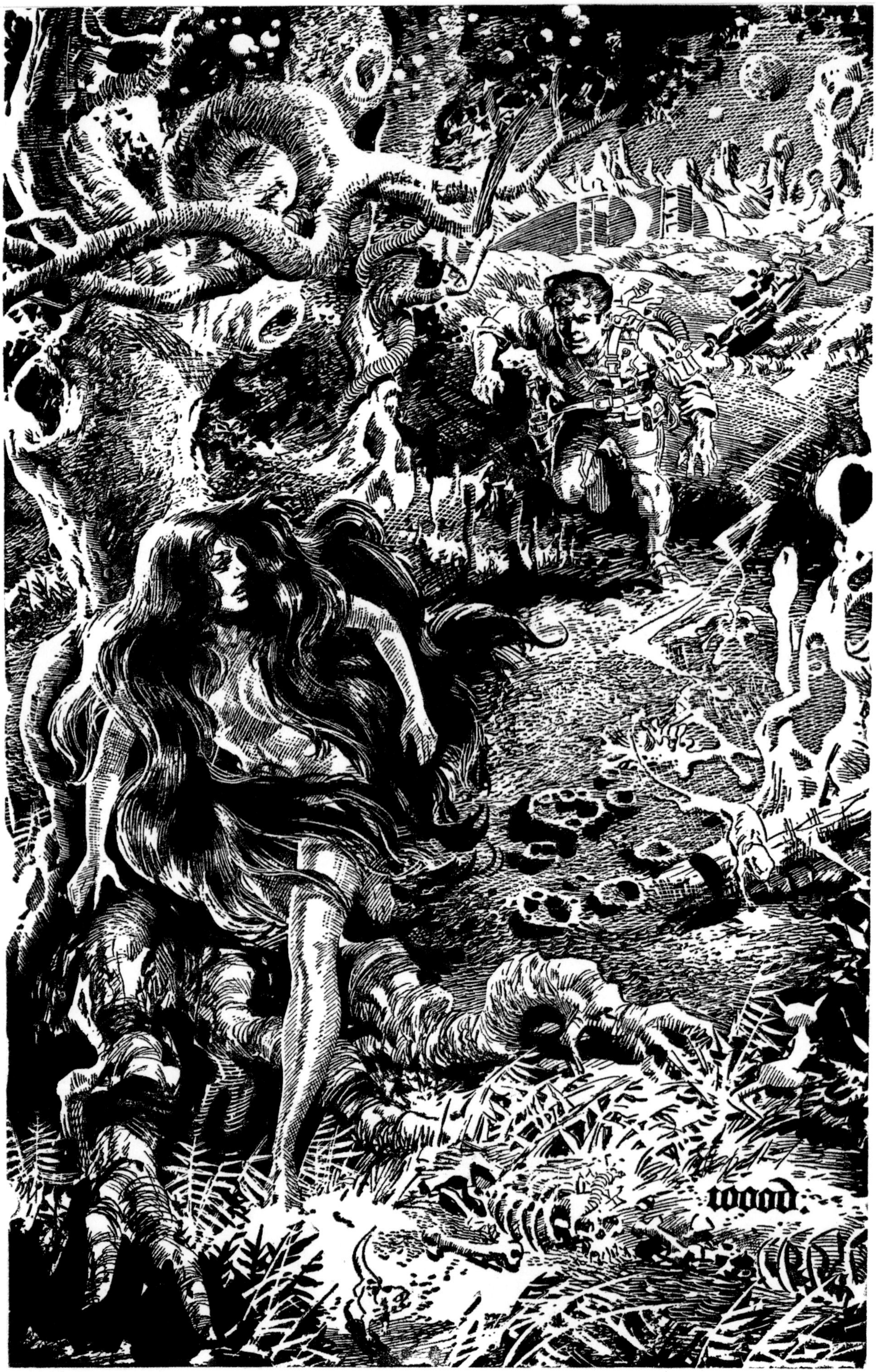

Planet Stories September 1953 Illustration for "The Un-Reconstructed Woman" by Hayden Howard

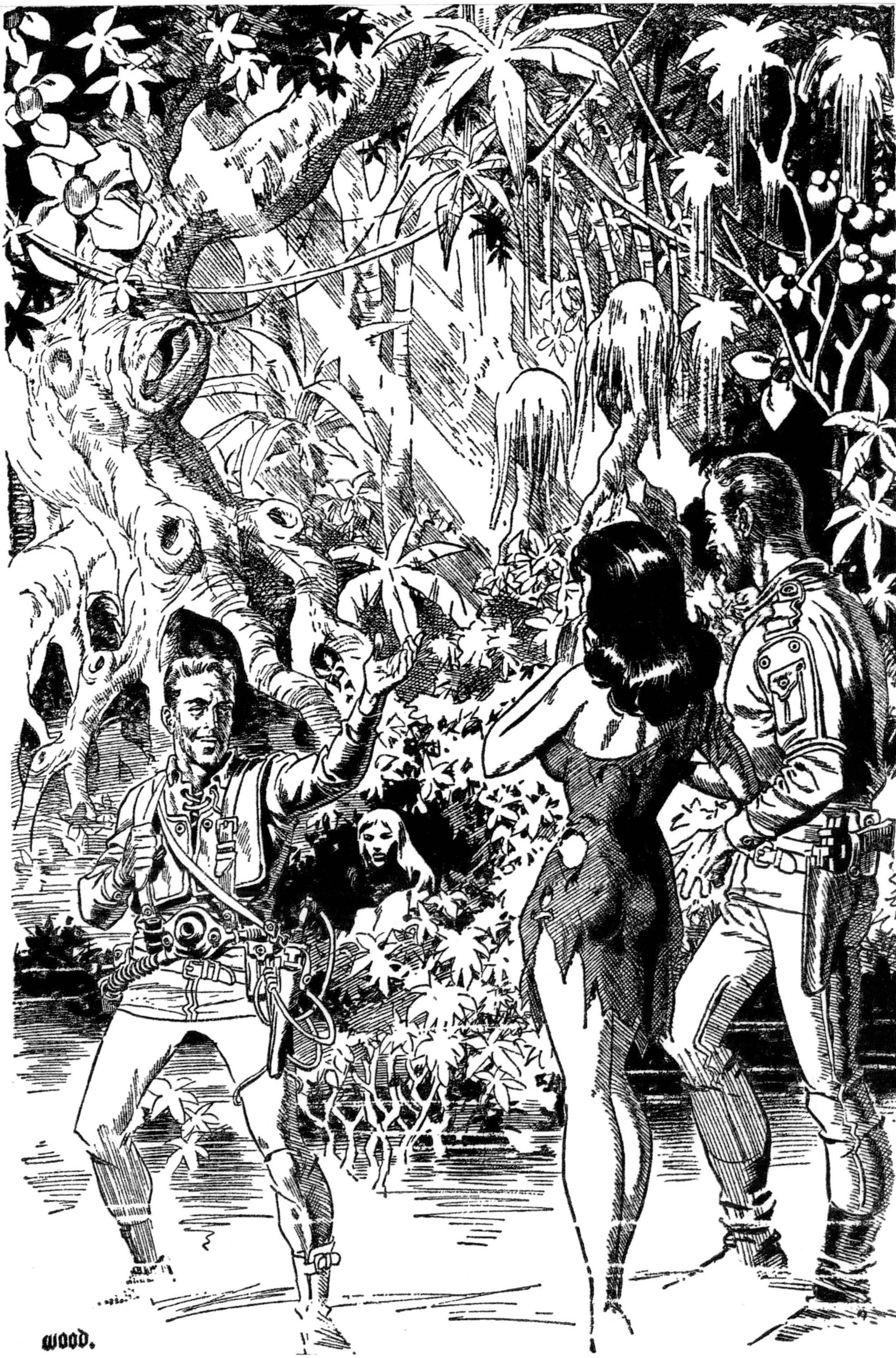

Planet Stories Summer 1954 Illustration for "Color Blind" by Charles A. Stearns

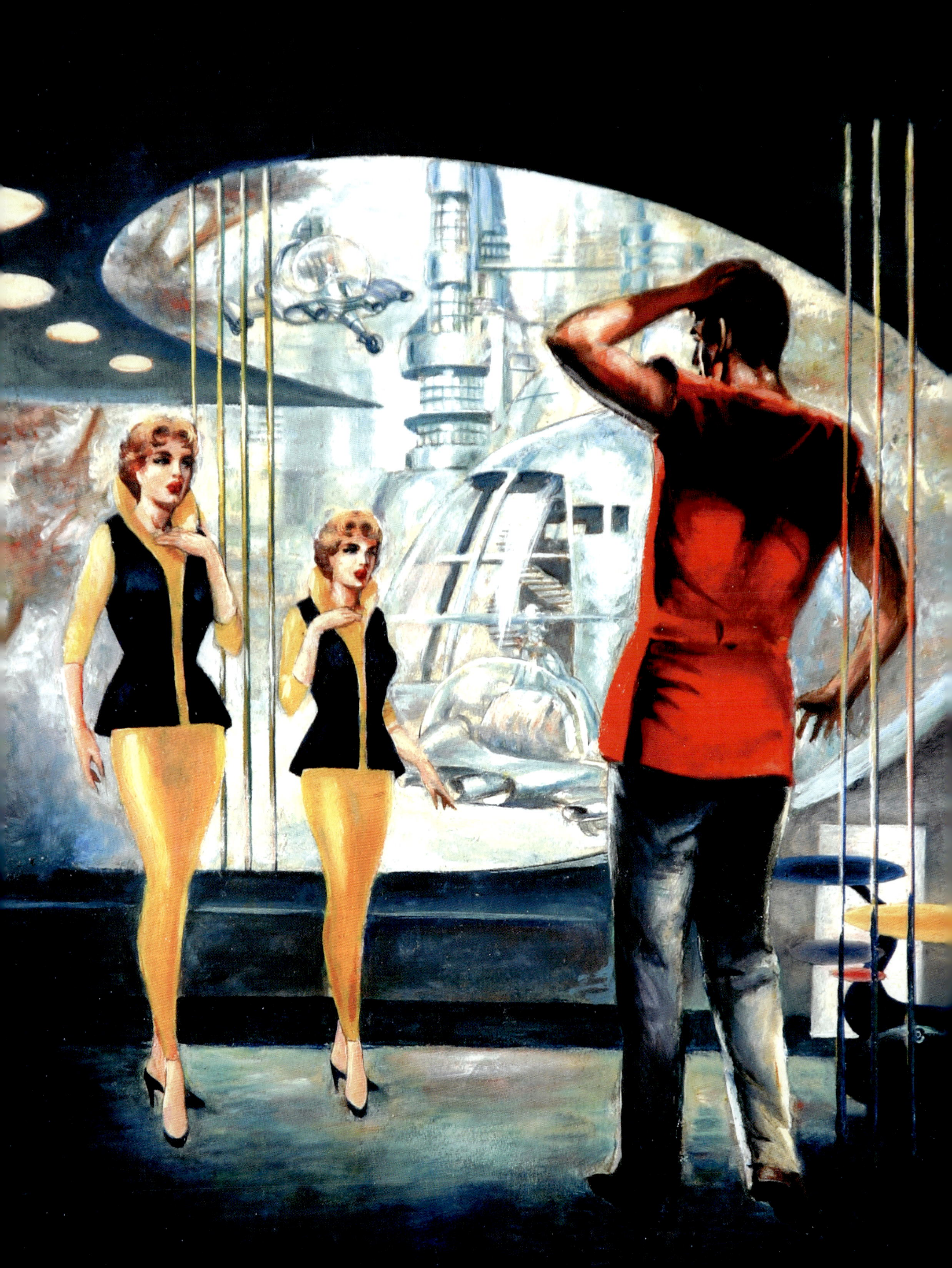

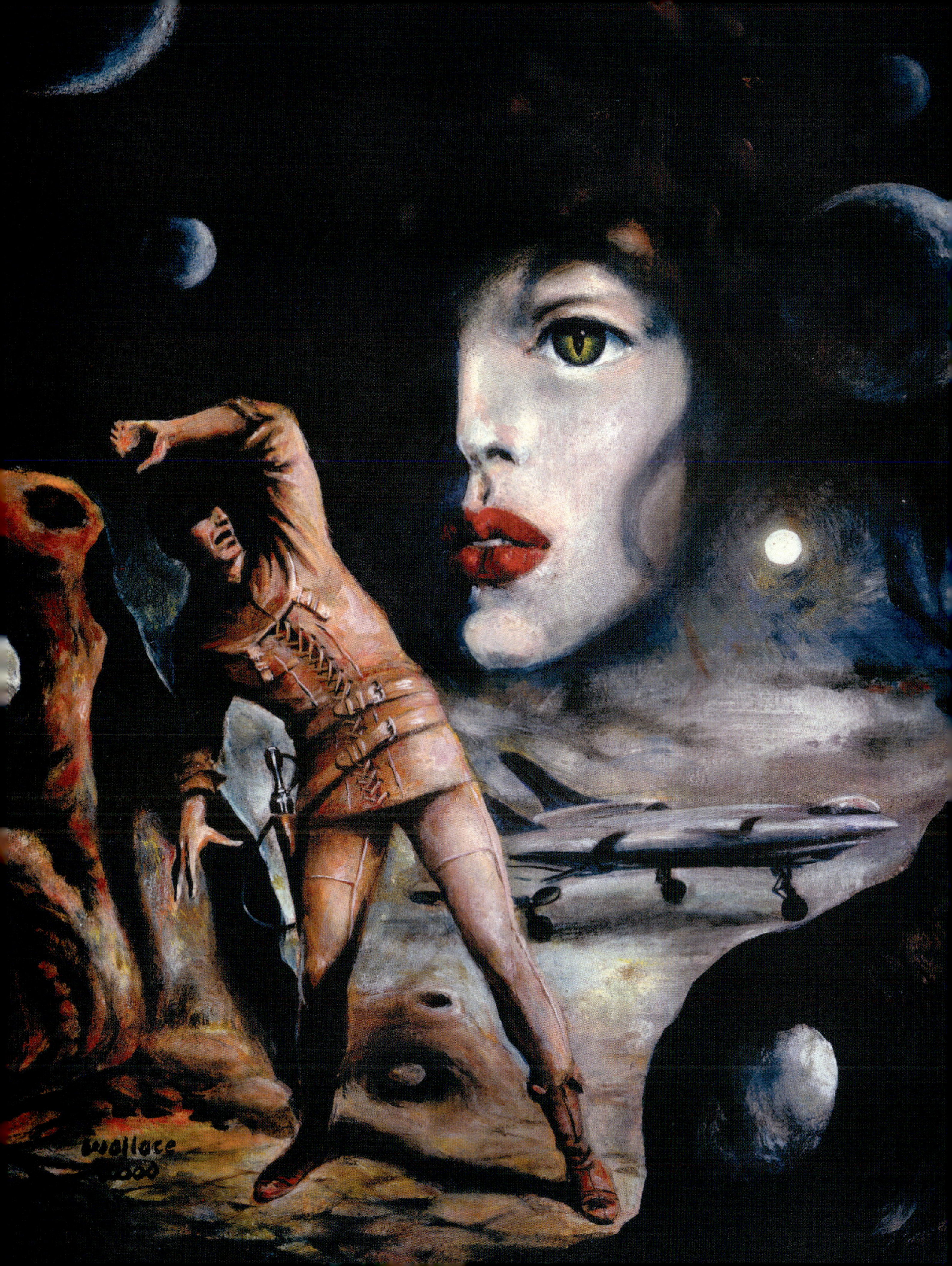
wallace

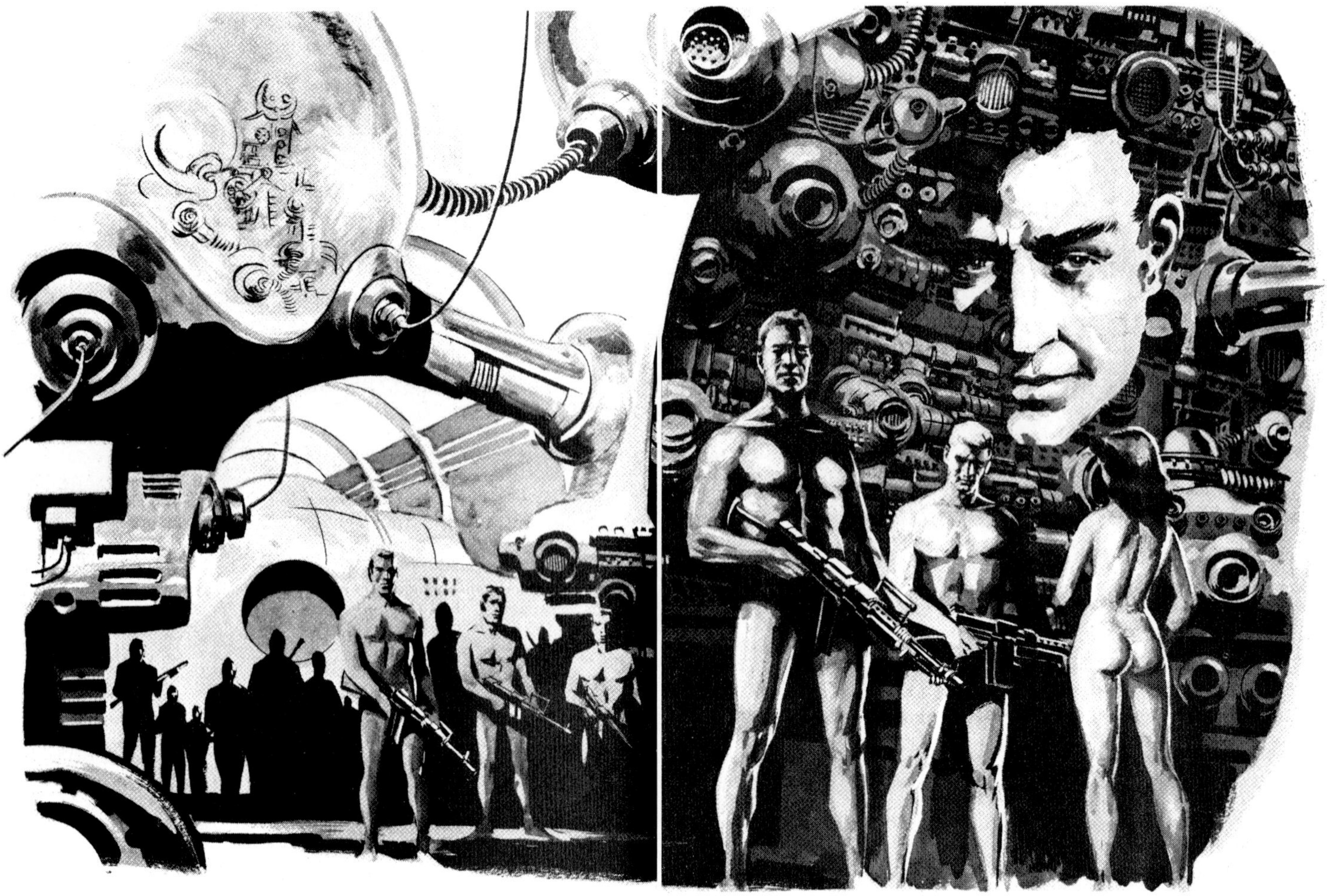

Previous spread:
Page 26: *Galaxy Science Fiction Novel* #30 1957 "Double Jeopardy" by Fletcher Pratt (Cover)

Page 27: *Galaxy Science Fiction Novel* #31 1958 "Shambleau" by C. L. Moore (Cover)

Above: *Galaxy Science Fiction* November 1957 Illustration for "Wolfbane" part 2 by Frederik Pohl and C. M. Kornbluth

Opposite: *Galaxy Science Fiction* October 1958 Illustration for "Lisbon Cubed" by William Tenn

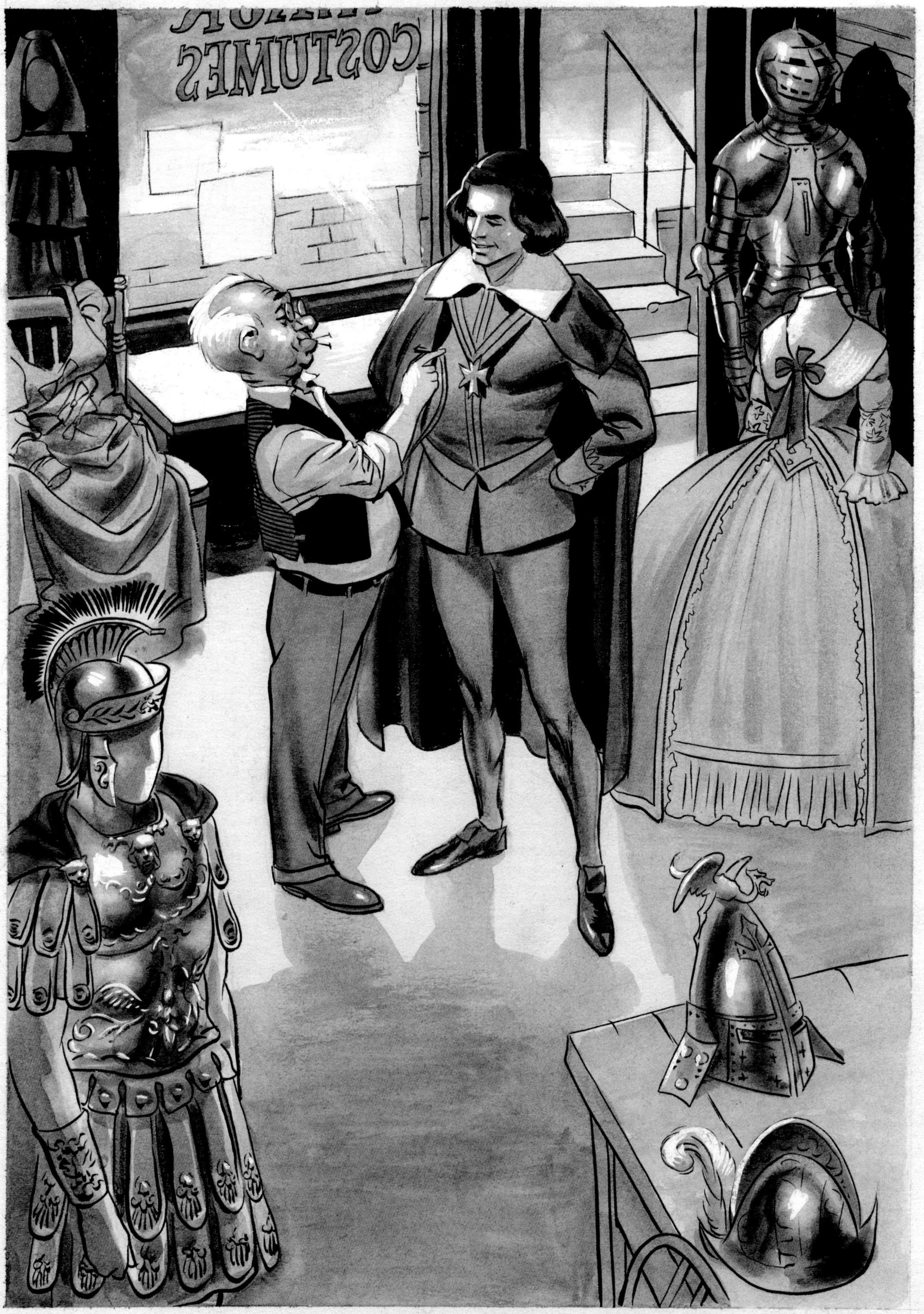
COSTUMES

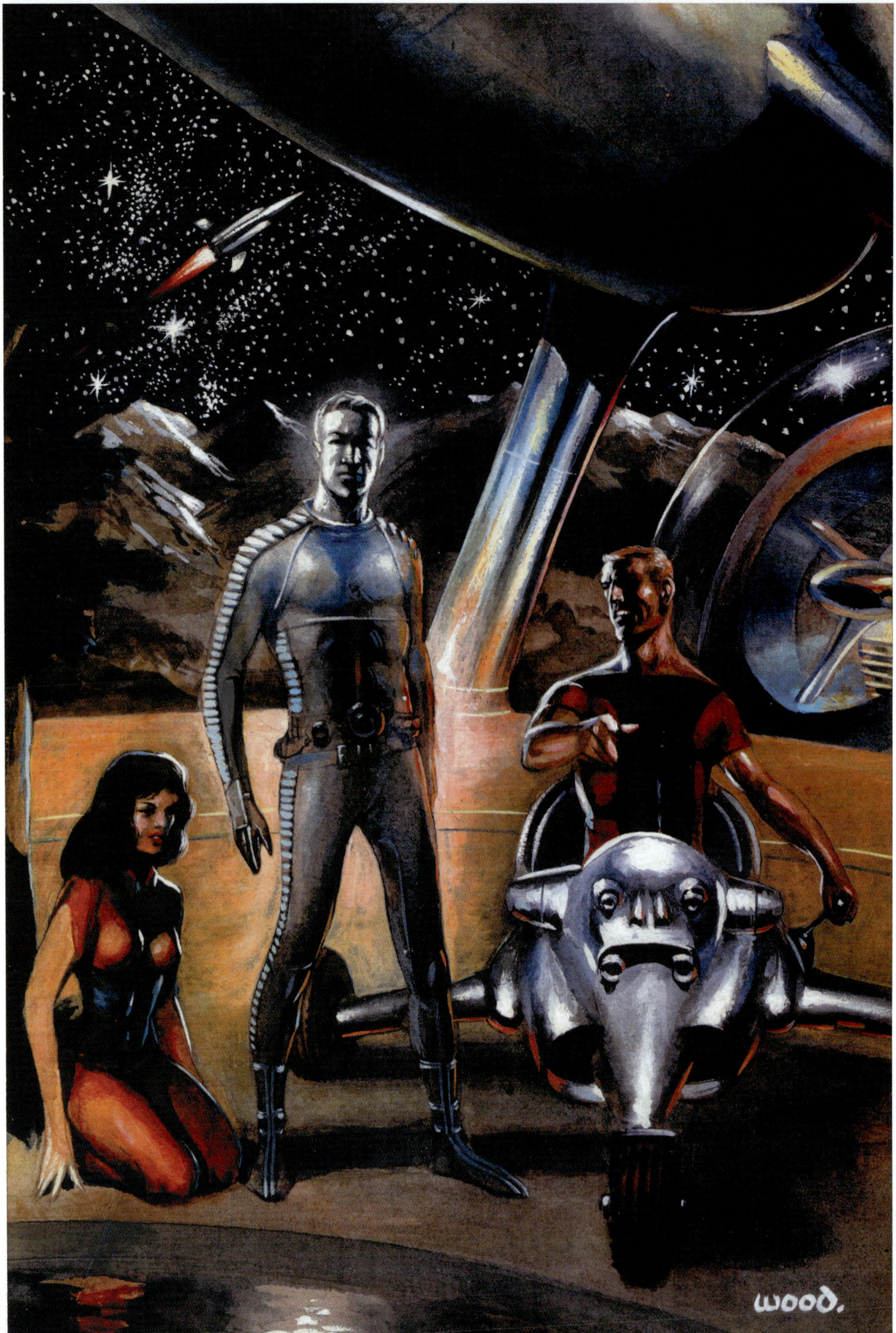

Galaxy Science Fiction Novel #32 1958 "Address Centauri" by F. L. Wallace (Cover)

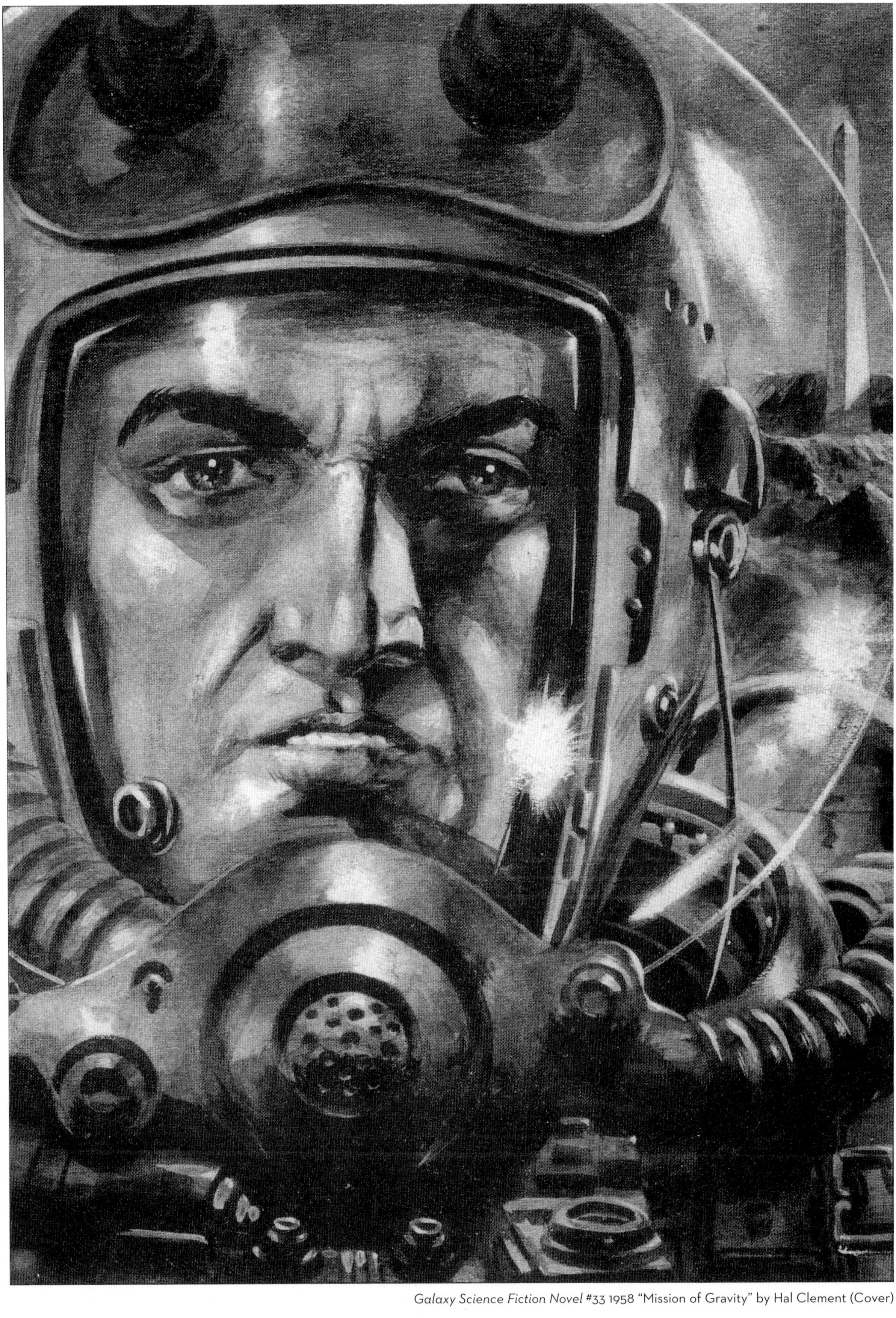

Galaxy Science Fiction Novel #33 1958 "Mission of Gravity" by Hal Clement (Cover)

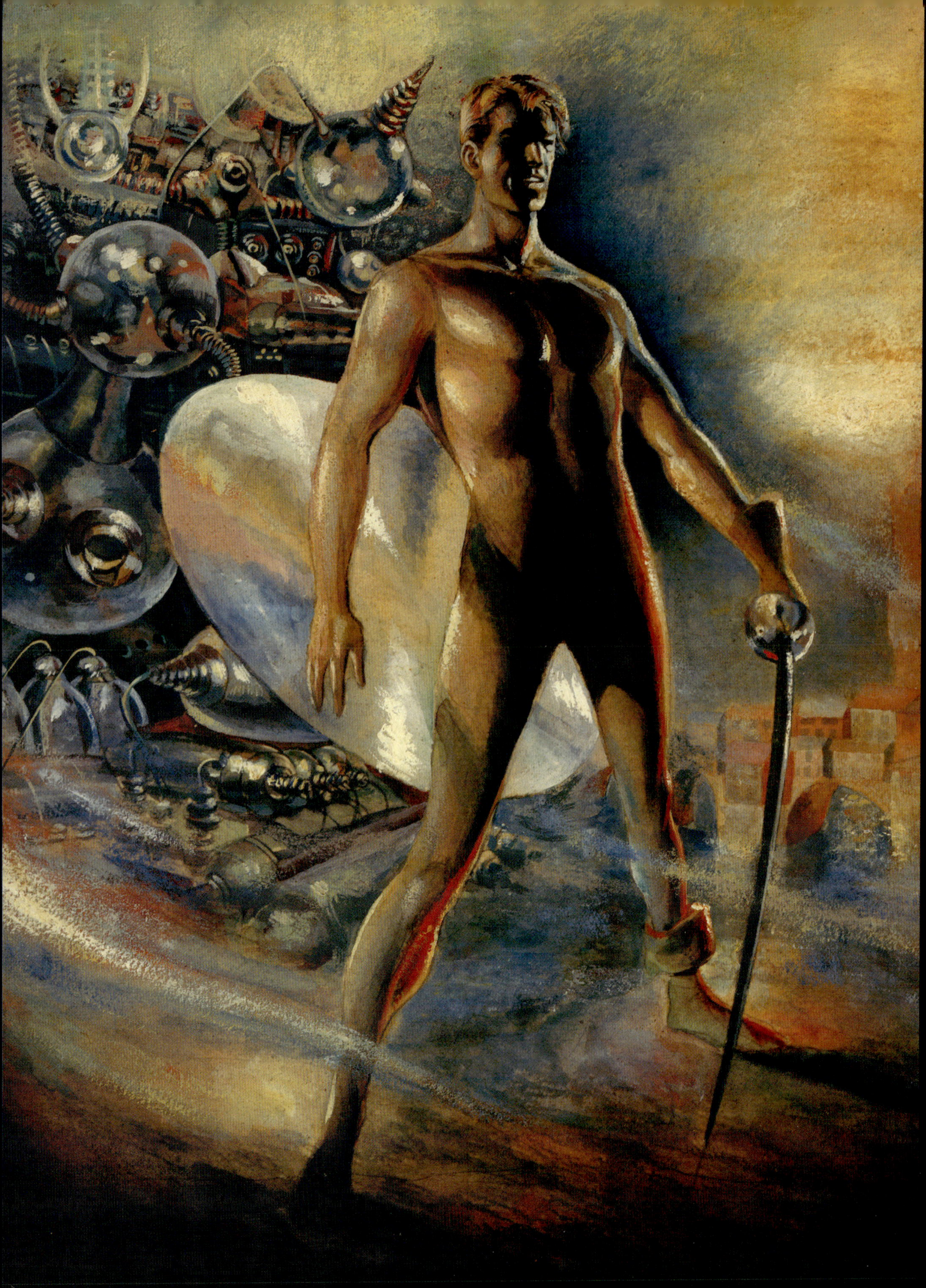

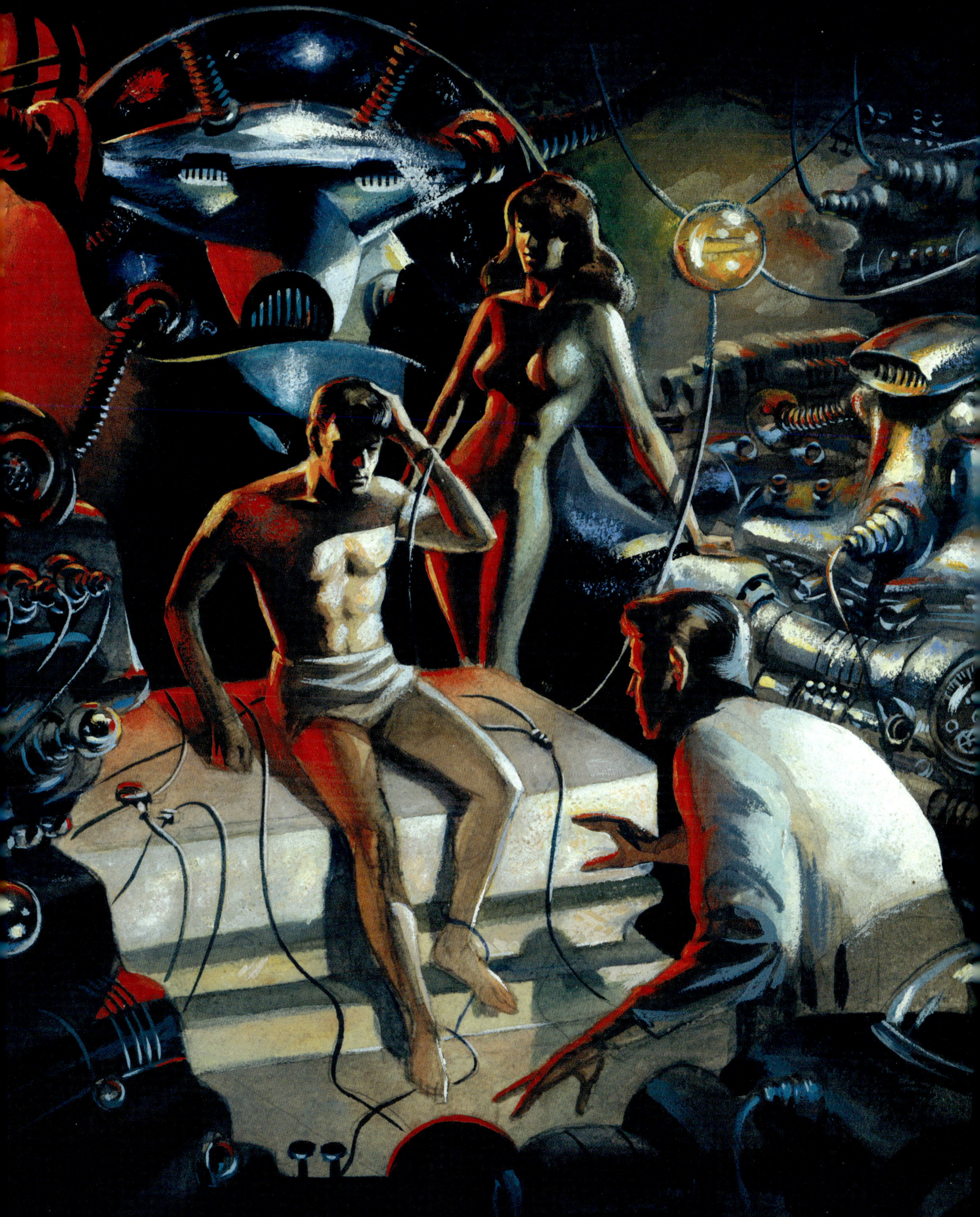

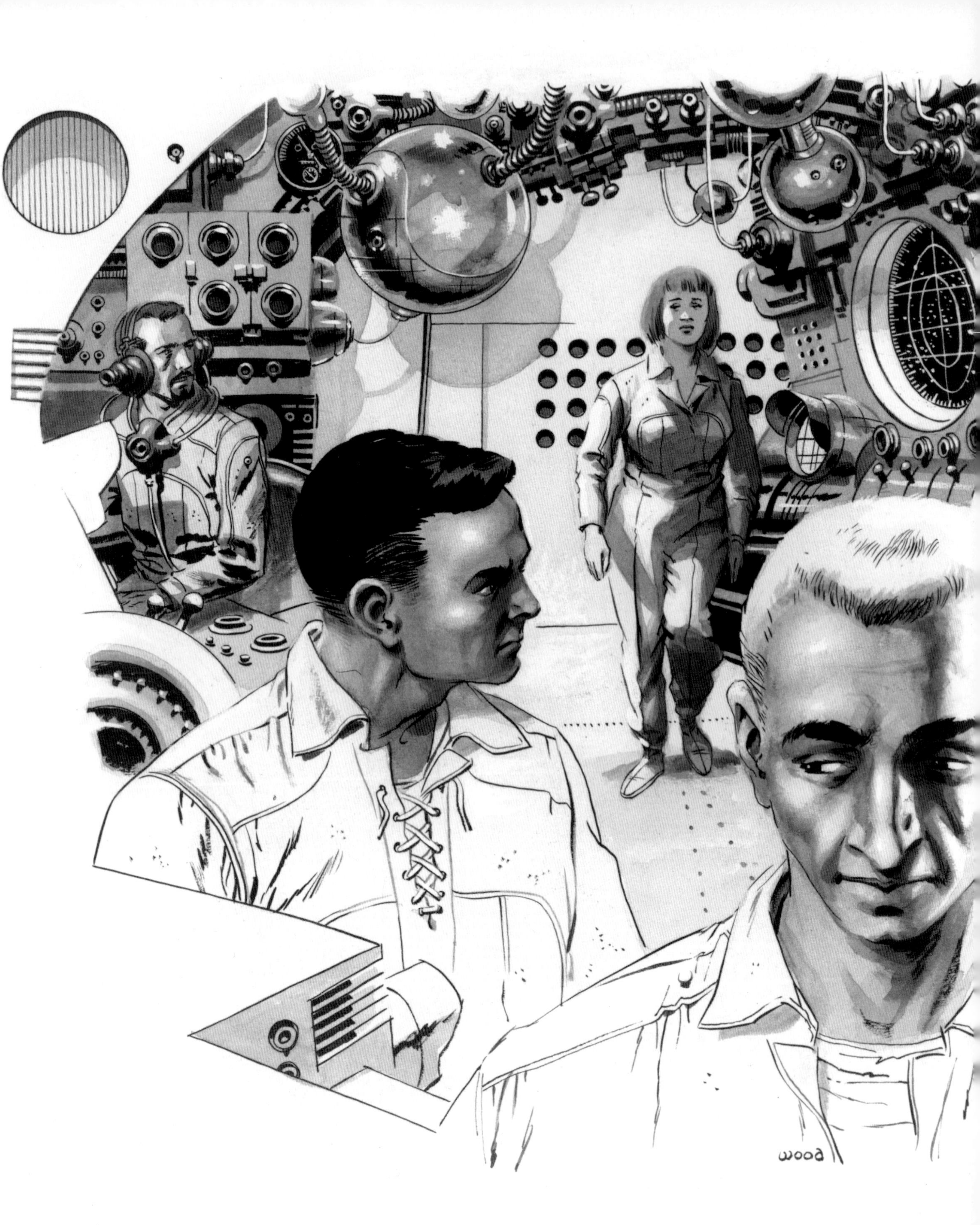

Previous spread:
Page 32: *Galaxy Science Fiction Novel* #34 1958 "Twice in Time" by Manly Wade Wellman (Cover)

Page 33: *Galaxy Science Fiction Novel* #35 1958 "The Forever Machine" by Mark Clifton and Frank Riley (Cover)

Above: *Galaxy Science Fiction* September 1957 Illustration for "The Pod in the Barrier" by Theodore Sturgeon

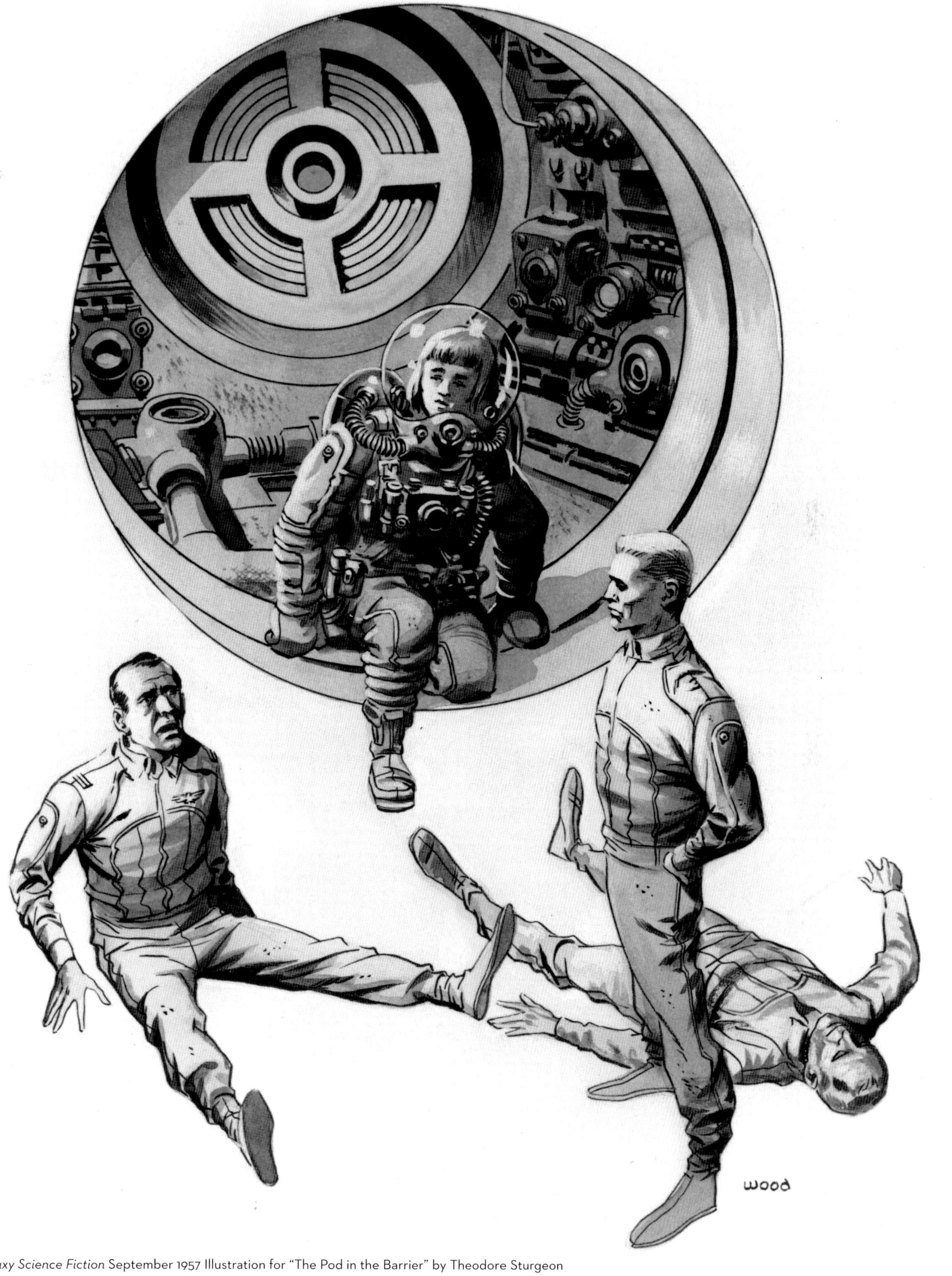

Galaxy Science Fiction September 1957 Illustration for "The Pod in the Barrier" by Theodore Sturgeon

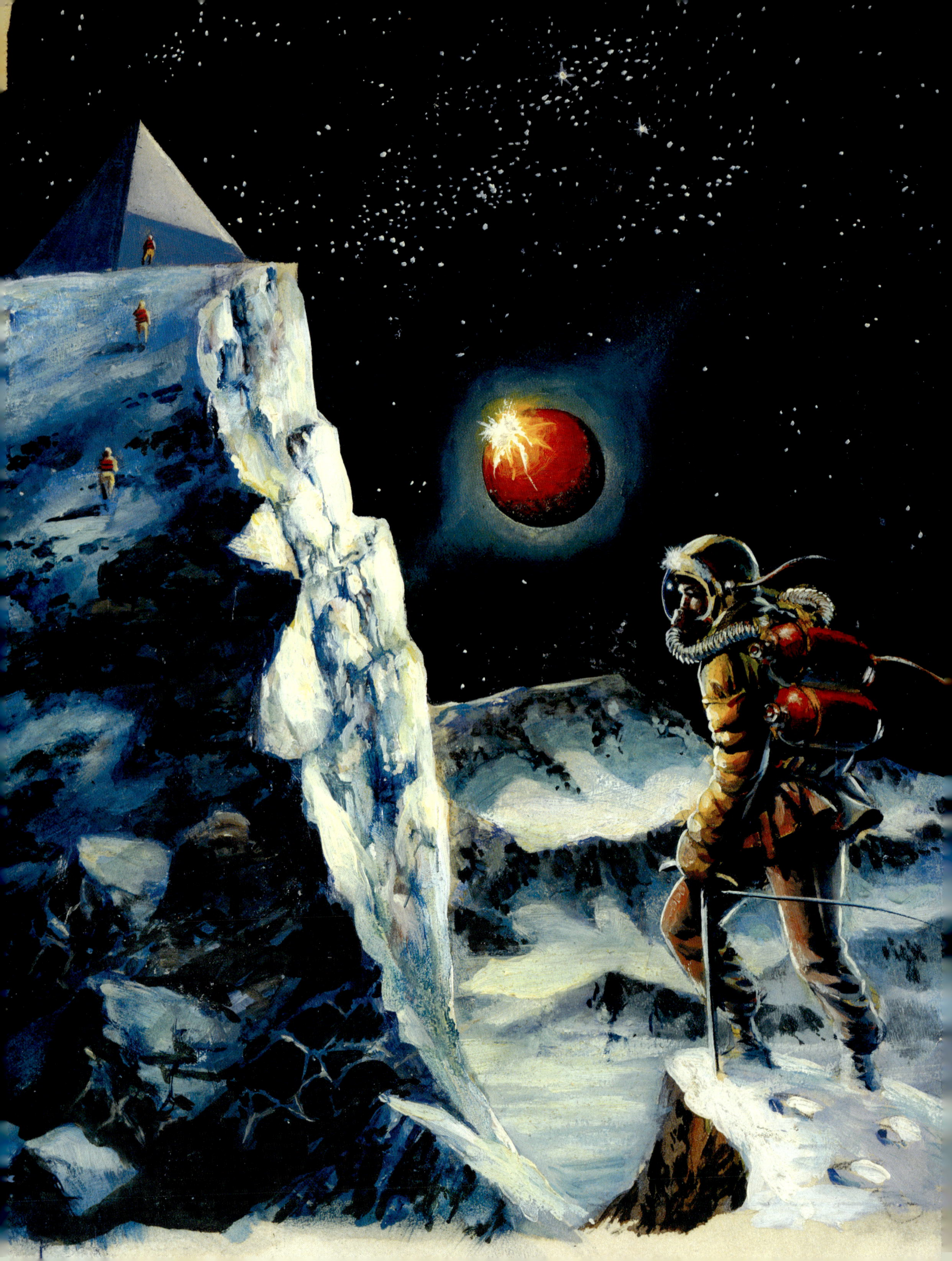

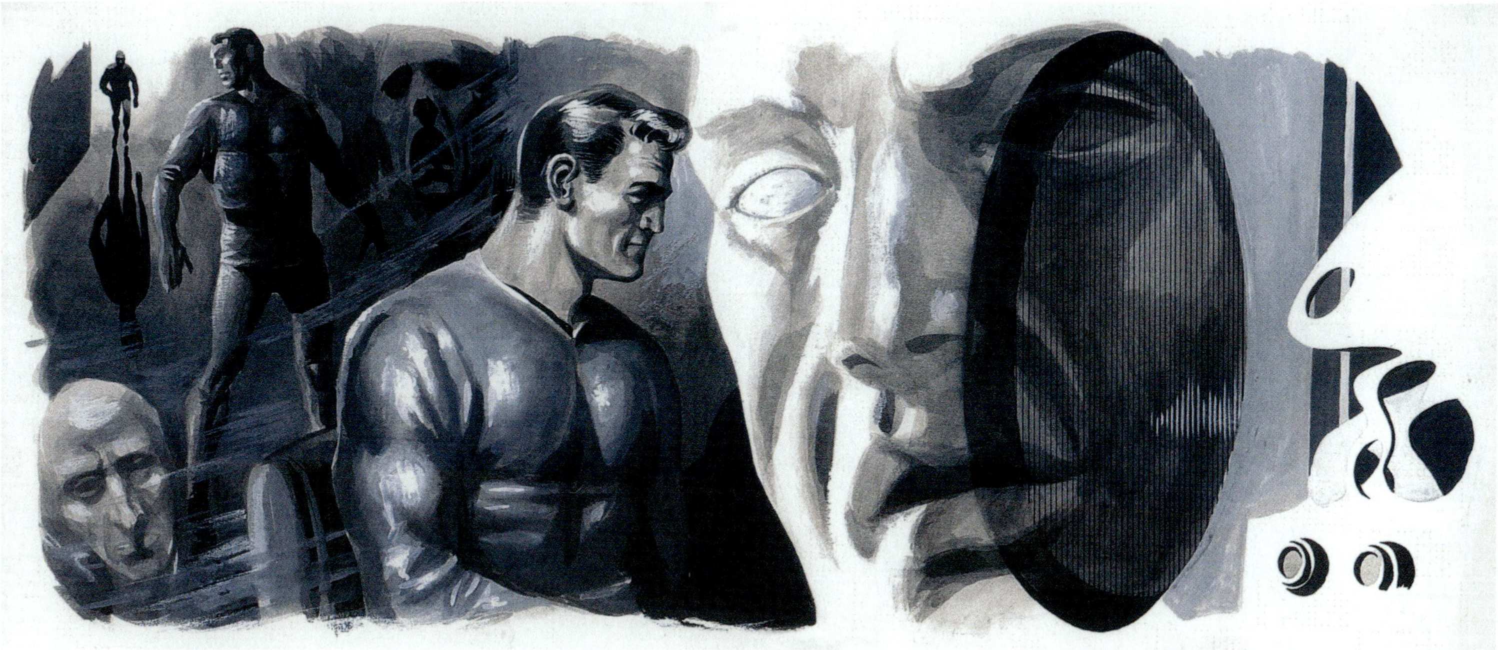

Top: *Galaxy Science Fiction* November 1958
Illustration for "Time Killer" part 2 by Robert Sheckley

Bottom: *Galaxy Science Fiction* November 1957
Illustration for "Wolfbane" part 2 by Frederik Pohl and C. M. Kornbluth

Opposite: *Galaxy Science Fiction* October 1957 (Cover)
Illustration for "Wolfbane" by Frederik Pohl and C. M. Kornbluth

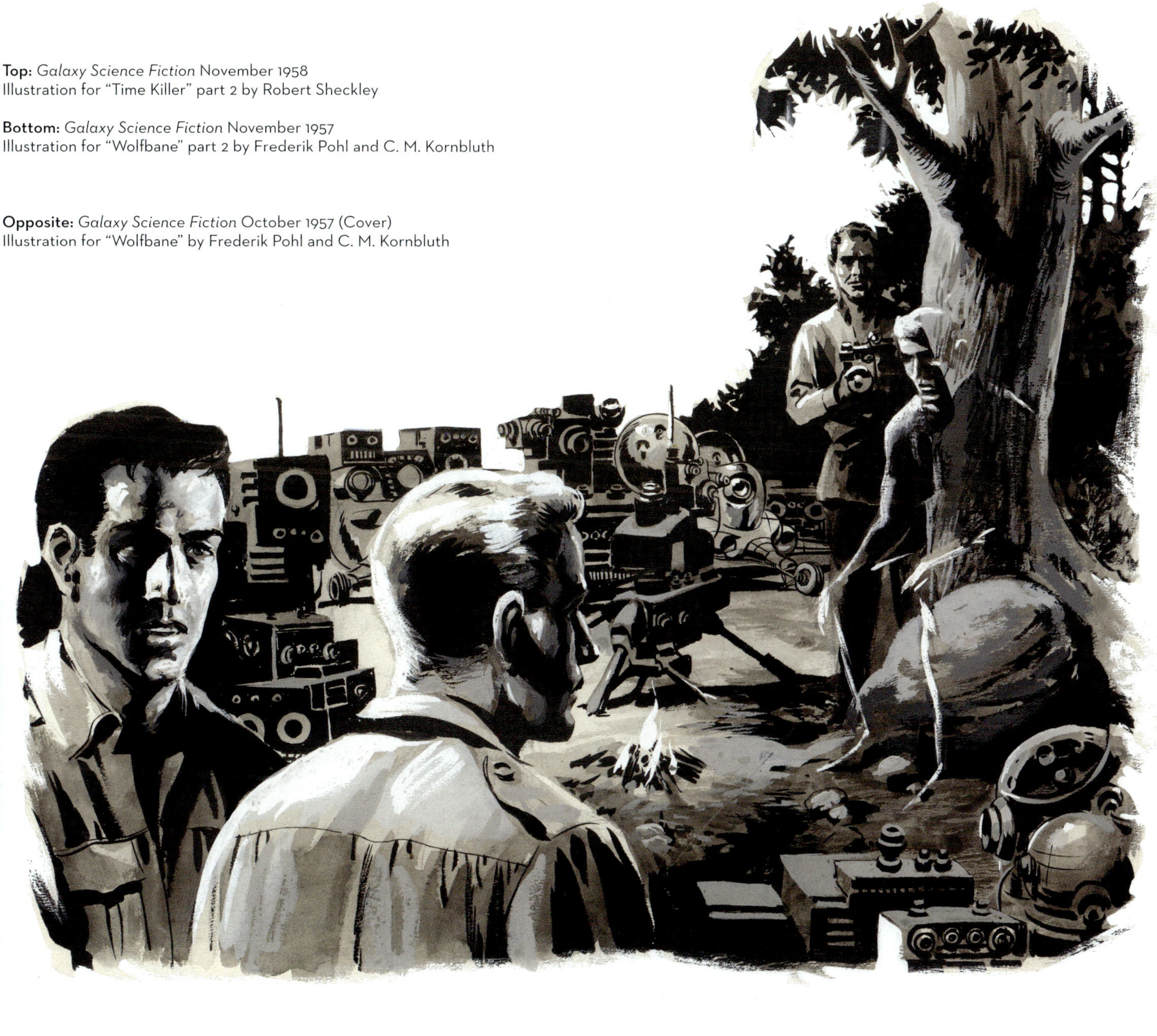

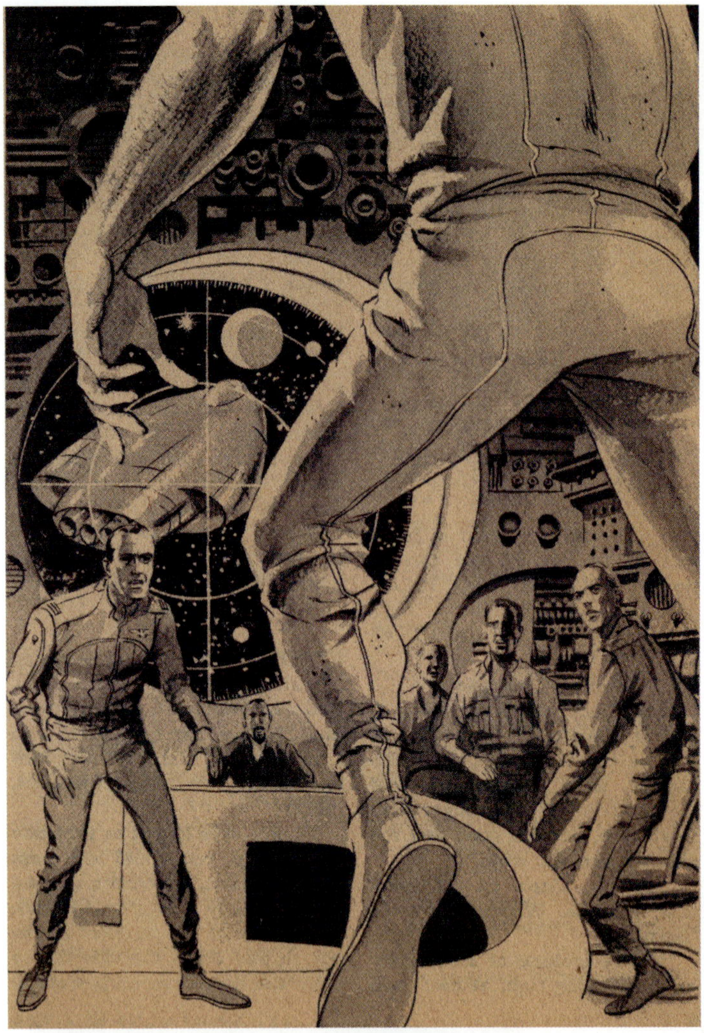

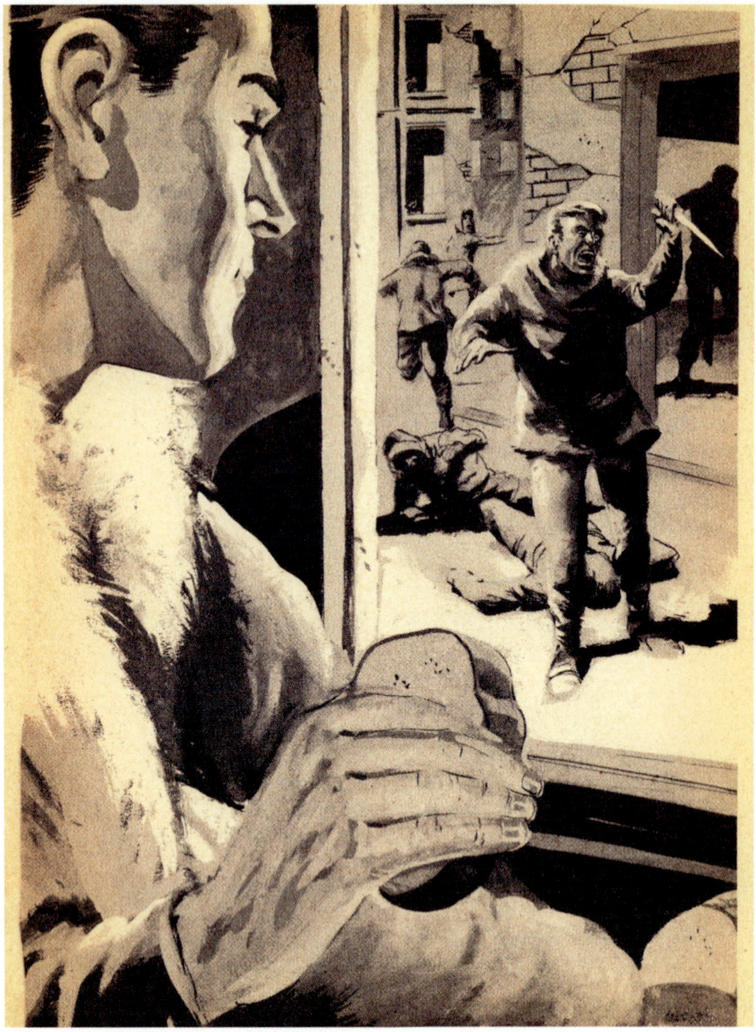

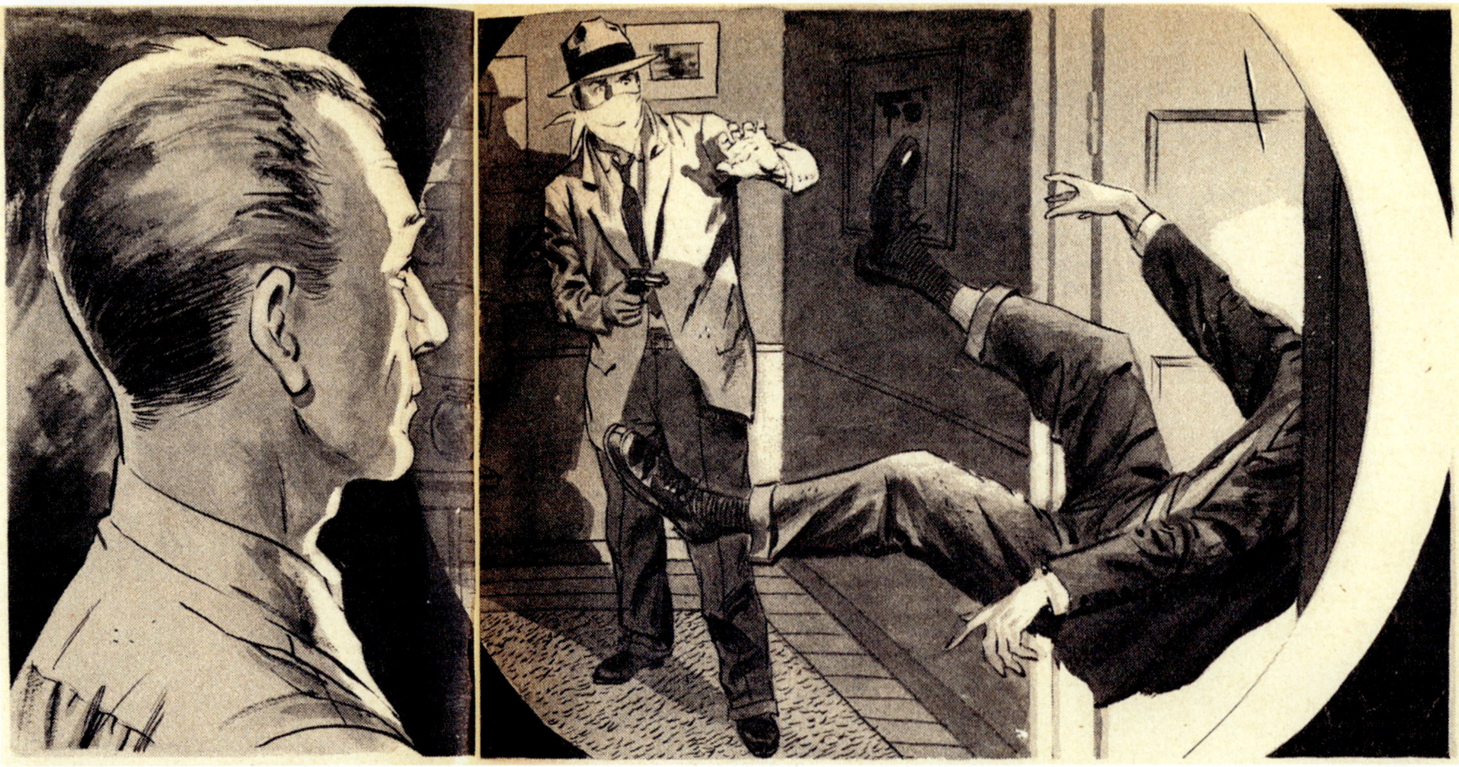

Clockwise from upper left:
Galaxy Science Fiction September 1957 Illustration for "The Pod in the Barrier" by Theodore Sturgeon

Galaxy Science Fiction October 1957 Illustration for "Wolfbane" part 1 by Frederik Pohl and C. M. Kornbluth

Galaxy Science Fiction February 1958 Illustration for "The Rule of the Door" by Lloyd Biggle Jr.

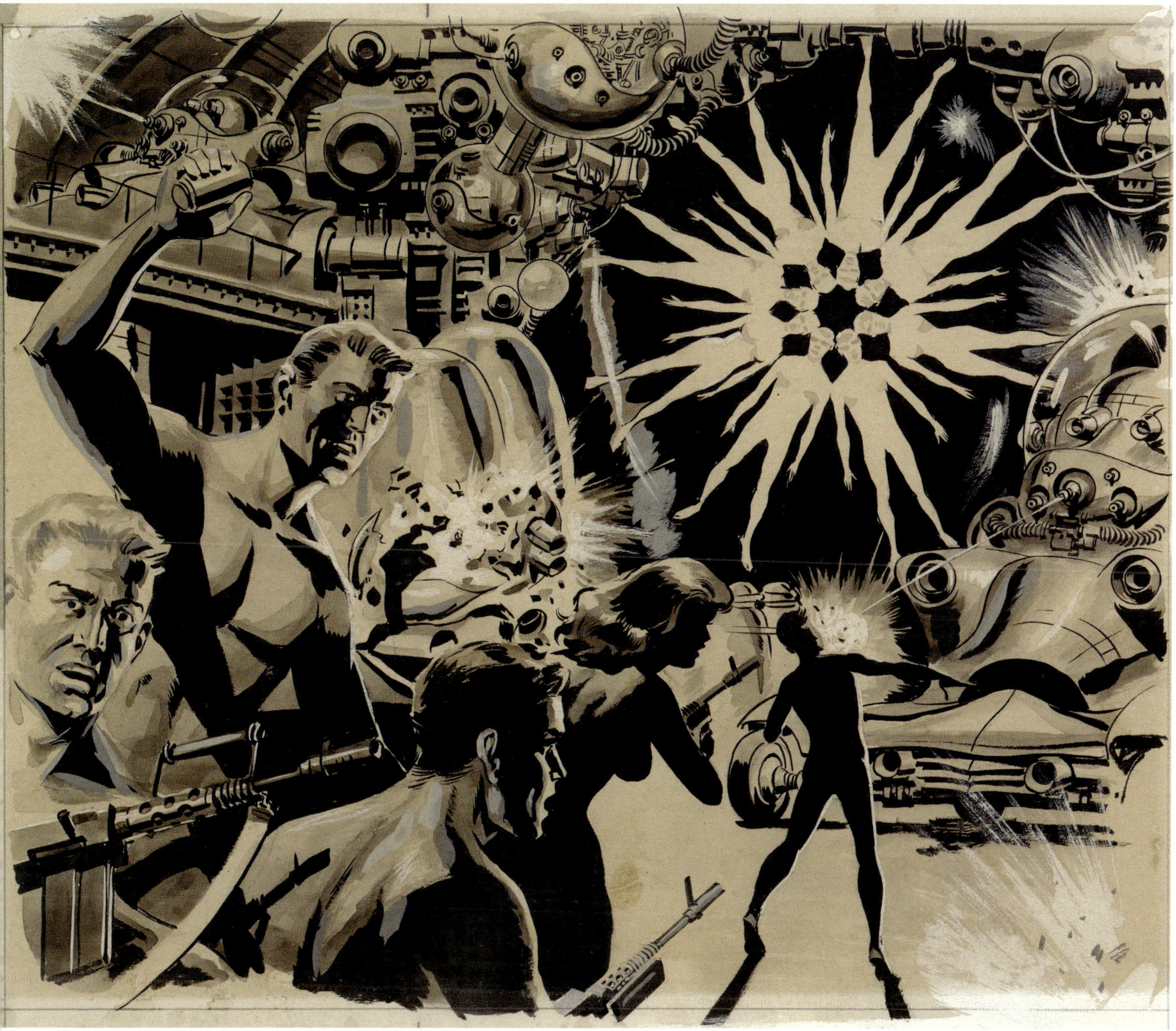

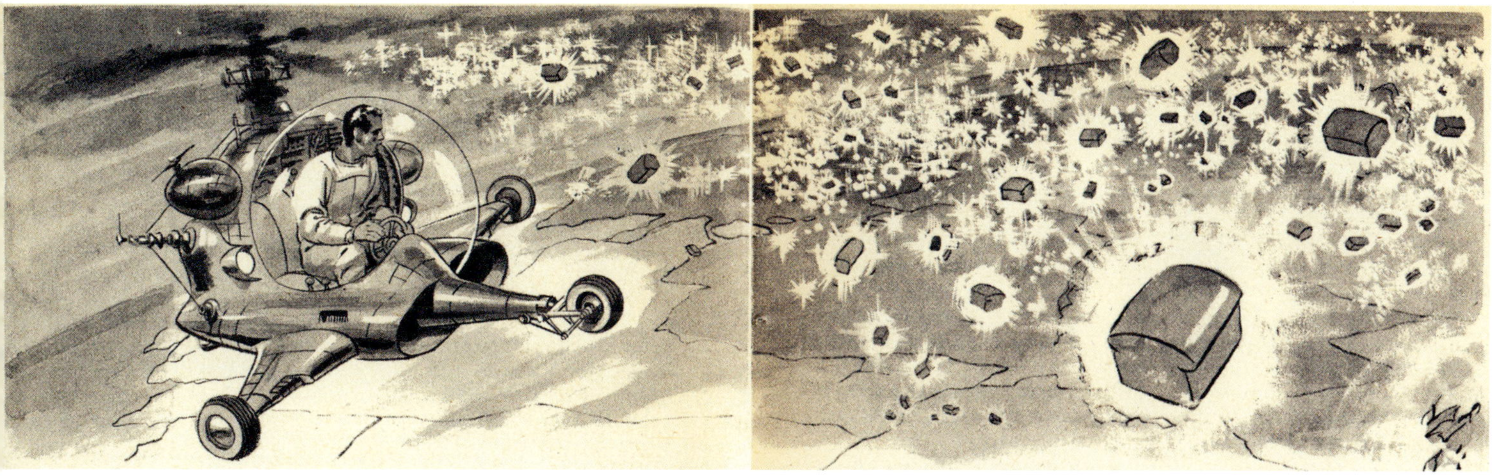

Top: *Galaxy Science Fiction* November 1957 Illustration for "Wolfbane" part 2 by Frederik Pohl and C. M. Kornbluth

Bottom: *Galaxy Science Fiction* February 1958 Illustration for "Bread Overhead" by Fritz Leiber

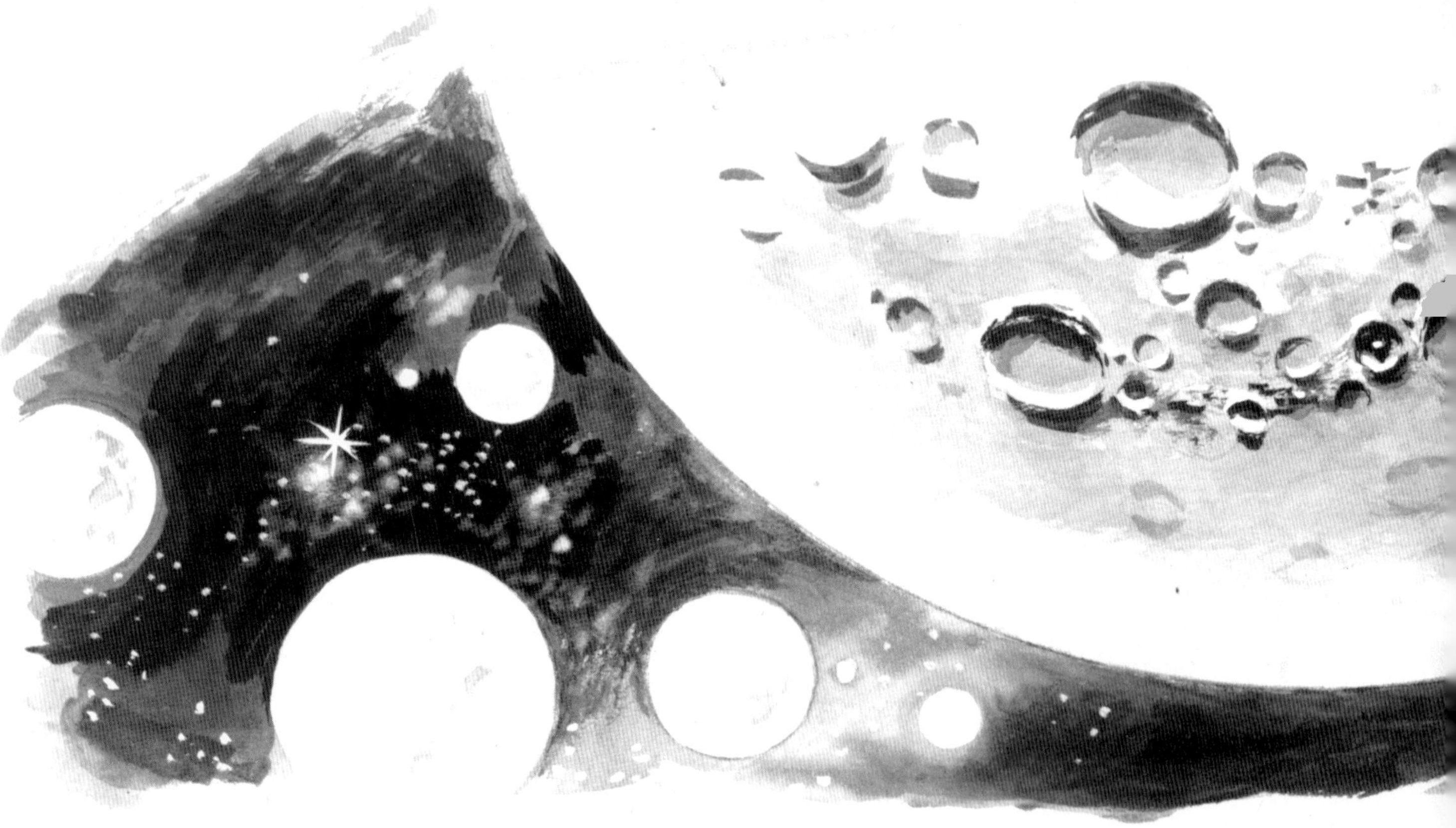

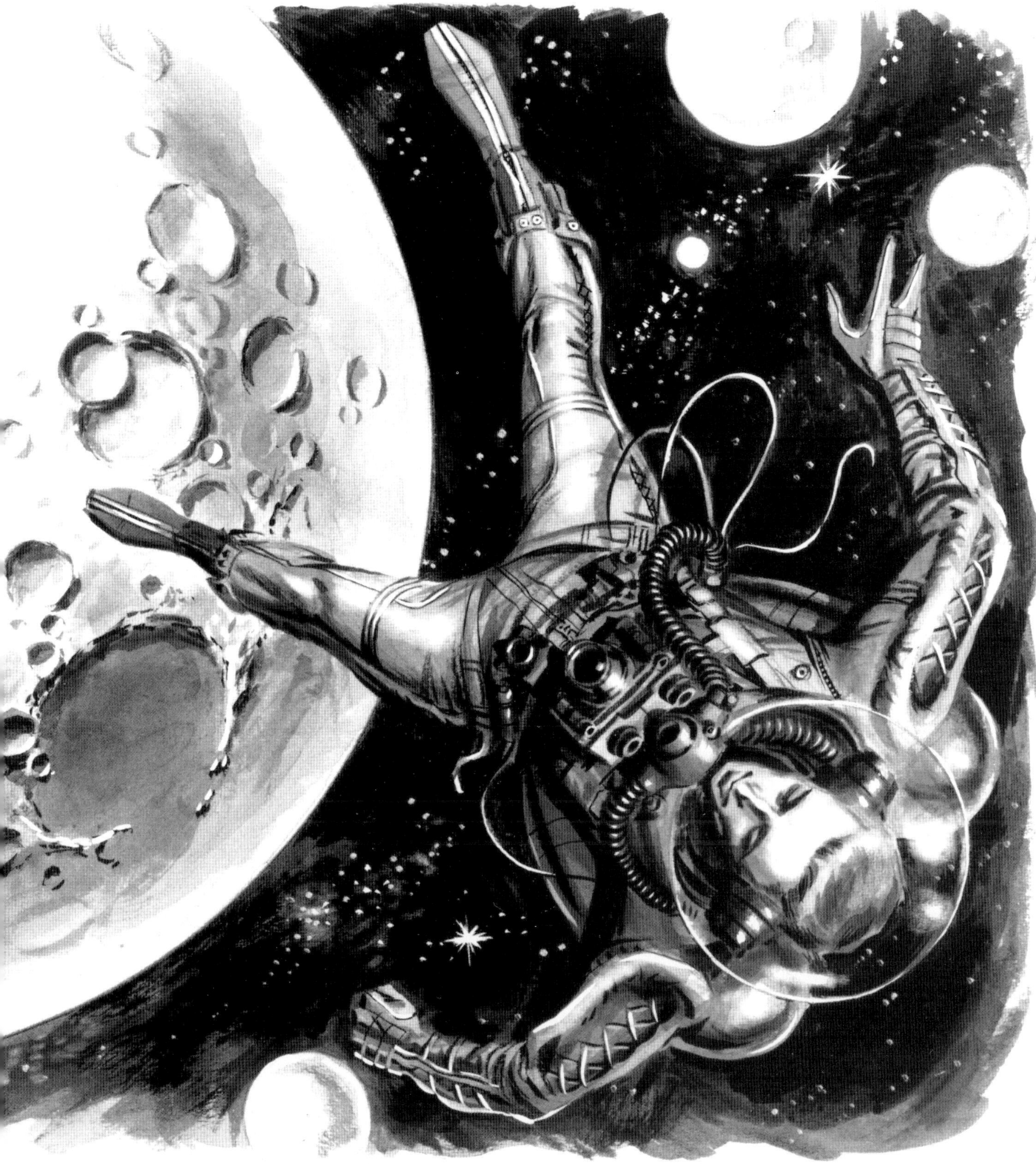

Galaxy Science Fiction December 1957 Illustration for “Payload” by Raymond F. Banks

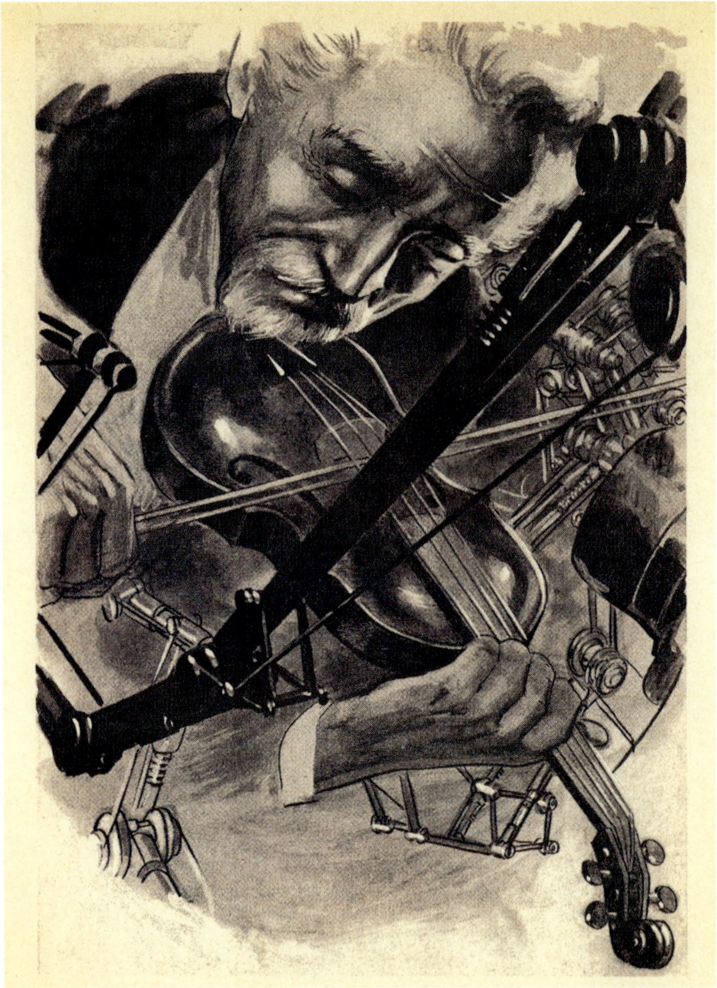

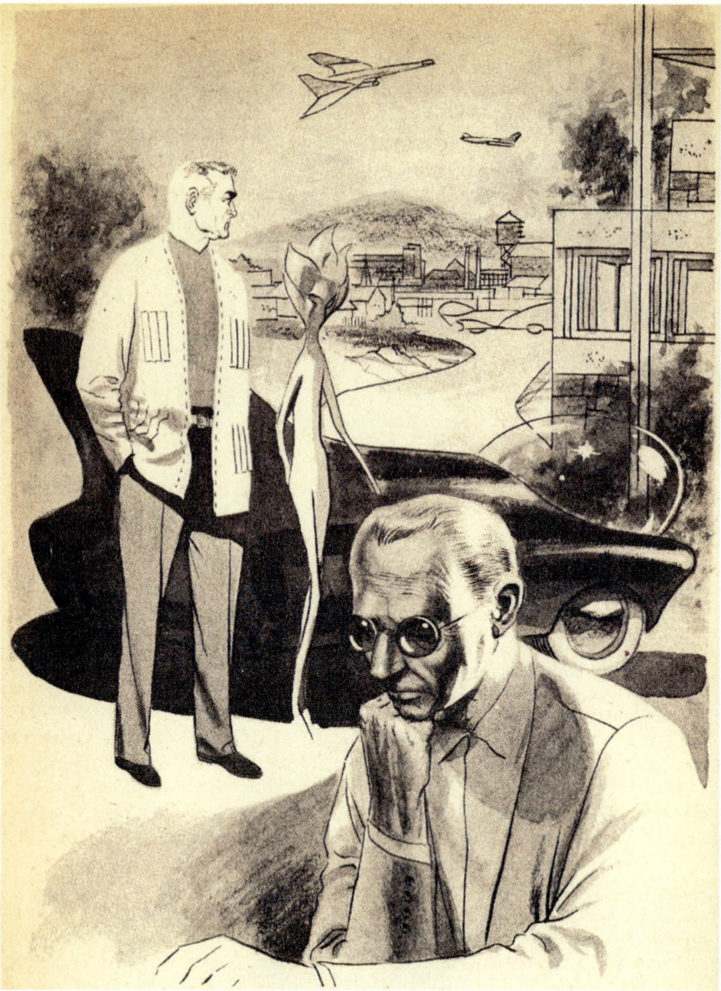

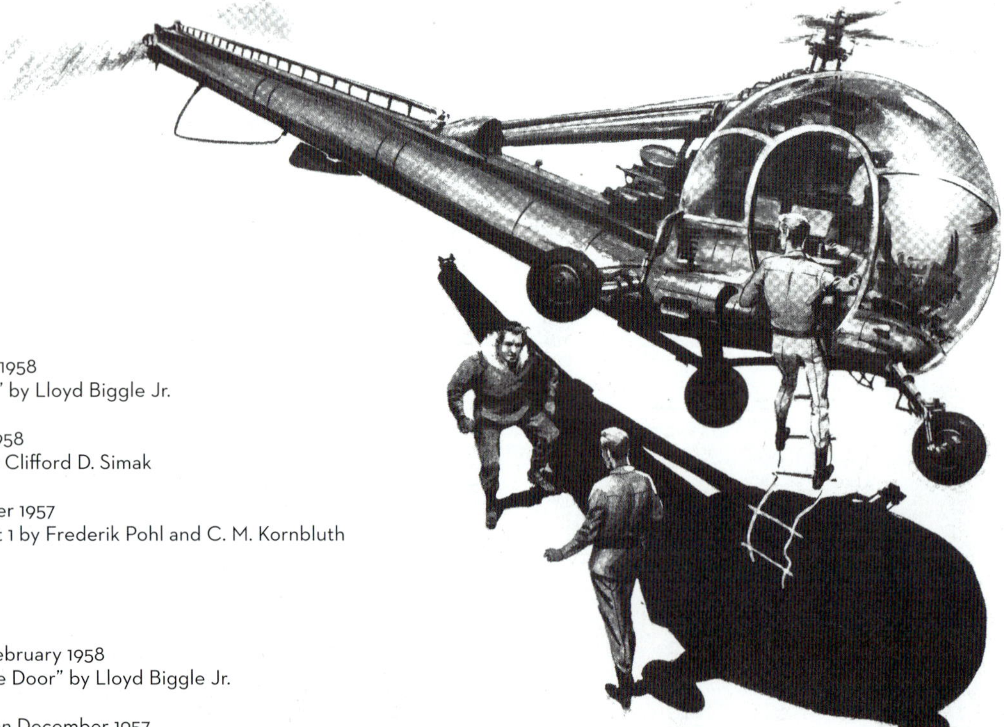

Clockwise from upper left:
Galaxy Science Fiction March 1958
Illustration for "Spare the Rod" by Lloyd Biggle Jr.

Galaxy Science Fiction April 1958
Illustration for "The Sitters" by Clifford D. Simak

Galaxy Science Fiction October 1957
Illustration for "Wolfbane" part 1 by Frederik Pohl and C. M. Kornbluth

Opposite:
Top: *Galaxy Science Fiction* February 1958
Illustration for "The Rule of the Door" by Lloyd Biggle Jr.

Bottom: *Galaxy Science Fiction* December 1957
Illustration for "Payload" by Raymond F. Banks

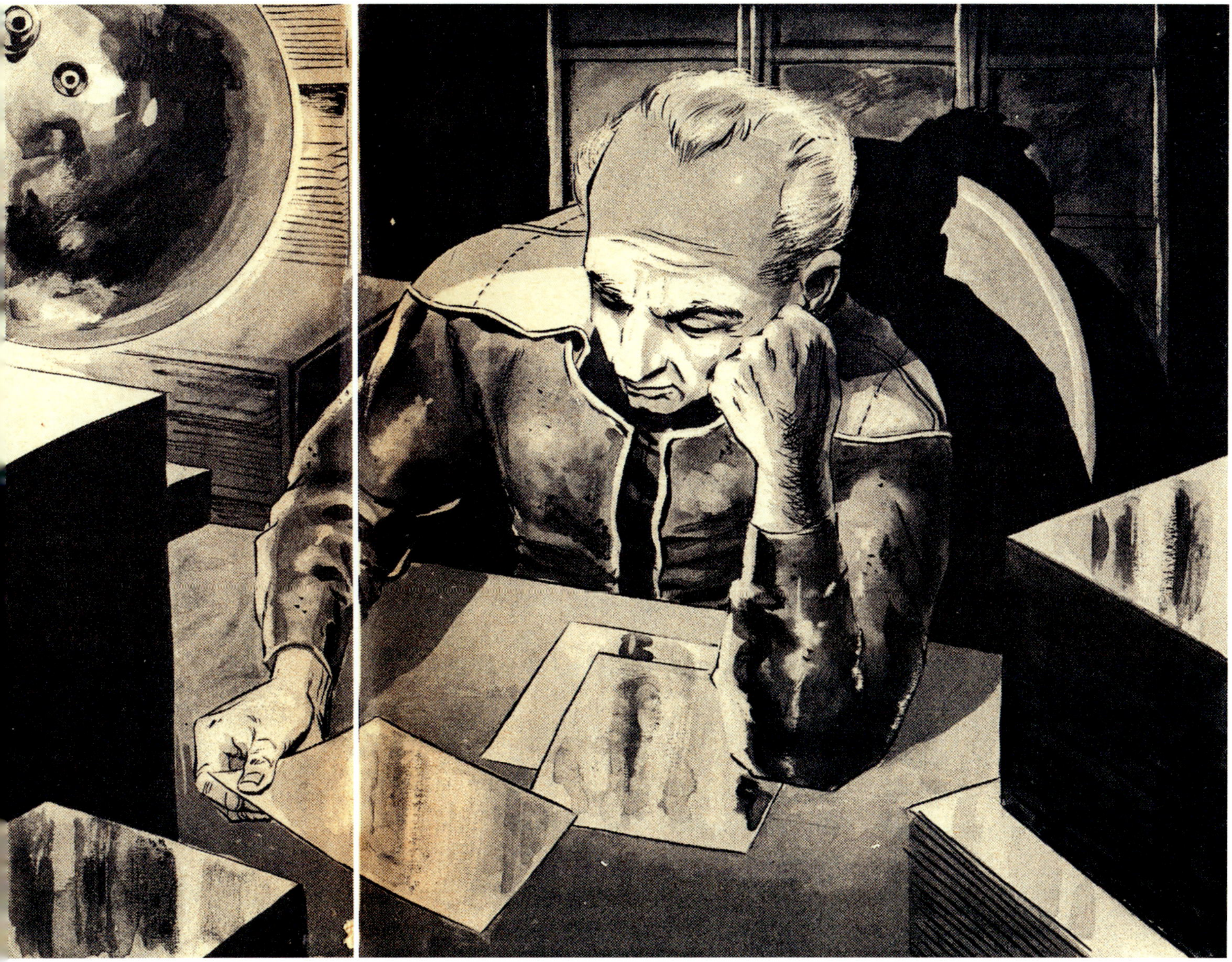

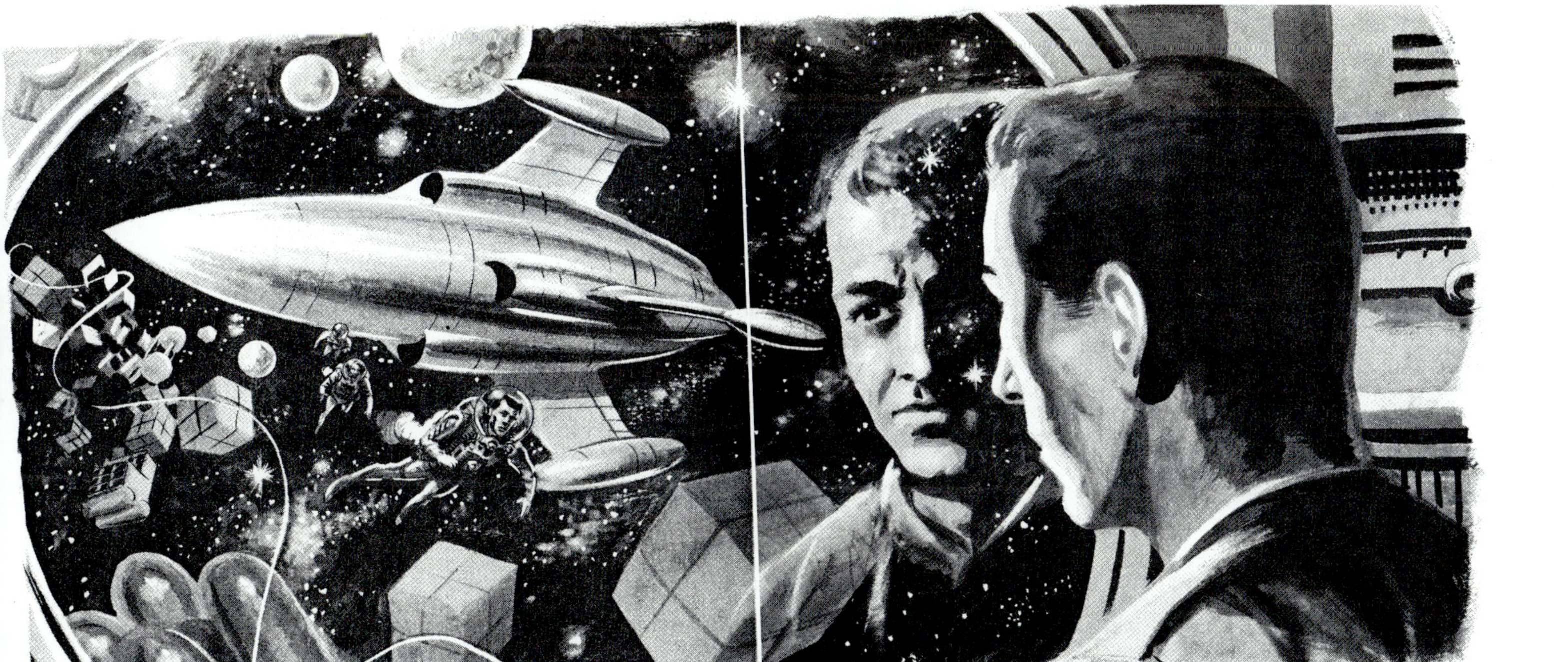

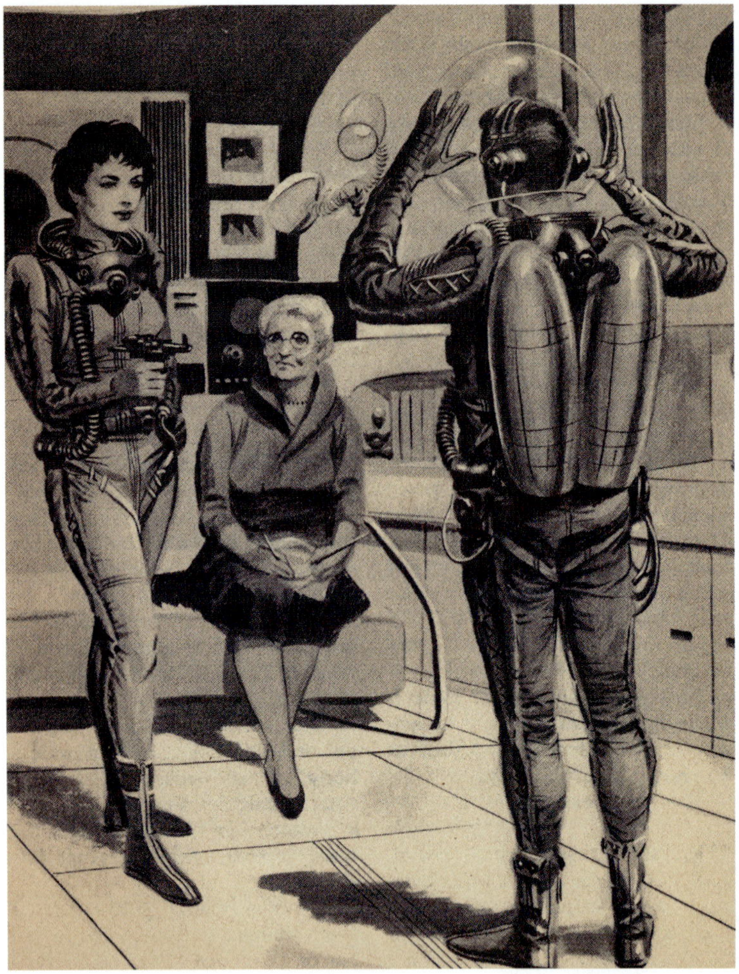

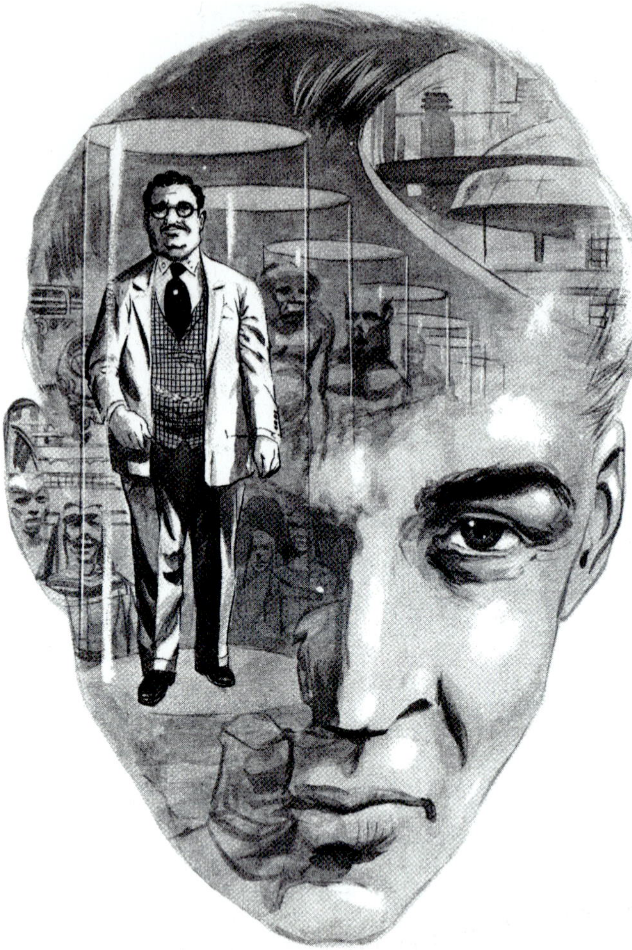

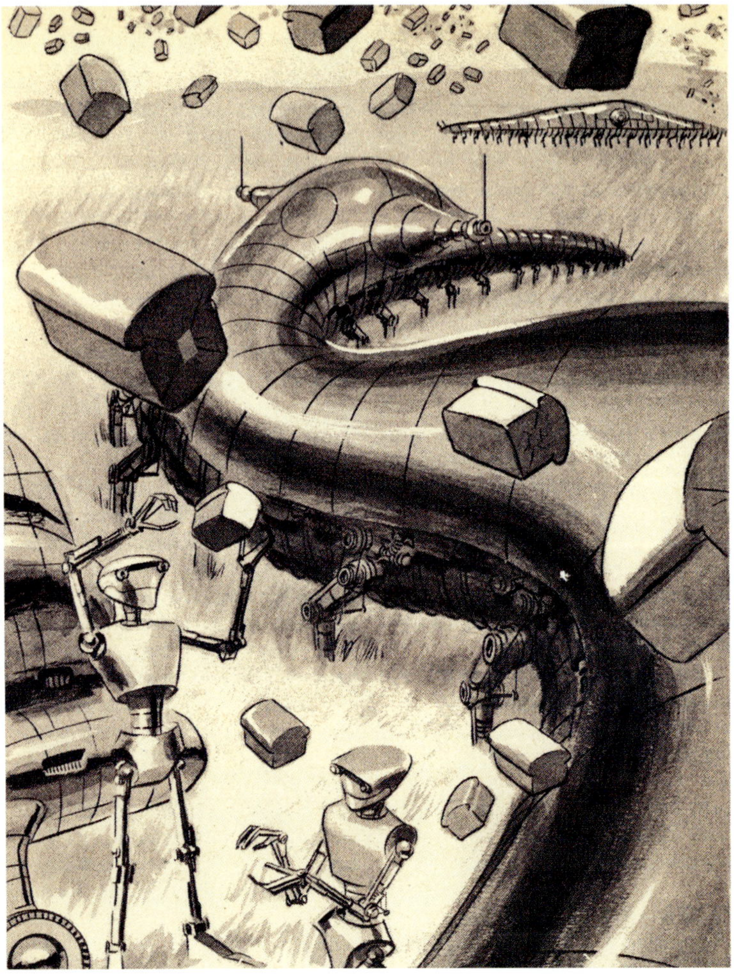

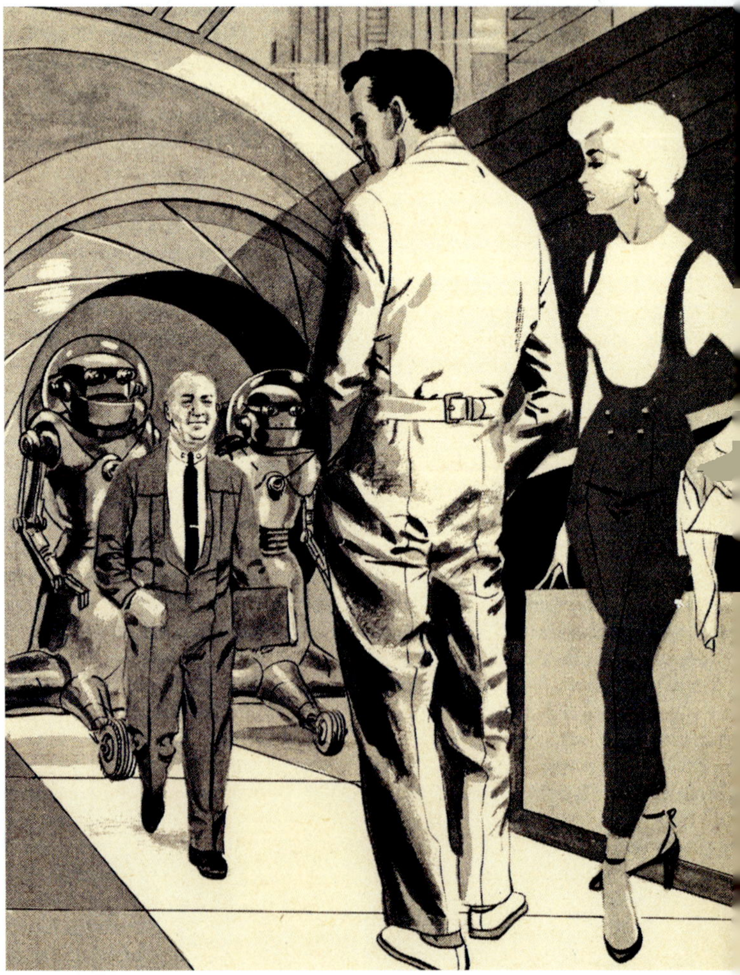

op:
alaxy Science Fiction October 1957
lustration for "Wolfbane" part 1 by Frederik Pohl
nd C. M. Kornbluth

ottom:
alaxy Science Fiction November 1958
lustration for "Birds of a Feather" by Robert Silverberg

)pposite:
:lockwise from upper left:
alaxy Science Fiction December 1957
lustration for "Payload" by Raymond F. Banks

alaxy Science Fiction February 1958
lustration for "The Rule of the Door" by Lloyd Biggle Jr.

oth bottom illustrations:
alaxy Science Fiction February 1958
lustration for "Bread Overhead" by Fritz Leiber

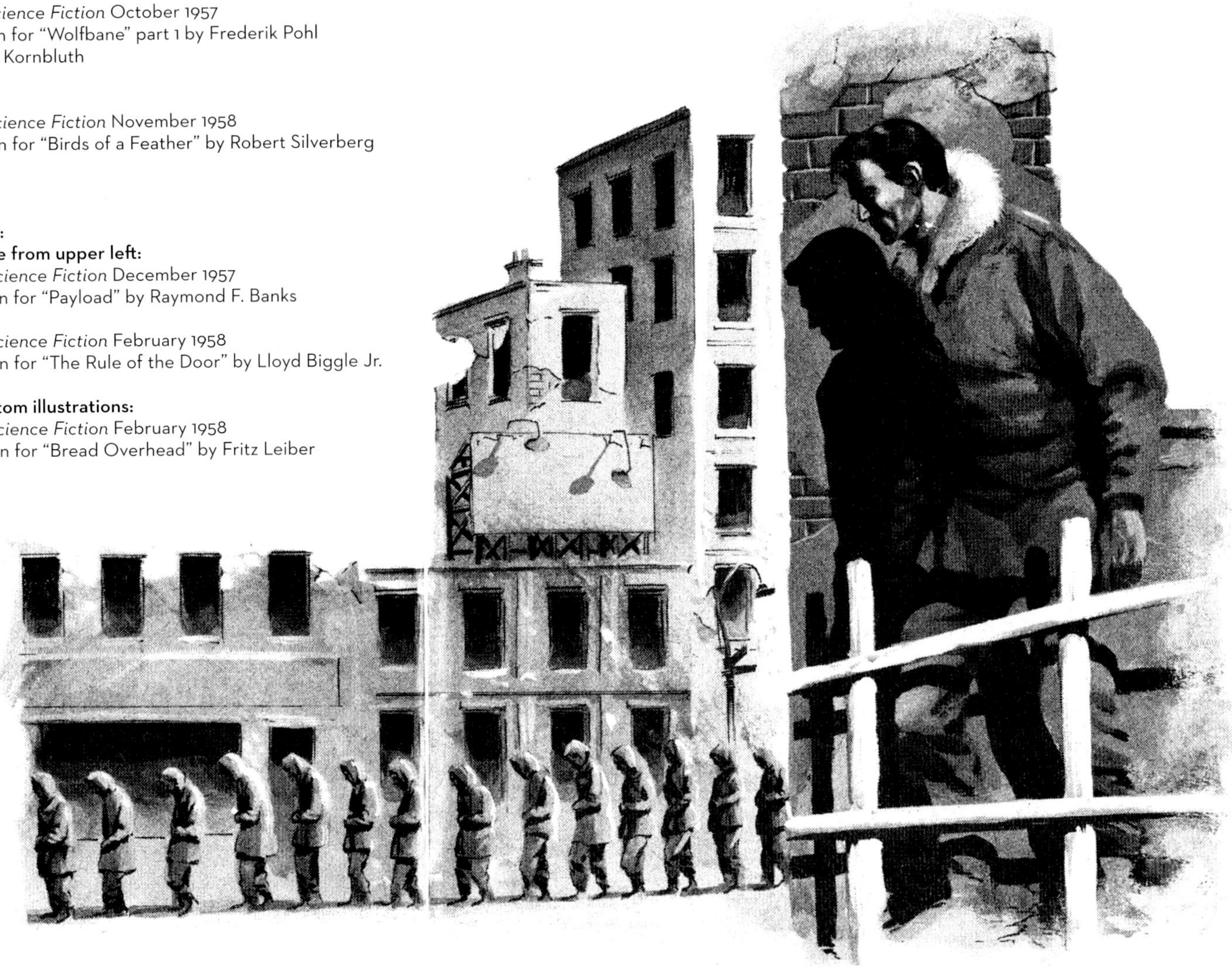

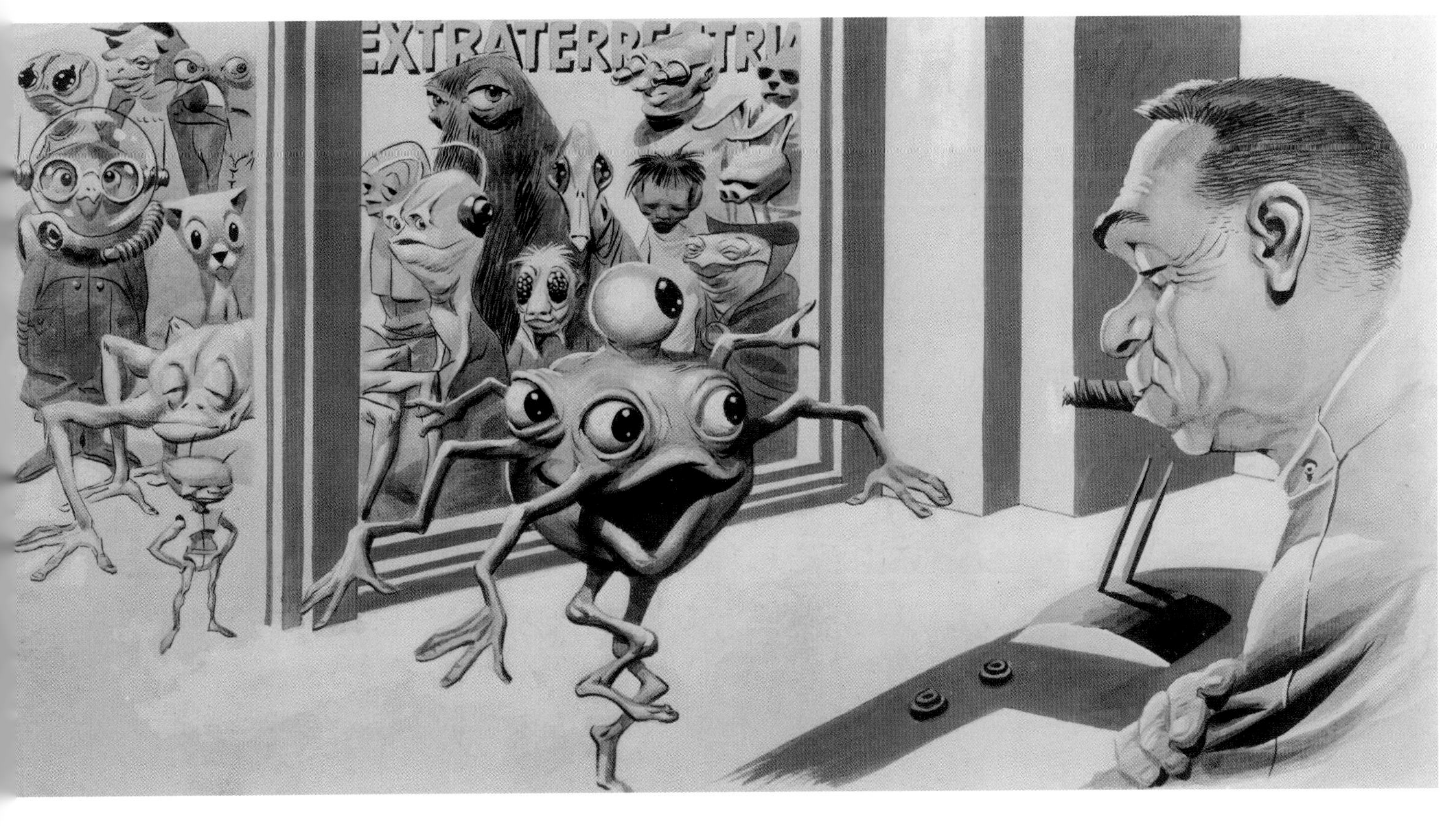

Clockwise from upper right:
Galaxy Science Fiction March 1958
Illustration for "Spare the Rod" by Lloyd Biggle Jr.

Galaxy Science Fiction April 1958
Illustration for "The Sitters" by Clifford D. Simak

Galaxy Science Fiction June 1958
Illustration for "Mars by Moonlight" by Paul Flehr

Opposite:
Galaxy Science Fiction March 1958
Illustration for "My Fair Planet" by Evelyn E. Smith

Next spread:
Page 48, clockwise from upper left:
Galaxy Science Fiction April 1958
Illustration for "First Man" by Clyde Brown

Galaxy Science Fiction May 1958
Illustration for "The Iron Chancellor" by Robert Silverberg

Galaxy Science Fiction July 1958
Illustration for "Innocent at Large" by Poul and Karen Anderson

Galaxy Science Fiction December 1962
Illustration for "The Creature from the Cleveland Depths" by Fritz Leiber

Page 49:
Galaxy Science Fiction April 1958
Illustration for "The Sitters" by Clifford D. Simak

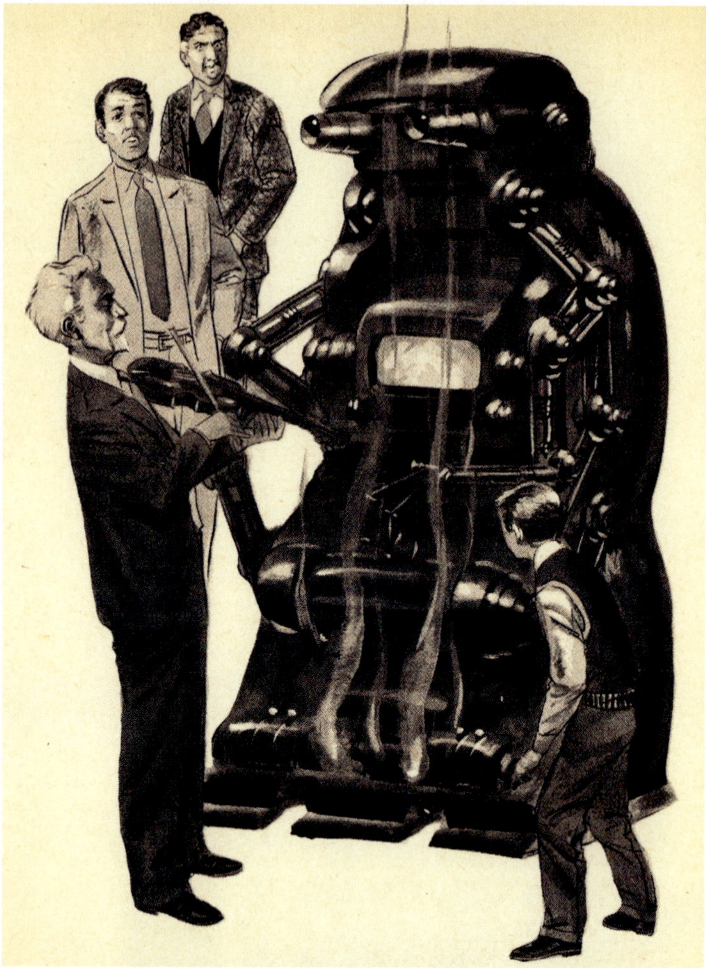

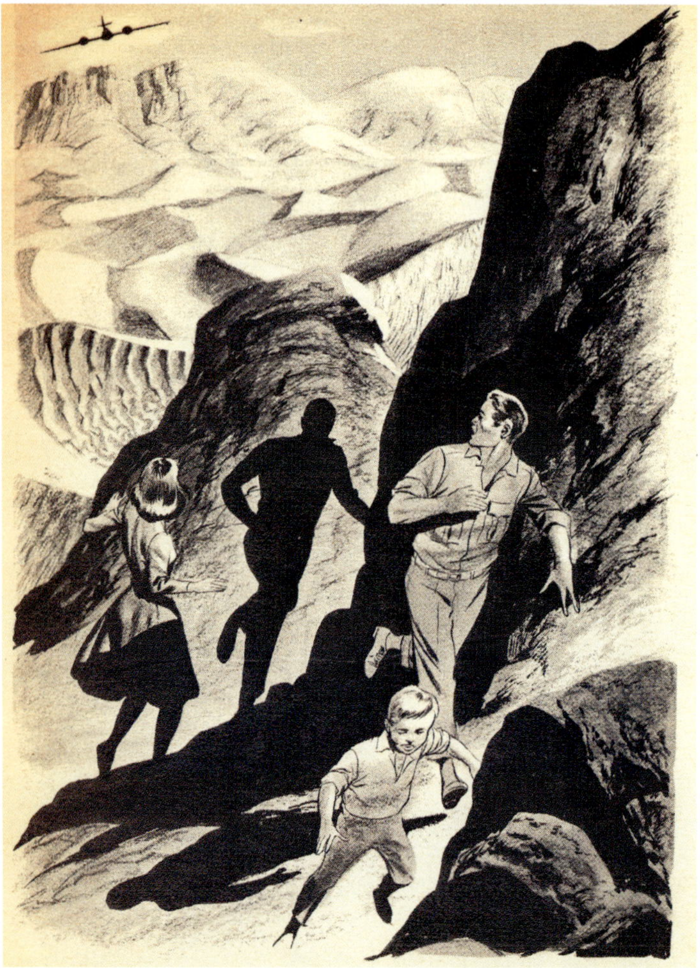

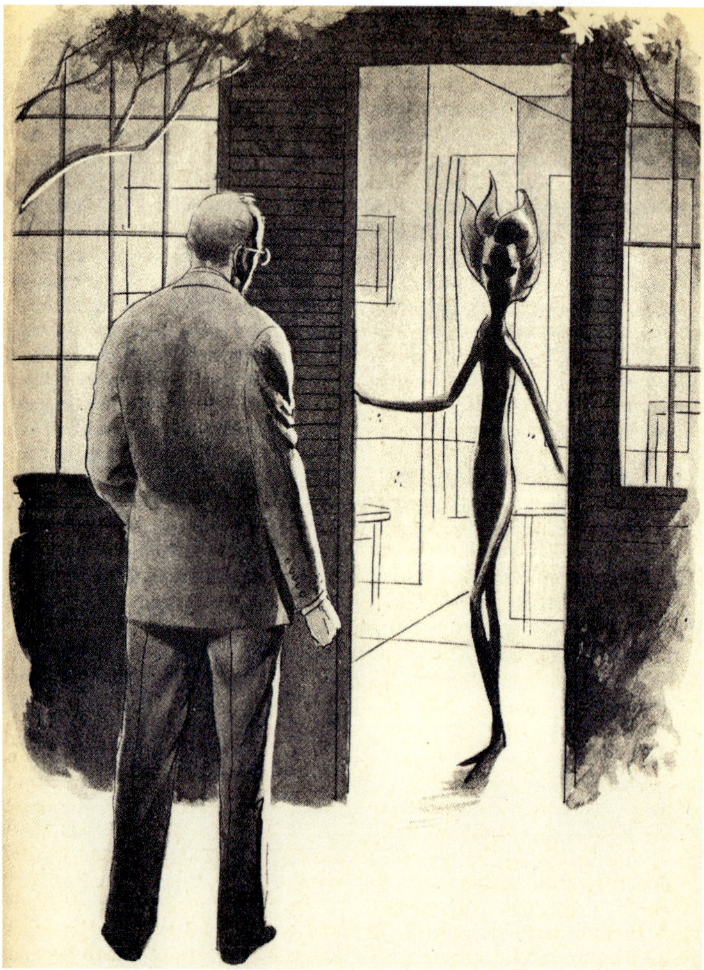

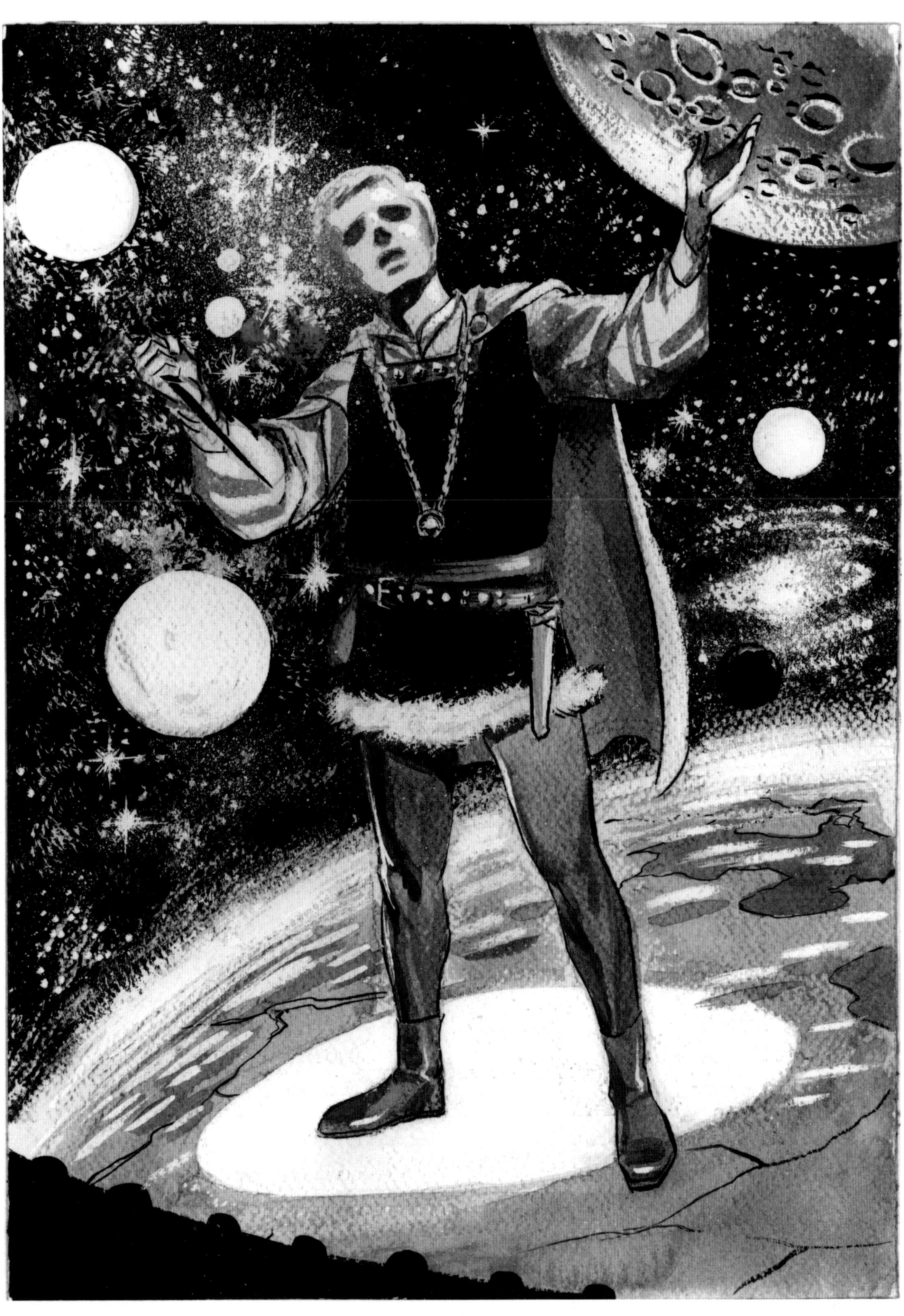

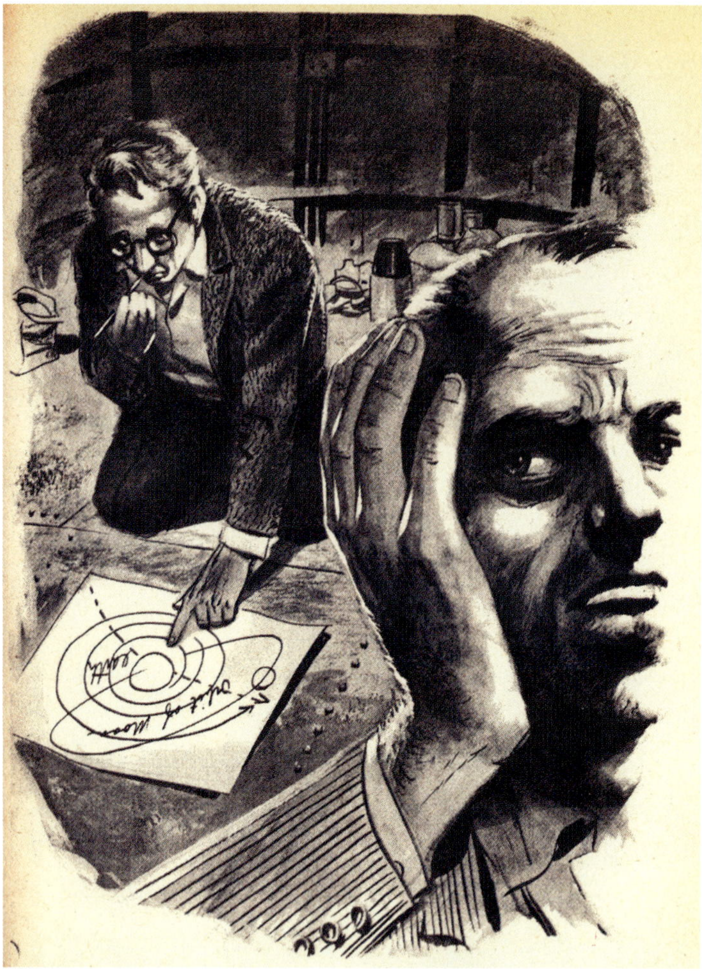

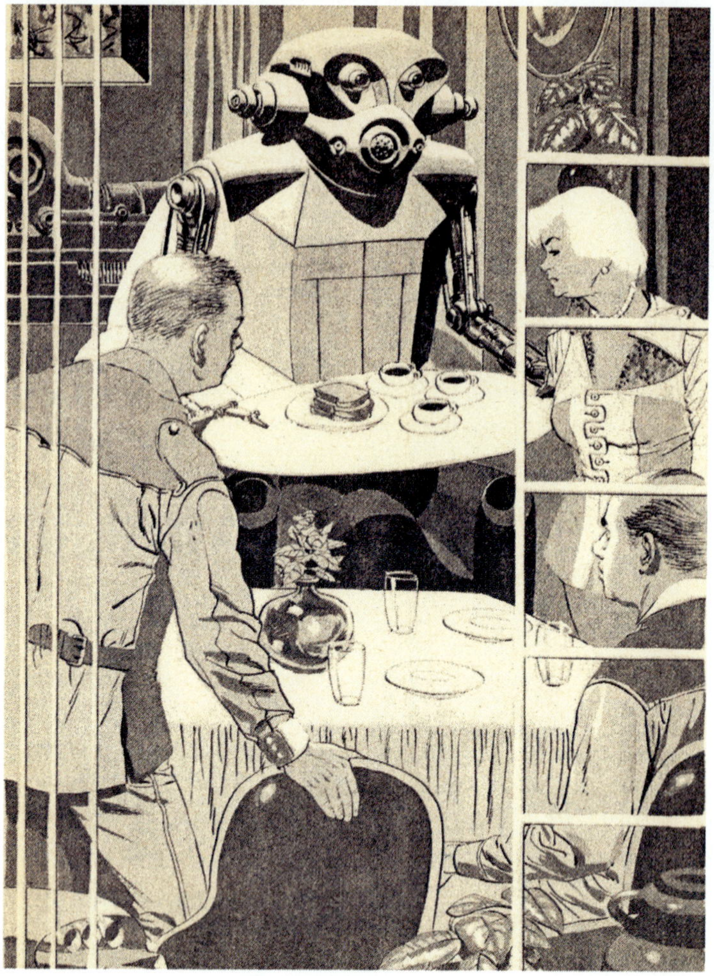

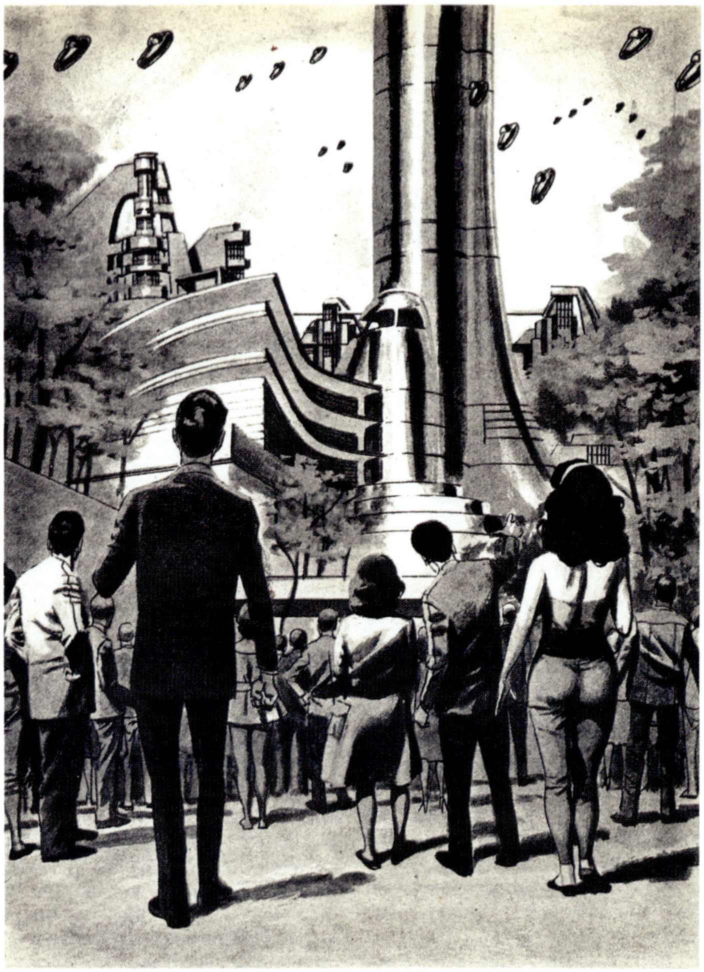

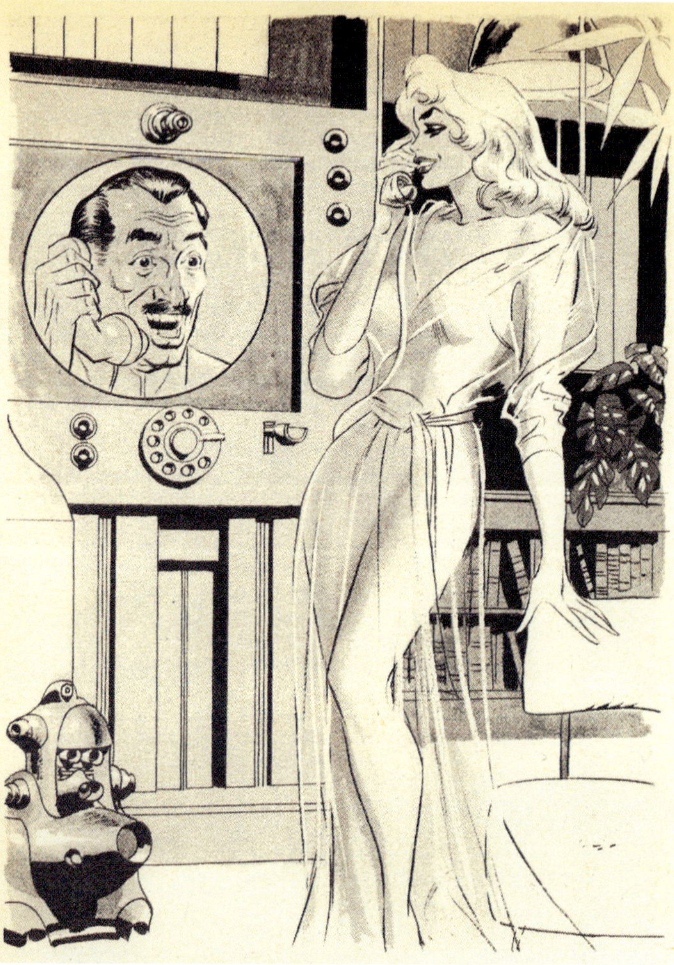

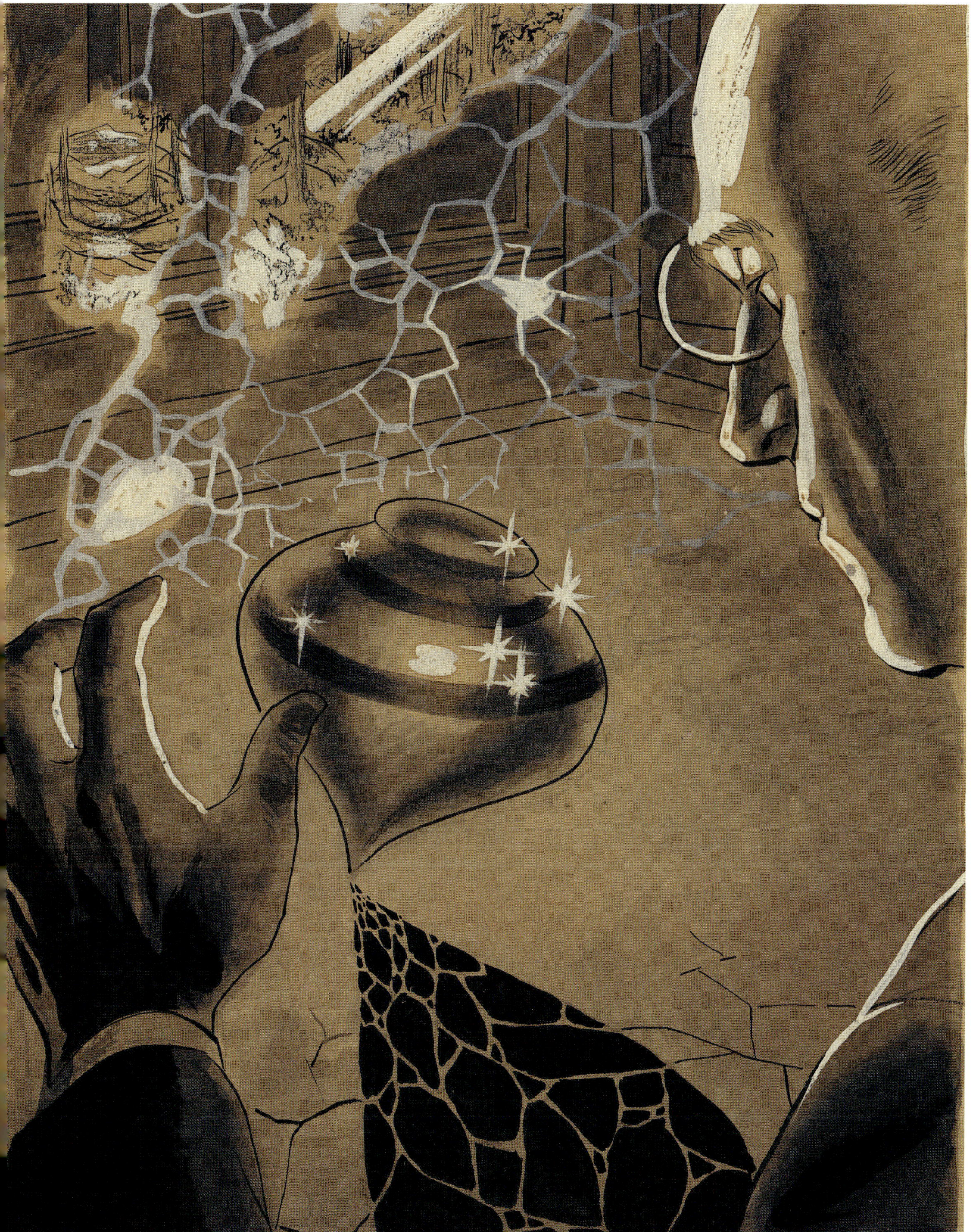

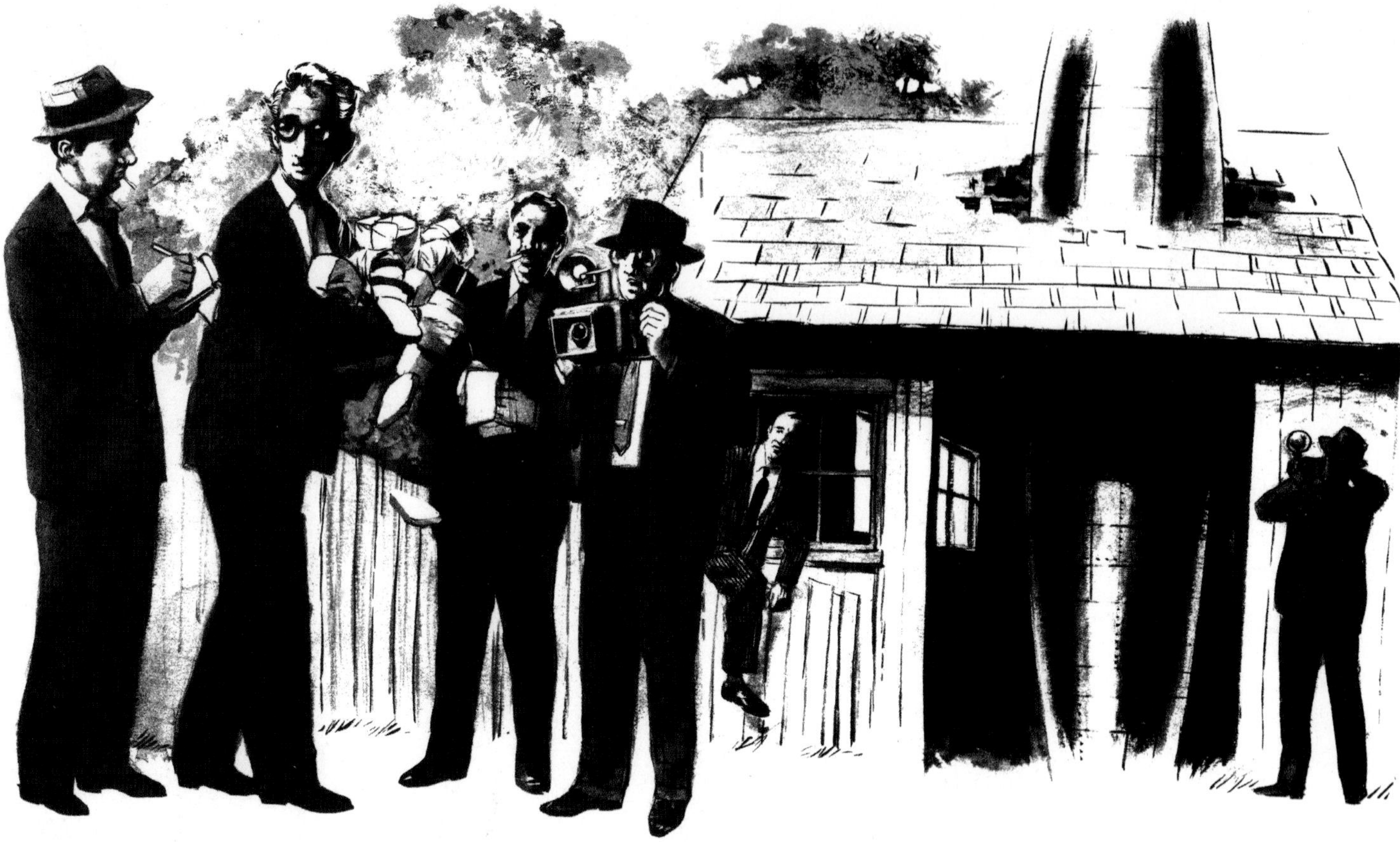

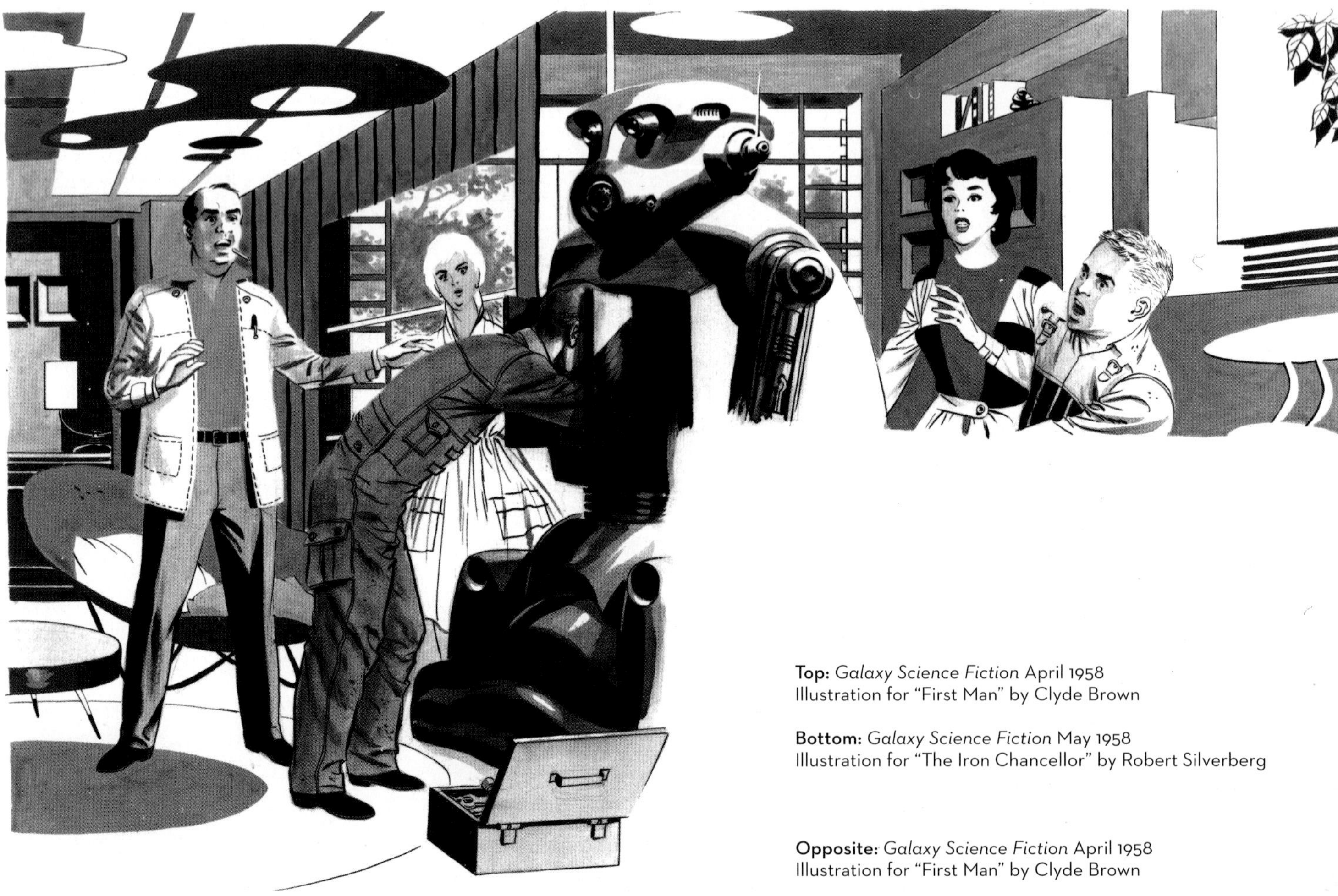

Top: *Galaxy Science Fiction* April 1958
Illustration for "First Man" by Clyde Brown

Bottom: *Galaxy Science Fiction* May 1958
Illustration for "The Iron Chancellor" by Robert Silverberg

Opposite: *Galaxy Science Fiction* April 1958
Illustration for "First Man" by Clyde Brown

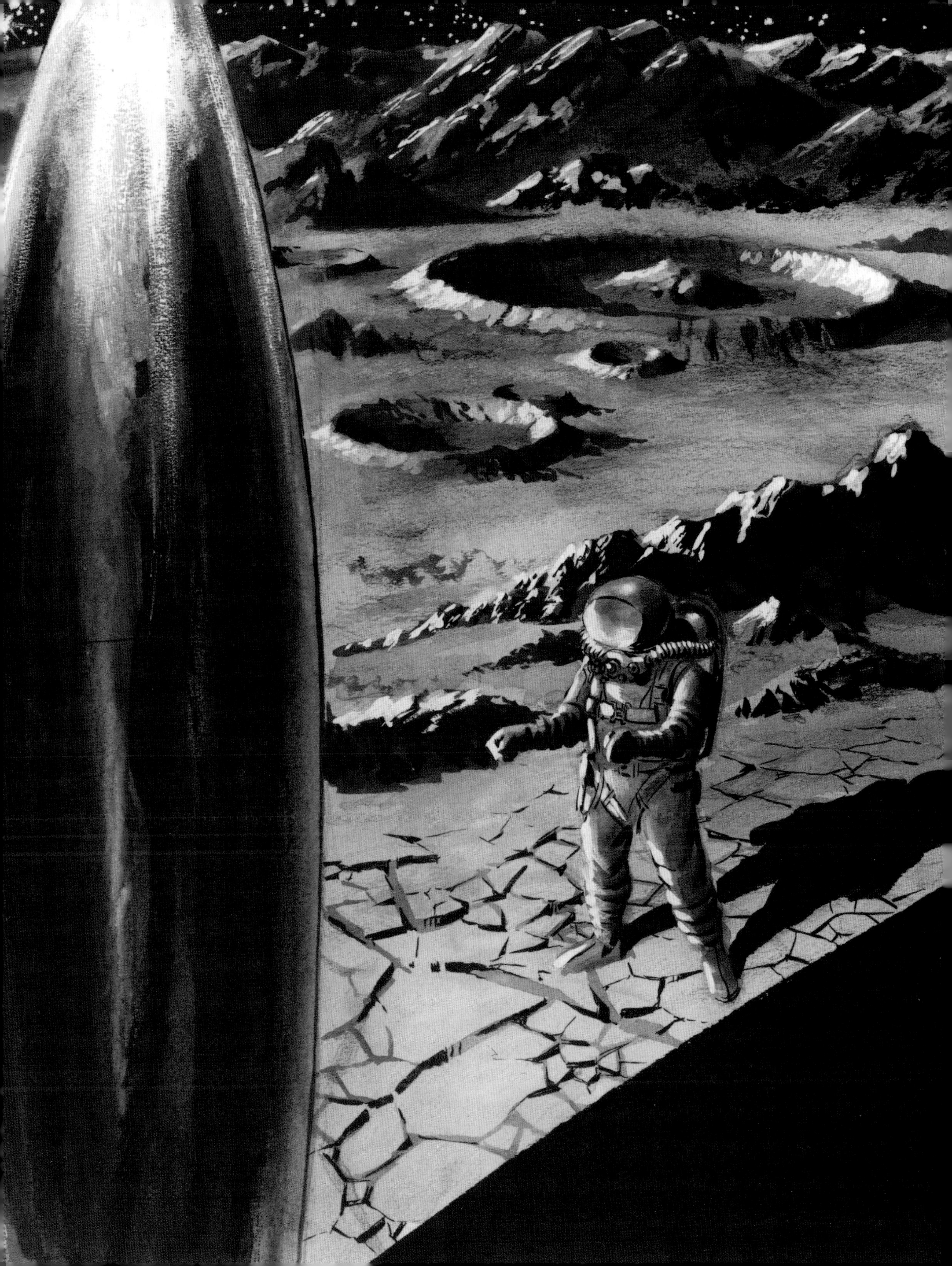

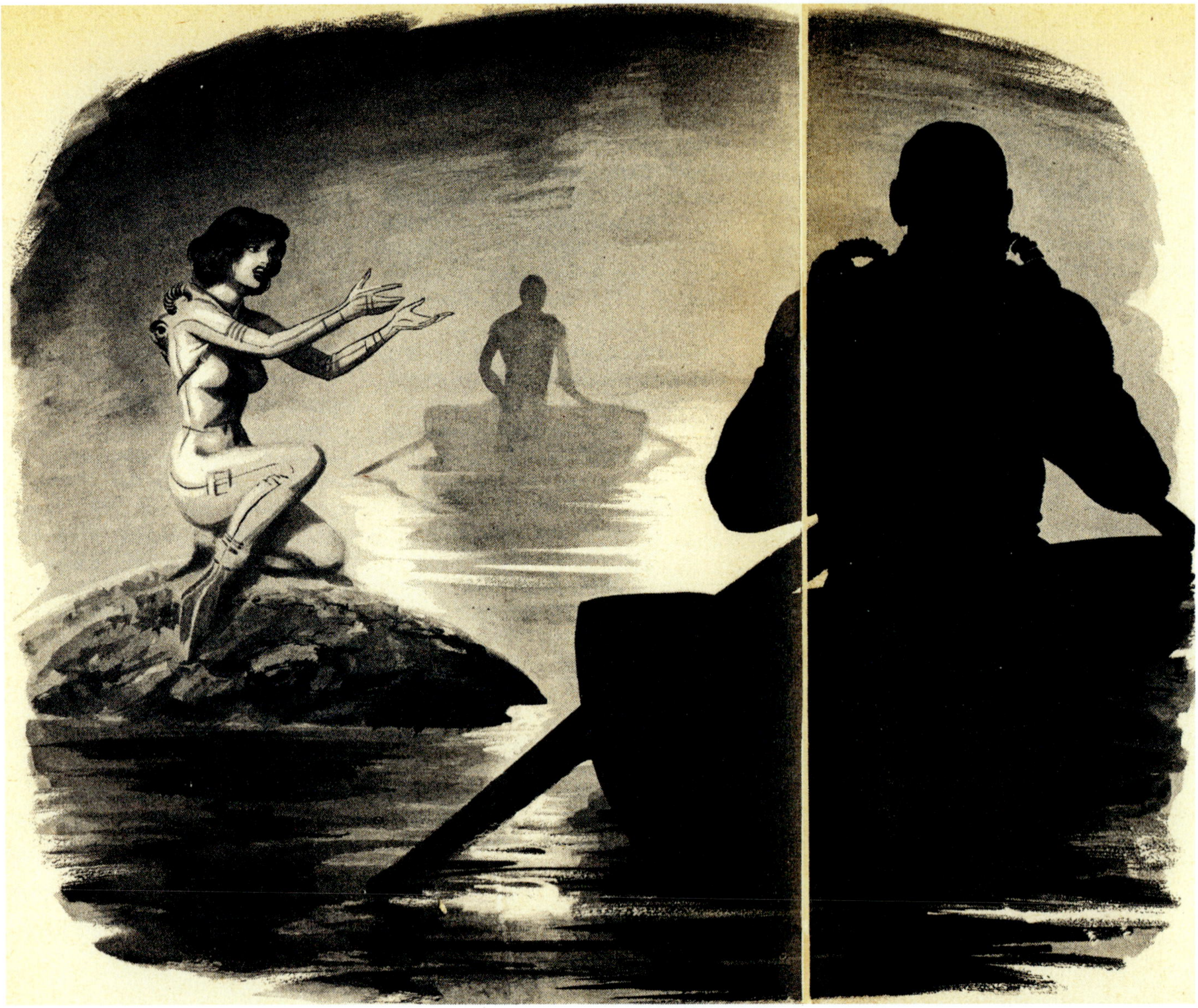

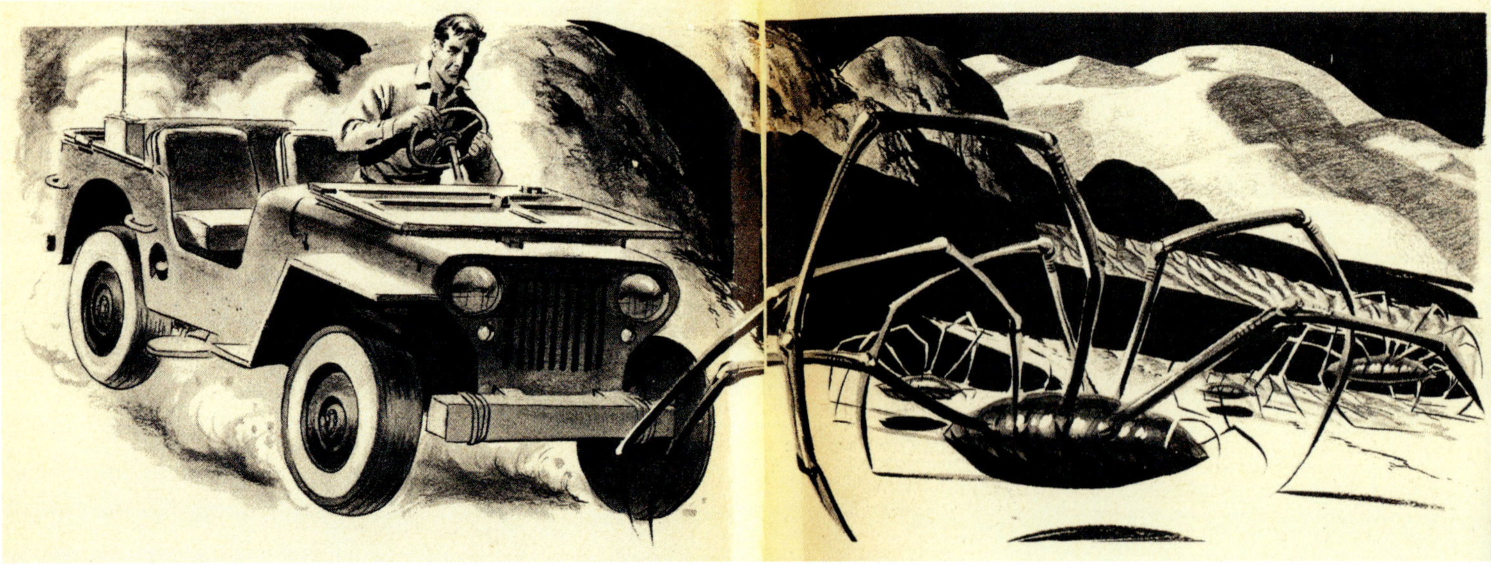

Top: *Galaxy Science Fiction* August 1958 Illustration for “Seven Deadly Virtues” by Paul Flehr

Bottom: *Galaxy Science Fiction* June 1958 Illustration for “Mars by Moonlight” by Paul Flehr

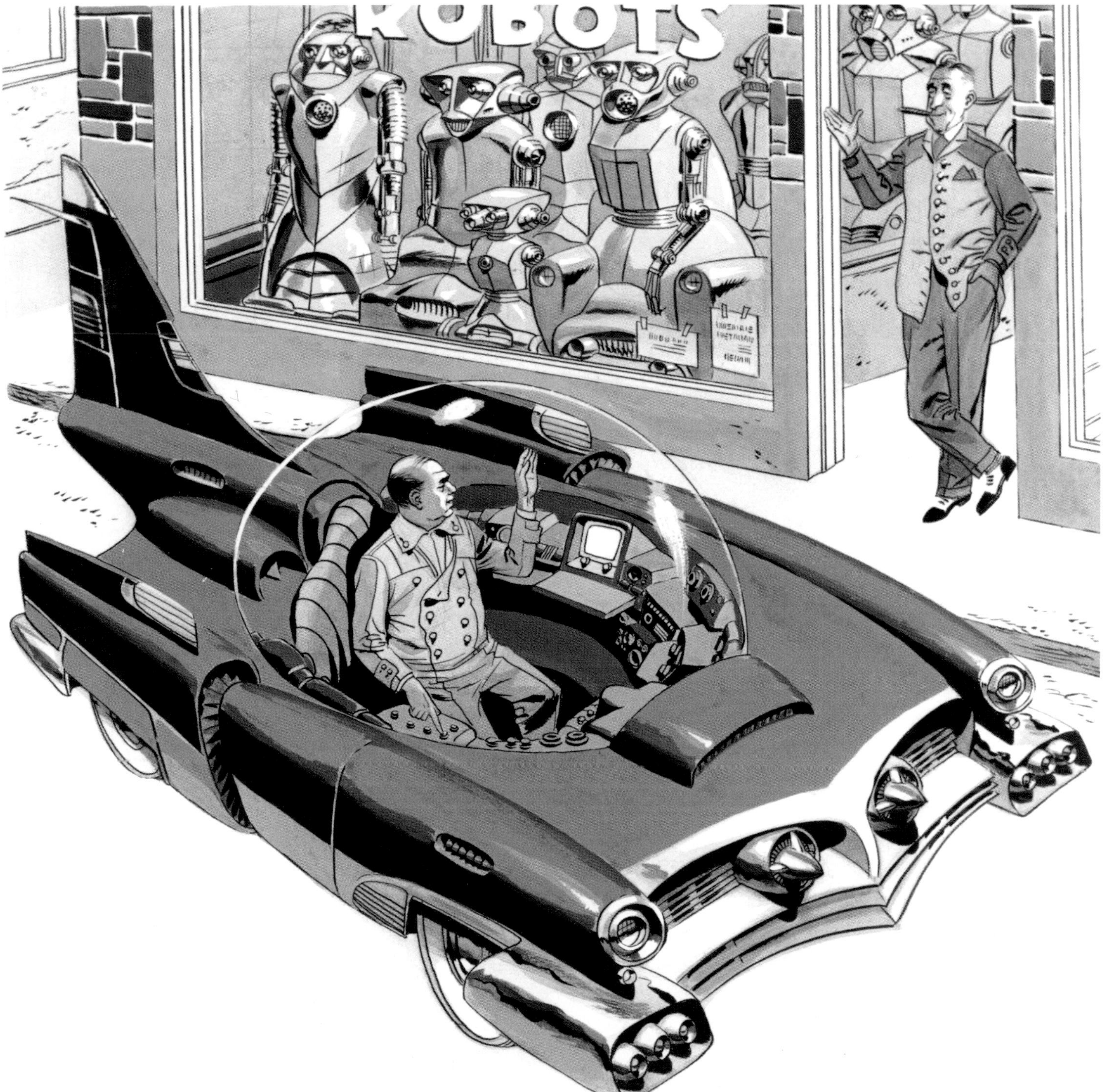

Galaxy Science Fiction May 1958 Illustration for "The Iron Chancellor" by Robert Silverberg

Top: *Galaxy Science Fiction* June 1958
Illustration for "The Gentlest Unpeople"
by Frederik Pohl

Bottom: *Galaxy Science Fiction* June 1958
Illustration for "Mars by Moonlight"
by Paul Flehr

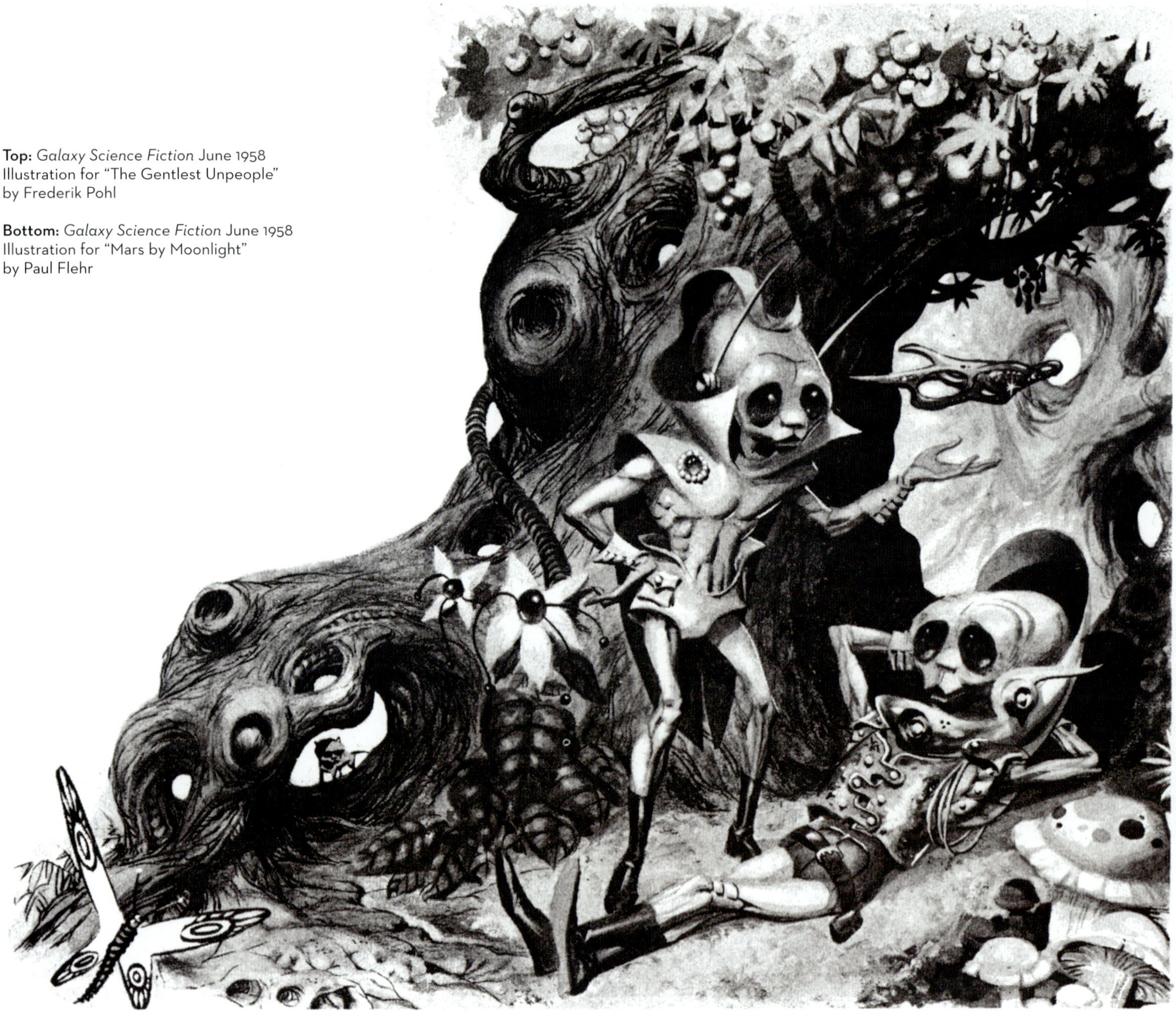

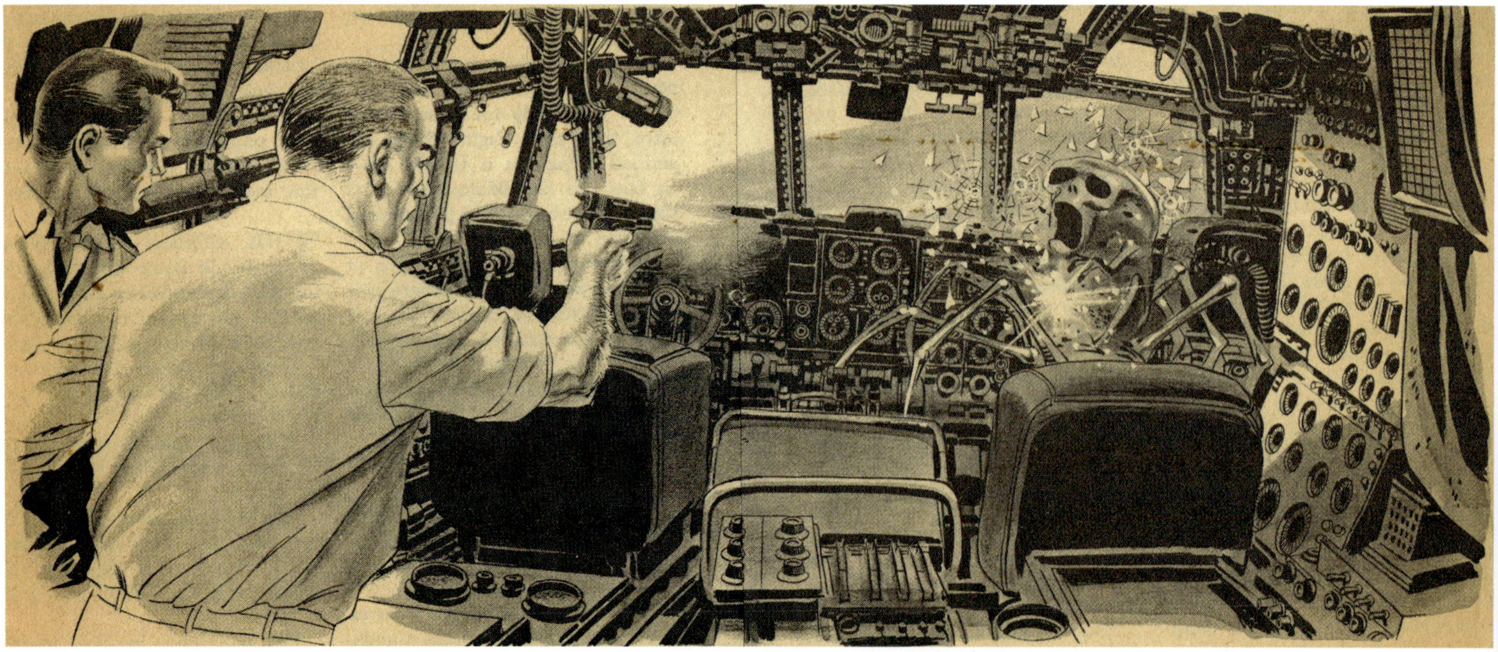

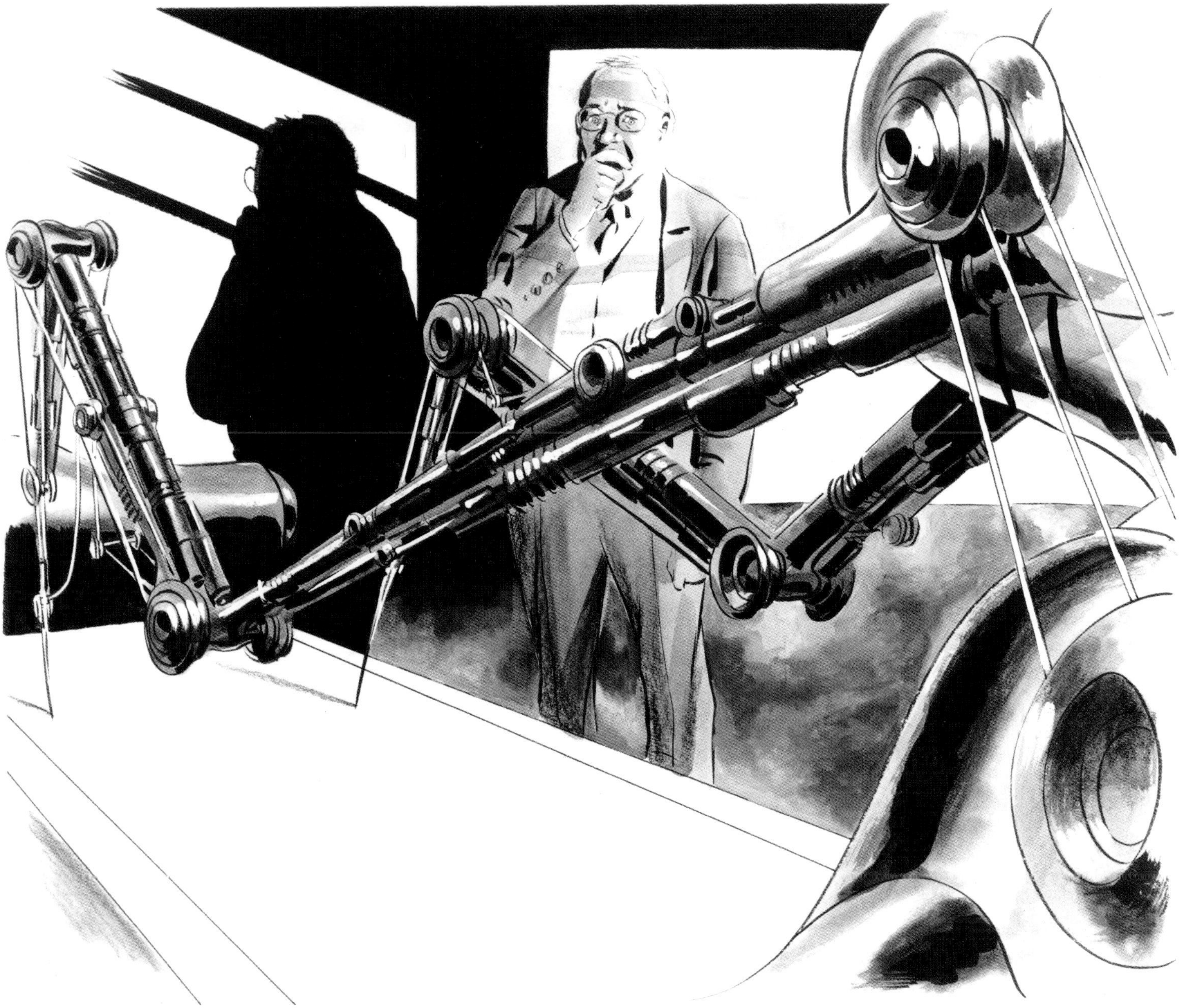

Galaxy Science Fiction September 1958 Illustration for "Thing of Beauty" by Damon Knight

Next spread:
Page 56, top: *Galaxy Science Fiction* September 1958 Illustration for "Lastborn" by Isaac Asimov

Page 56, bottom: *Galaxy Science Fiction* October 1958 Illustration for "Lisbon Cubed" by William Tenn

Page 57, upper left and bottom: *Galaxy Science Fiction* September 1958 Illustration for "Lastborn" by Isaac Asimov

Page 57, upper right: *Galaxy Science Fiction* September 1958 Illustration for "Thing of Beauty" by Damon Knight

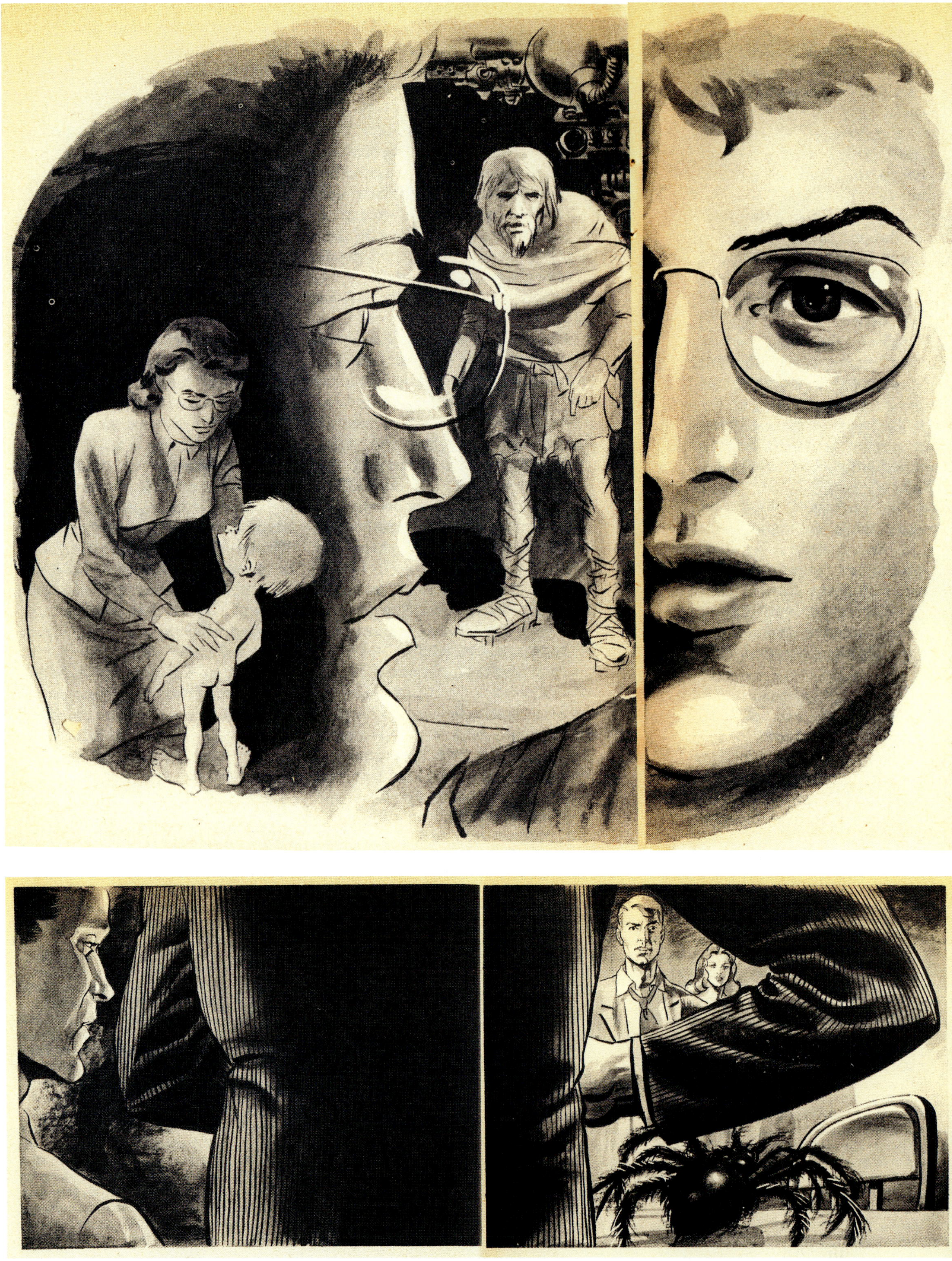

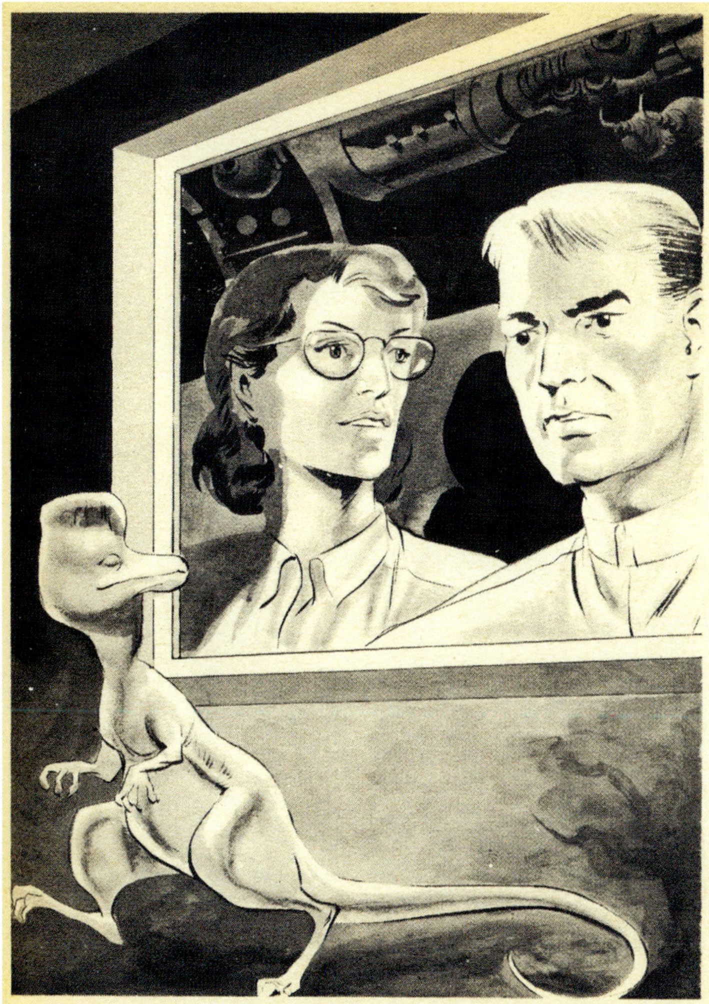

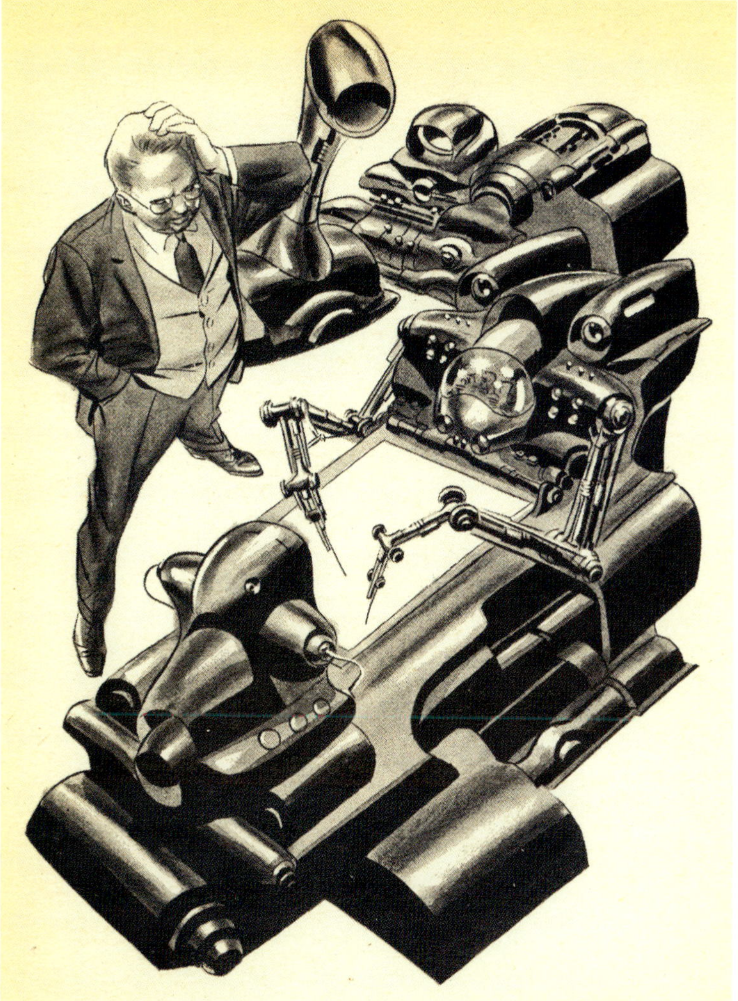

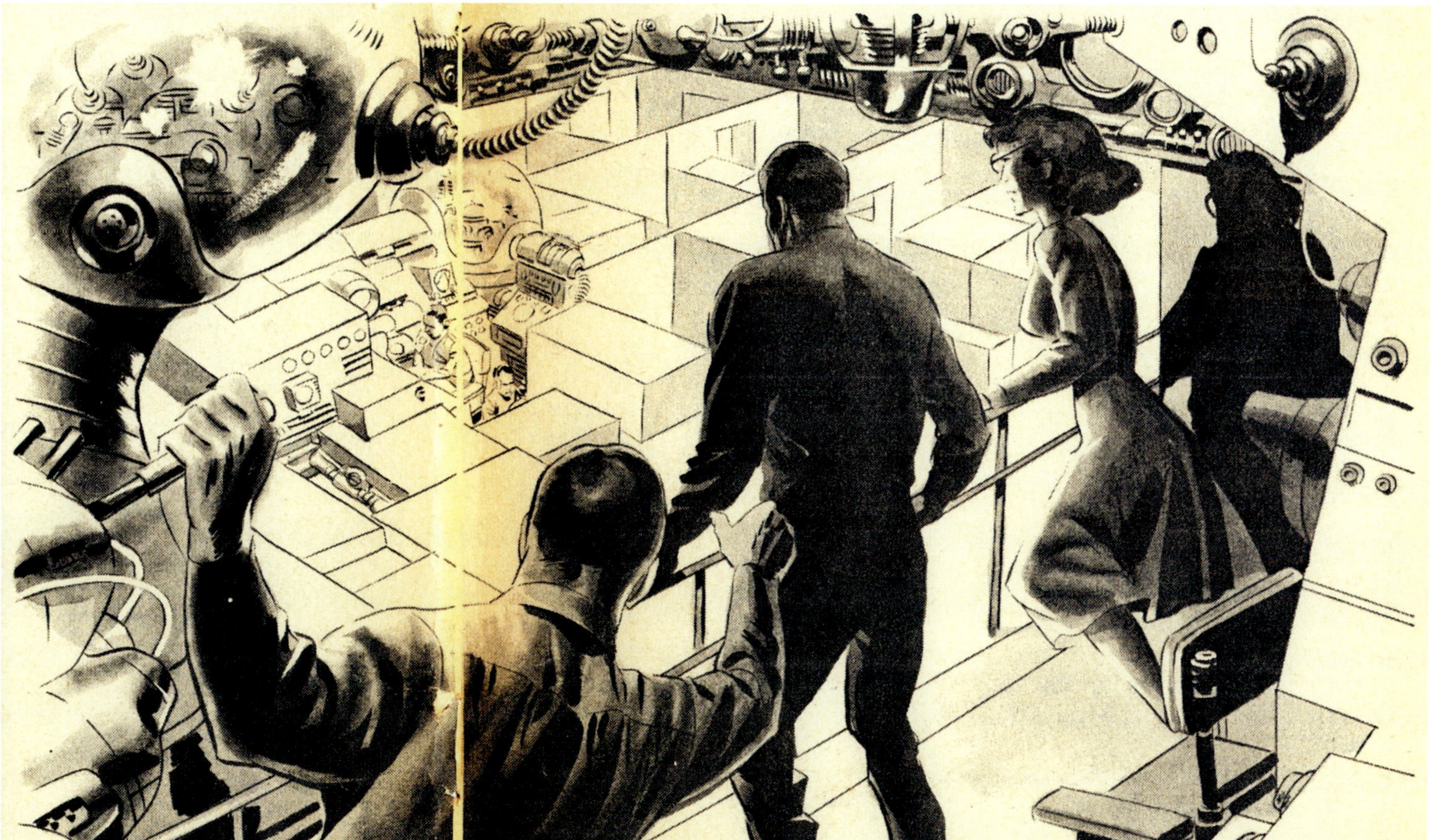

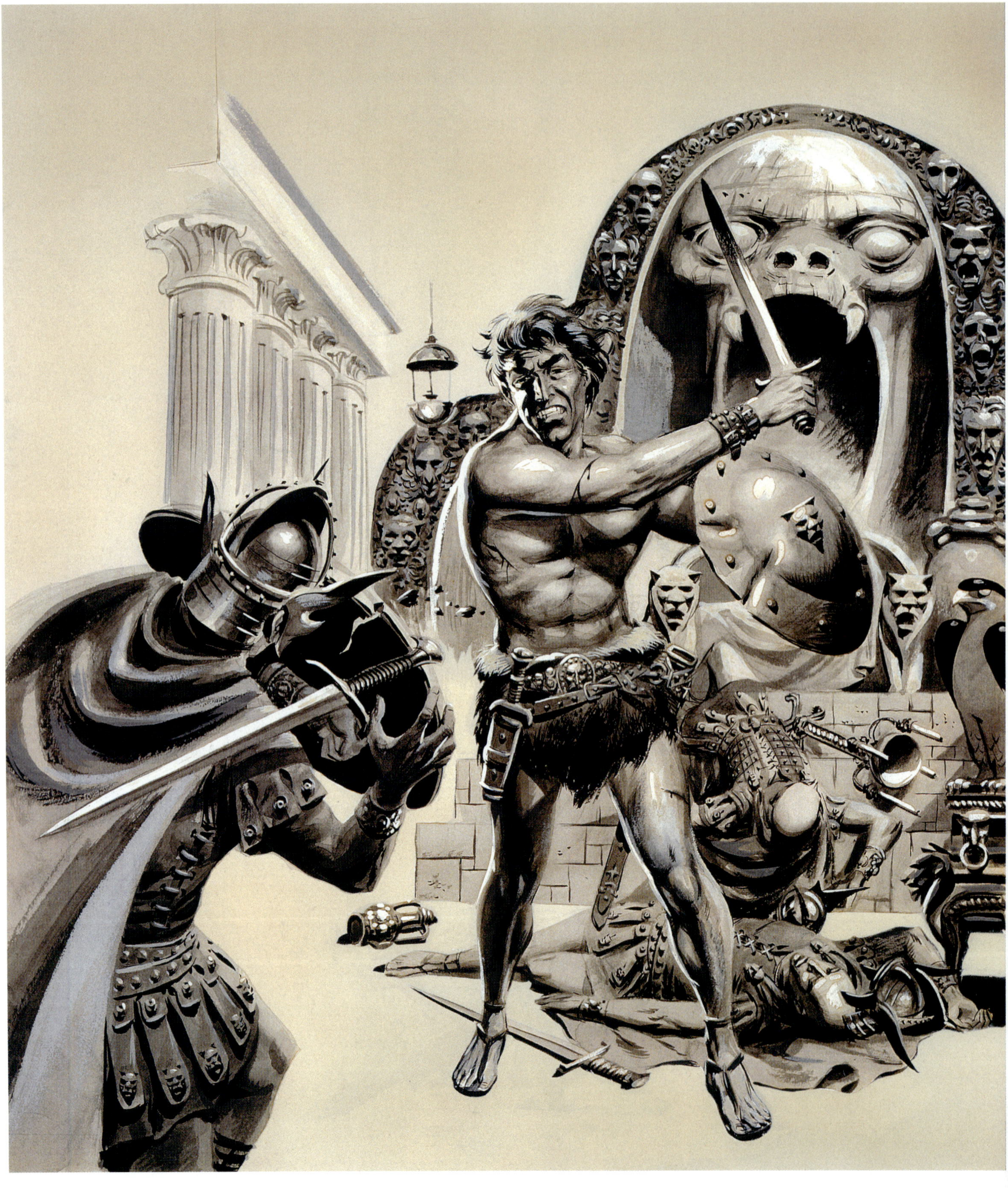

The Return of Conan by Björn Nyberg and L. Sprague de Camp, Gnome Press, 1957 (Cover)

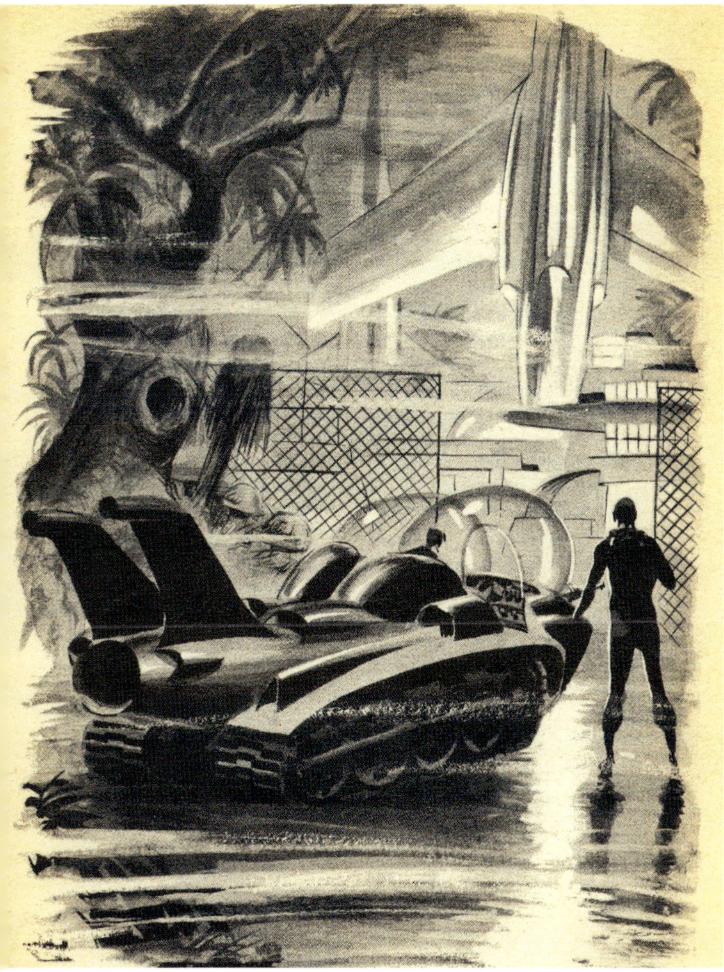

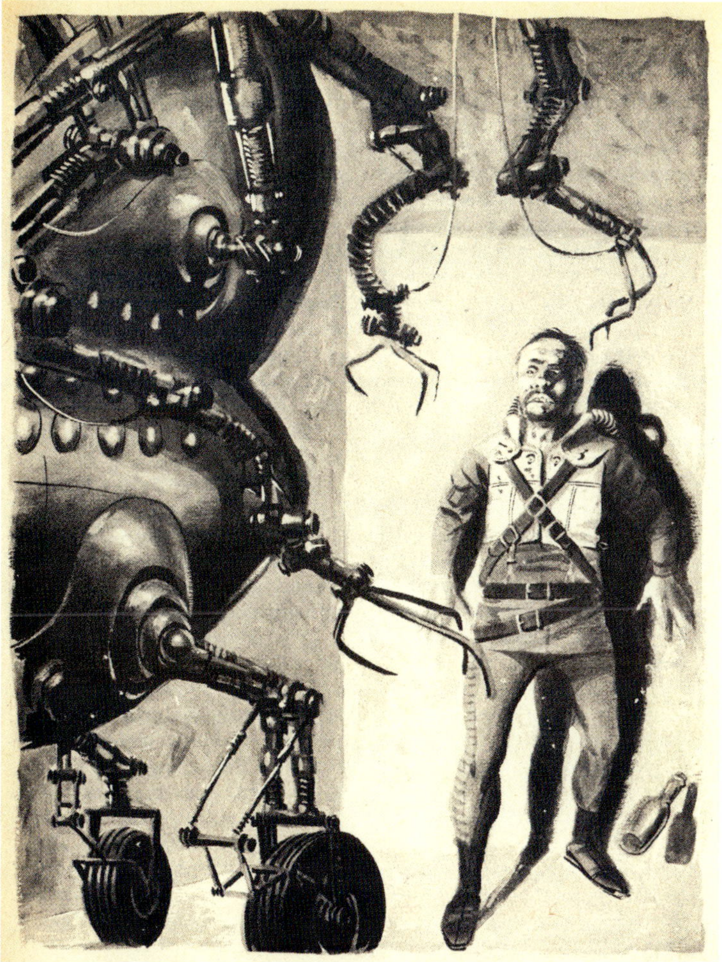

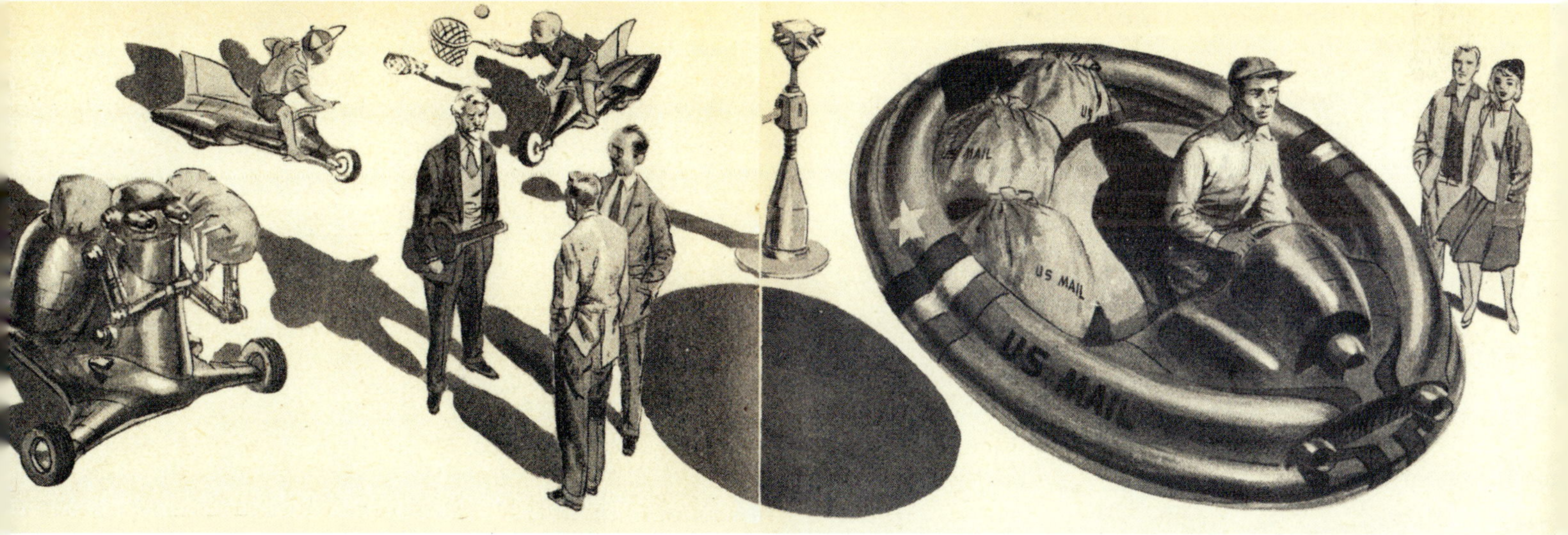

Clockwise from upper left:
Galaxy Science Fiction August 1958
Illustration for "Seven Deadly Virtues" by Paul Flehr

Galaxy Science Fiction June 1958
Illustration for "The Gentlest Unpeople" by Frederik Pohl

Galaxy Science Fiction March 1958
Illustration for "Spare the Rod" by Lloyd Biggle Jr.

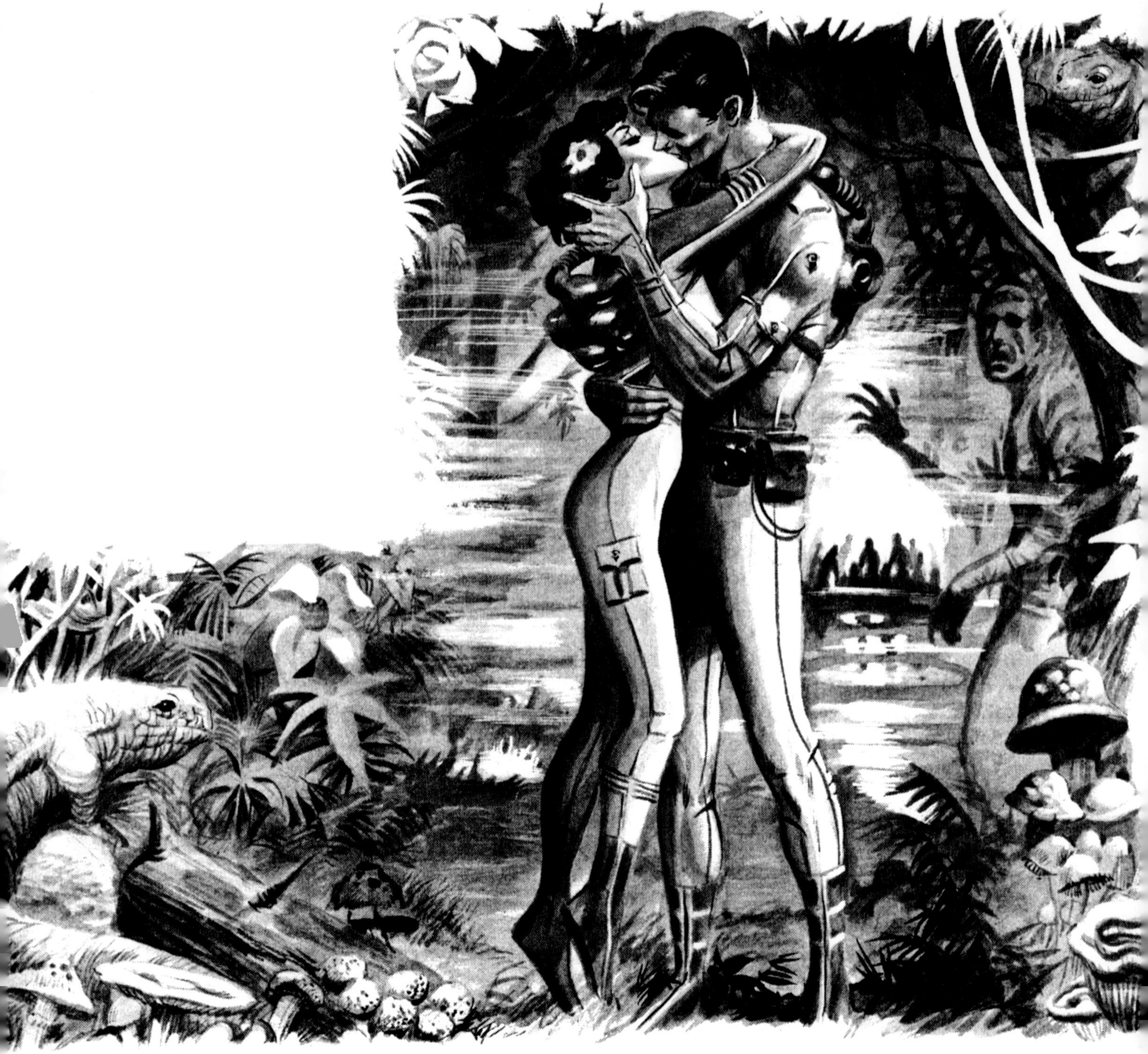

Galaxy Science Fiction August 1958
Illustration for "Seven Deadly Virtues" by Paul Flehr

Opposite: *Galaxy Science Fiction* July 1958
Illustration for "Innocent at Large" by Poul and Karen Anderson

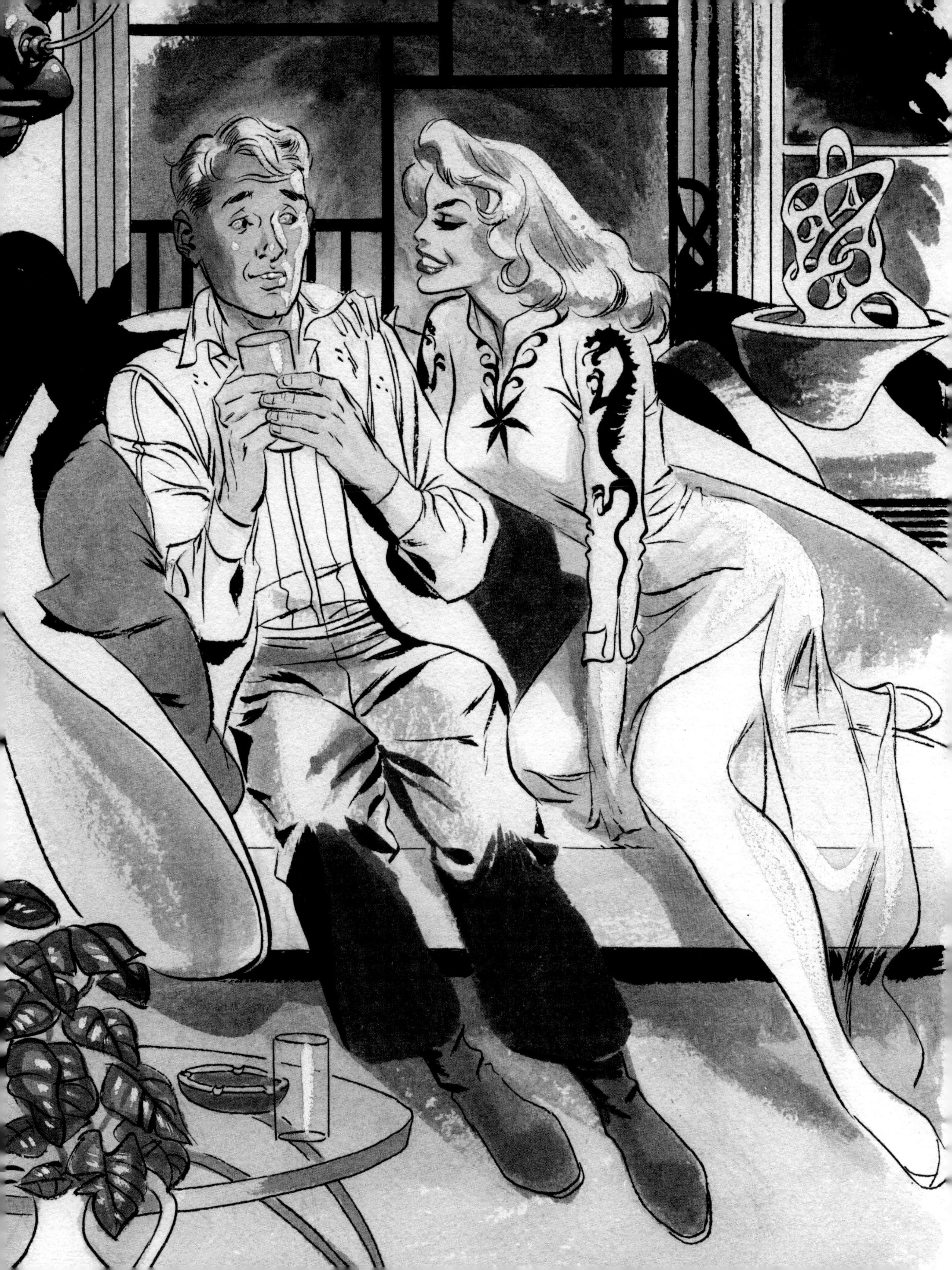

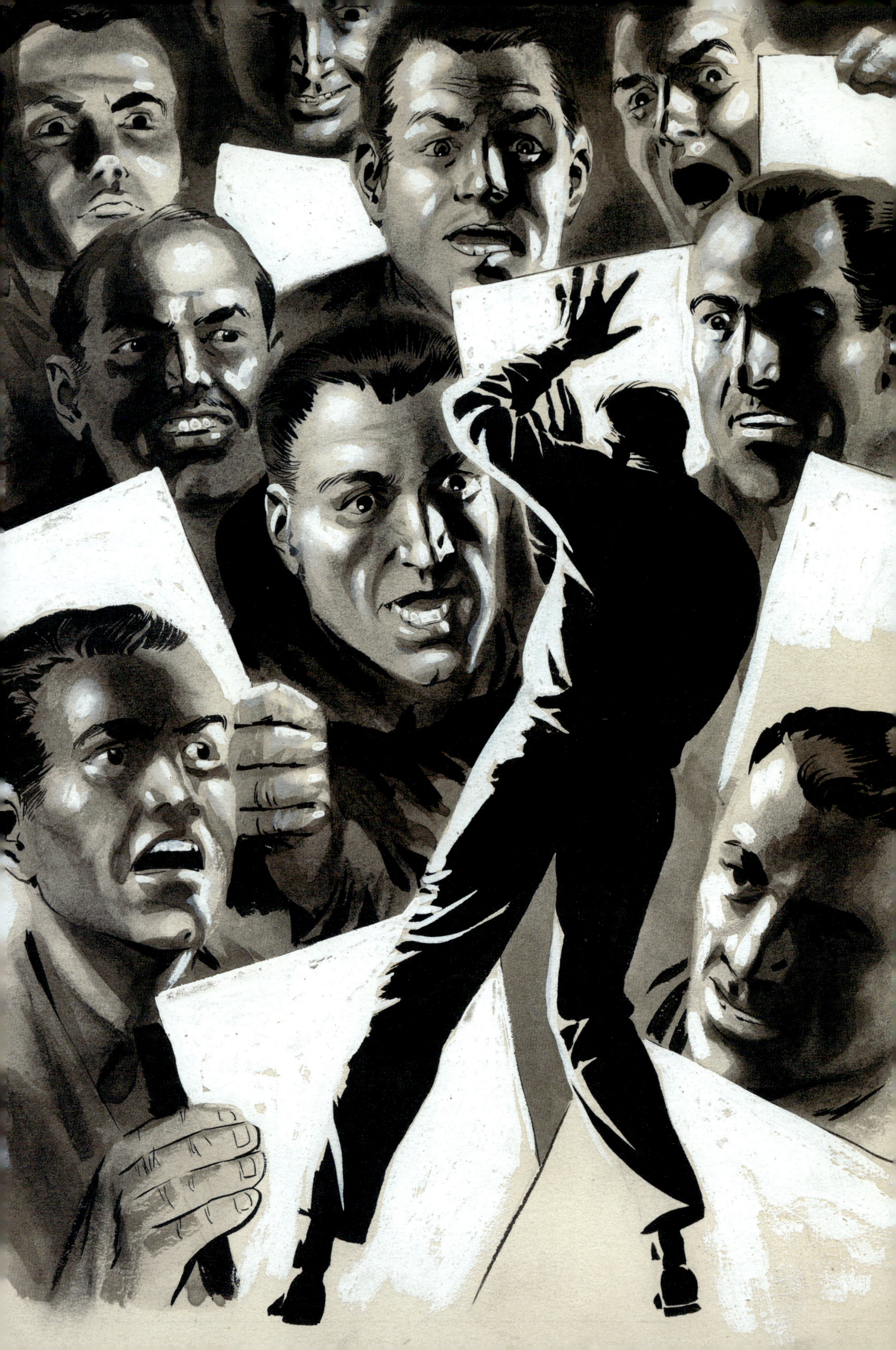

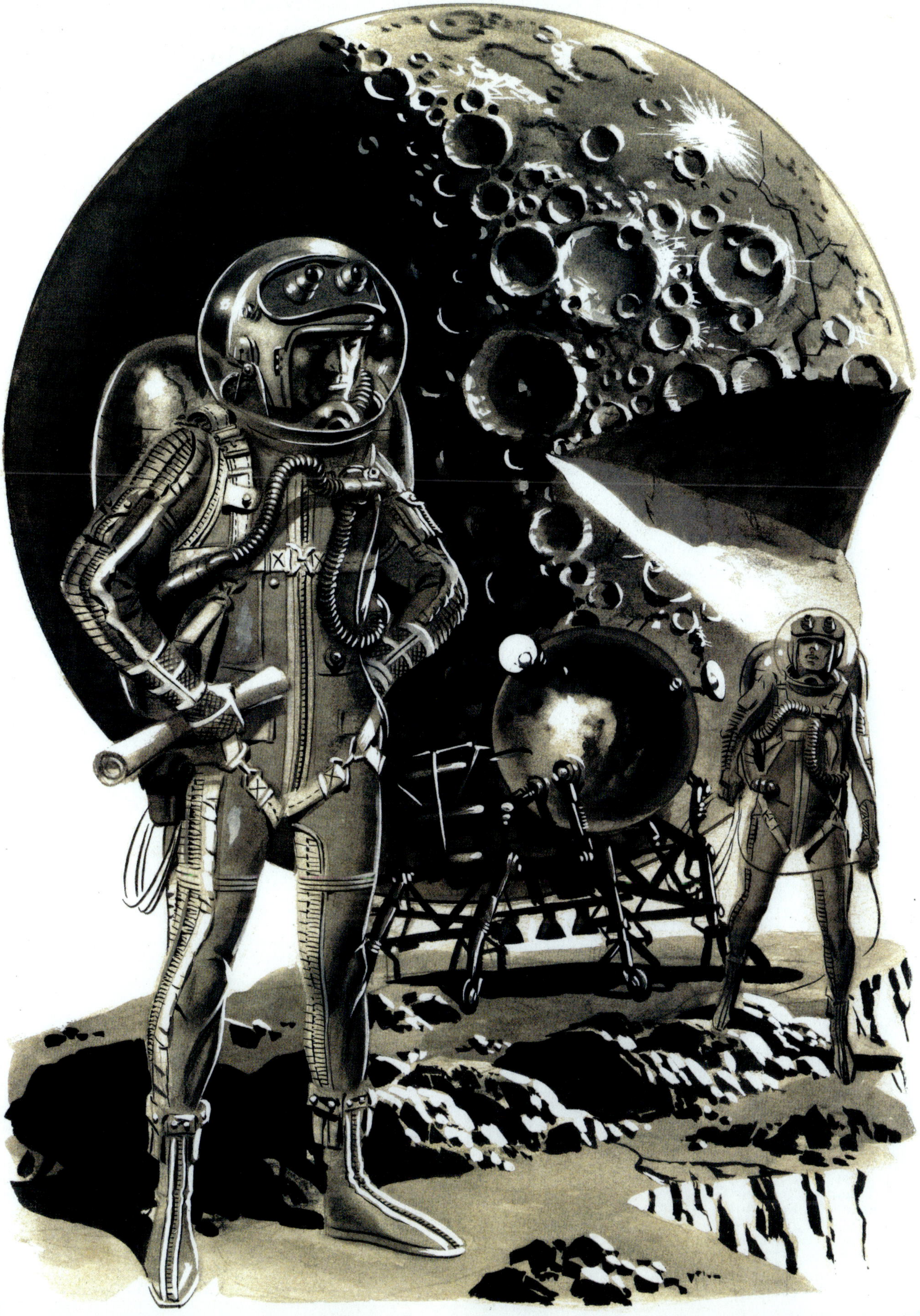

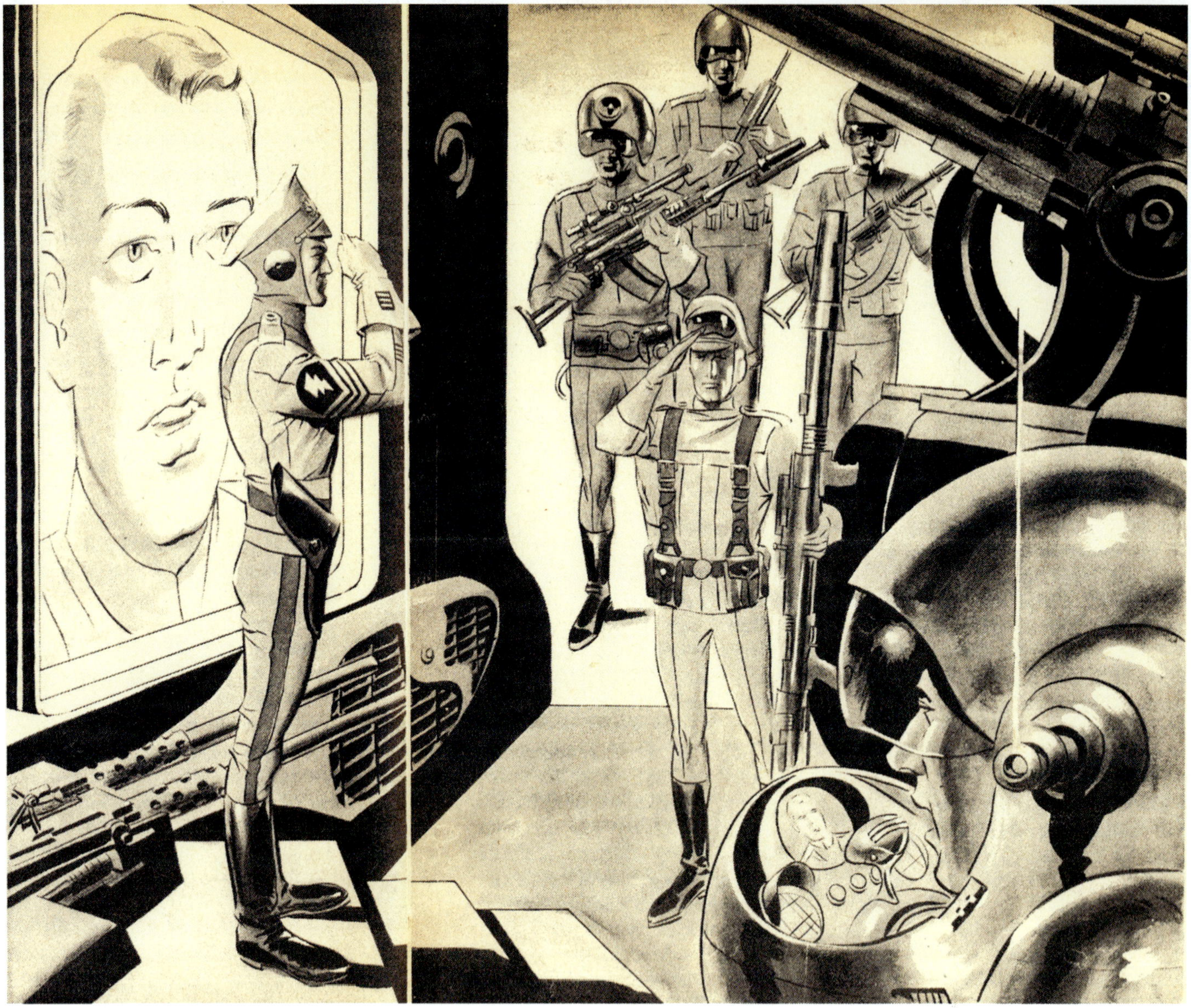

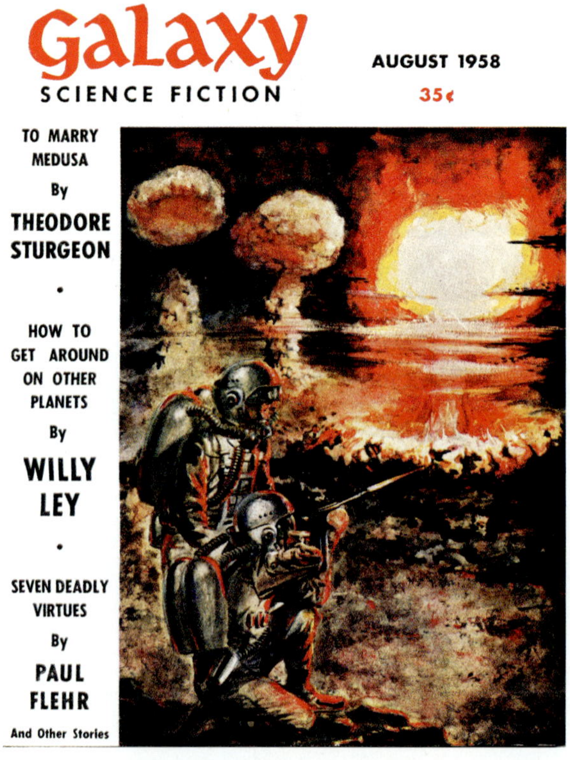

Top: *Galaxy Science Fiction* October 1958
Illustration for "The Wizards of Pung's Corners" by Frederik Pohl

Bottom: *Galaxy Science Fiction* August 1958 (Cover)
"Fireman of 2025 A.D. Fighting Small Industrial Blasts"

Opposite: *Galaxy Science Fiction* October 1958
Illustration for "The Wizards of Pung's Corners" by Frederik Pohl

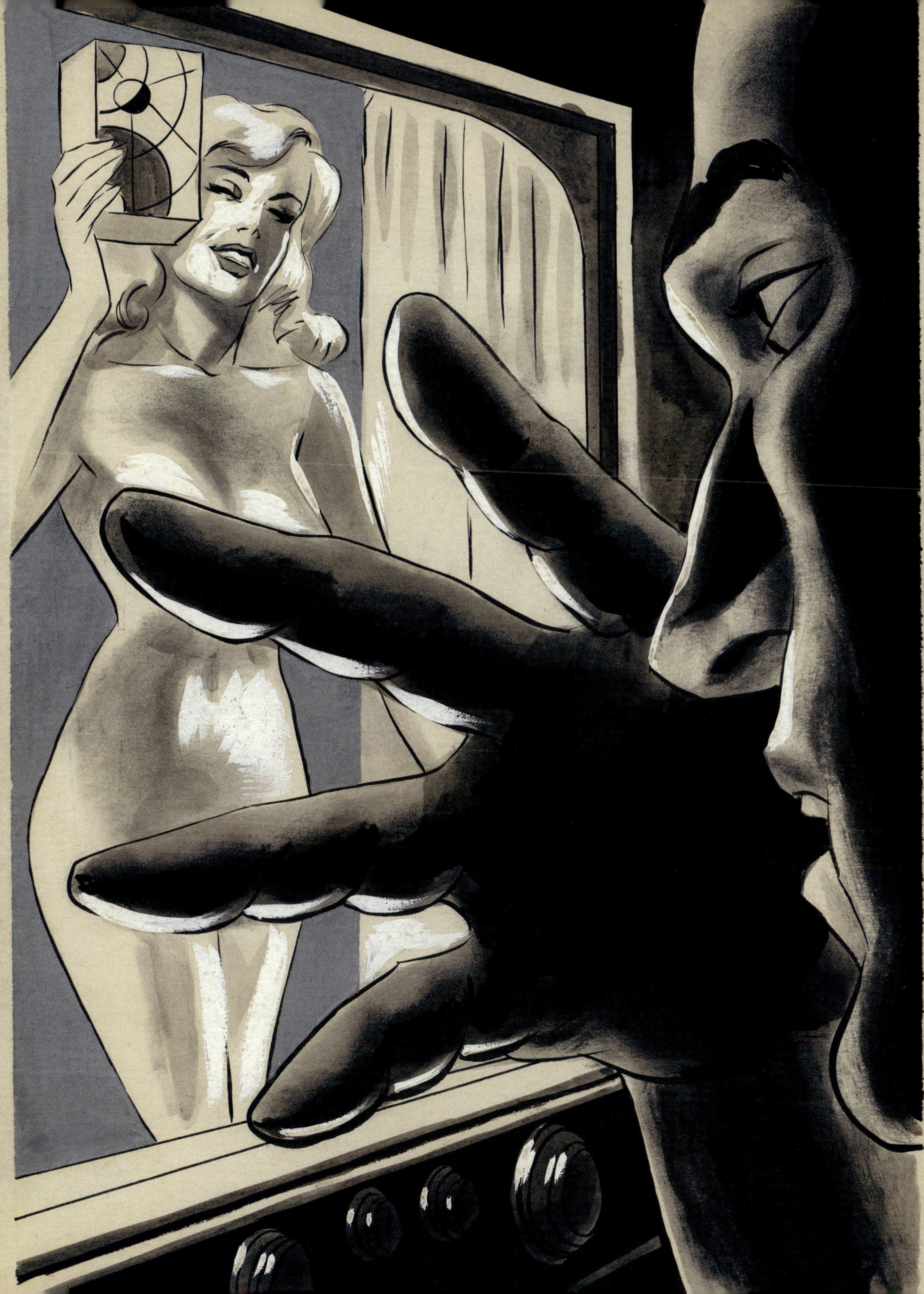

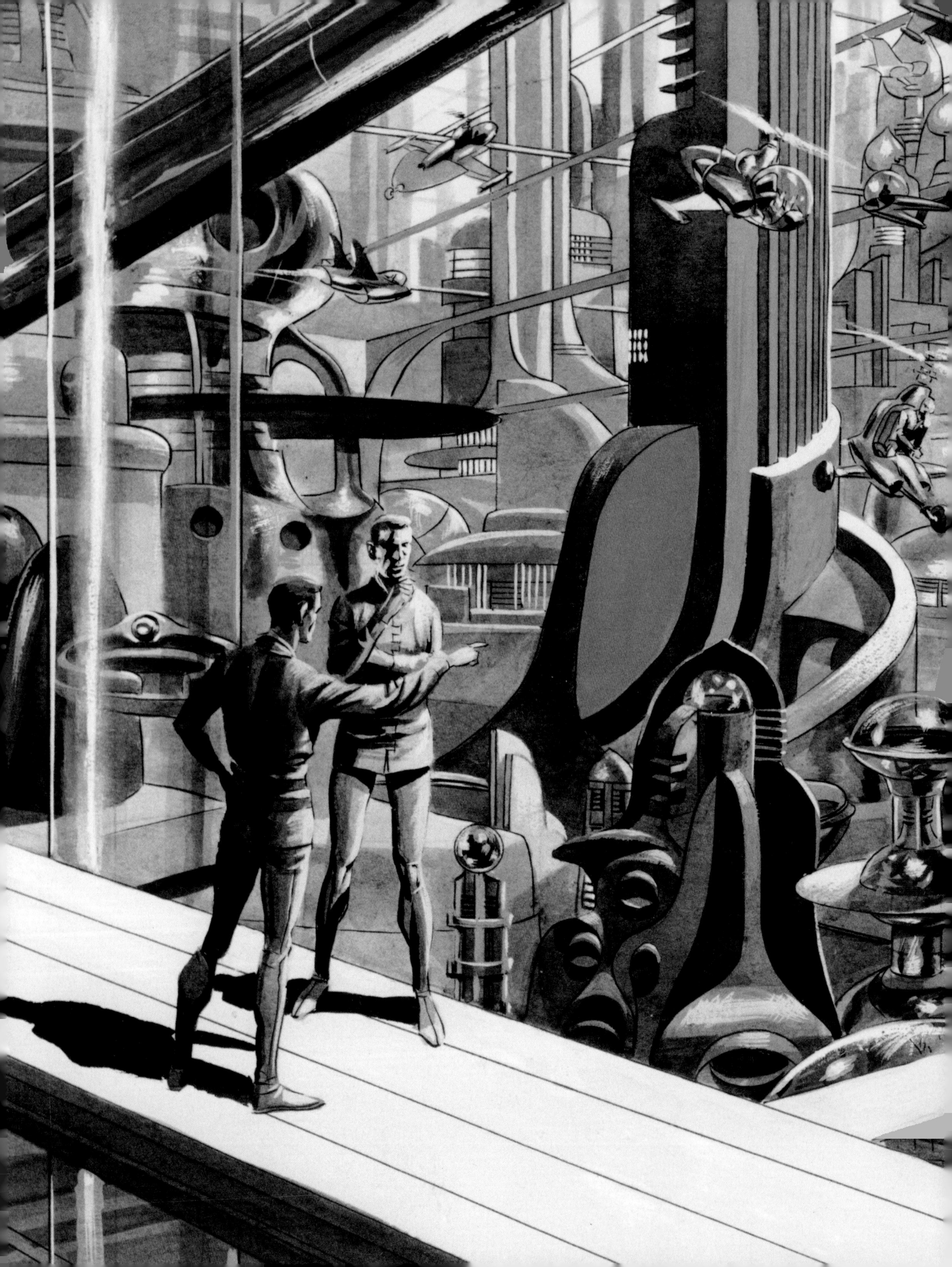

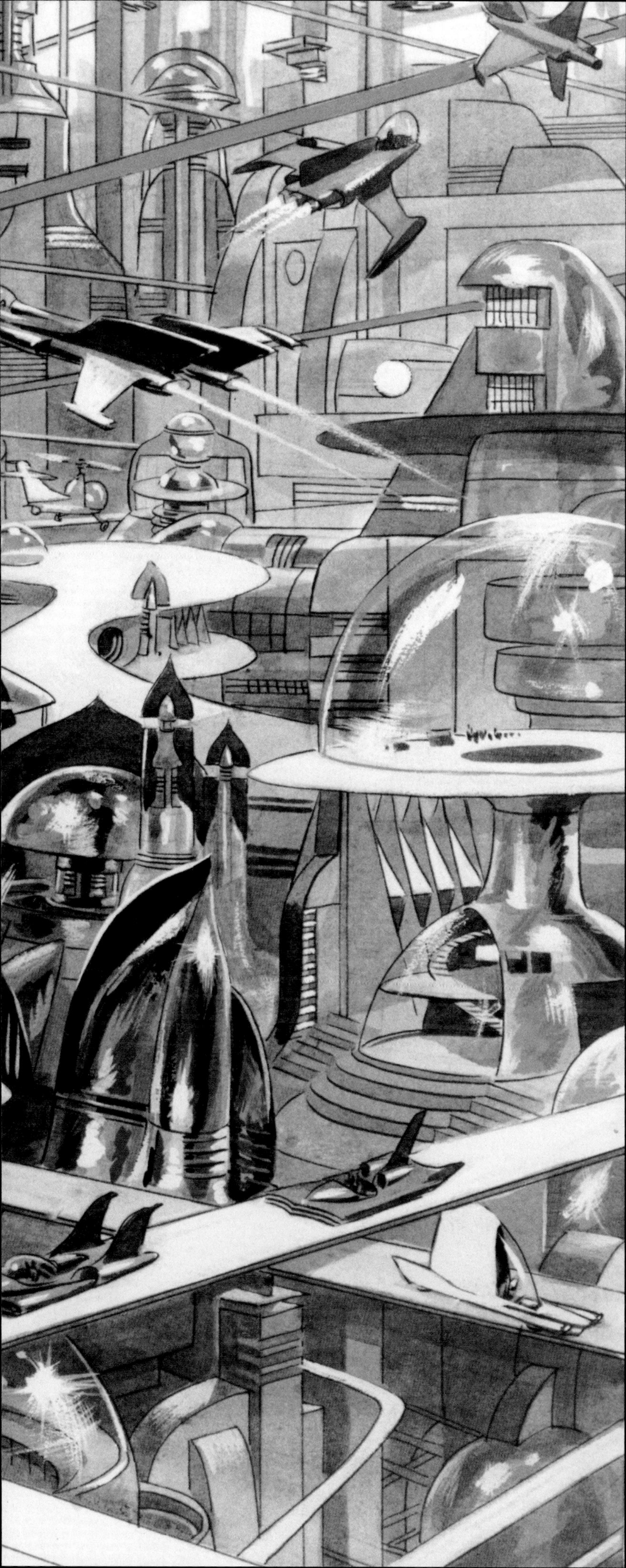

Galaxy Science Fiction October 1958
Illustration for "Time Killer" part 1 by Robert Sheckley

Next page, clockwise from upper left:
Galaxy Science Fiction December 1958
Illustration for "Ullward's Retreat" by Jack Vance

Galaxy Science Fiction October 1958
Illustration for "Time Killer" part 1 by Robert Sheckley

Galaxy Science Fiction September 1958
Illustration for "Thing of Beauty" by Damon Knight

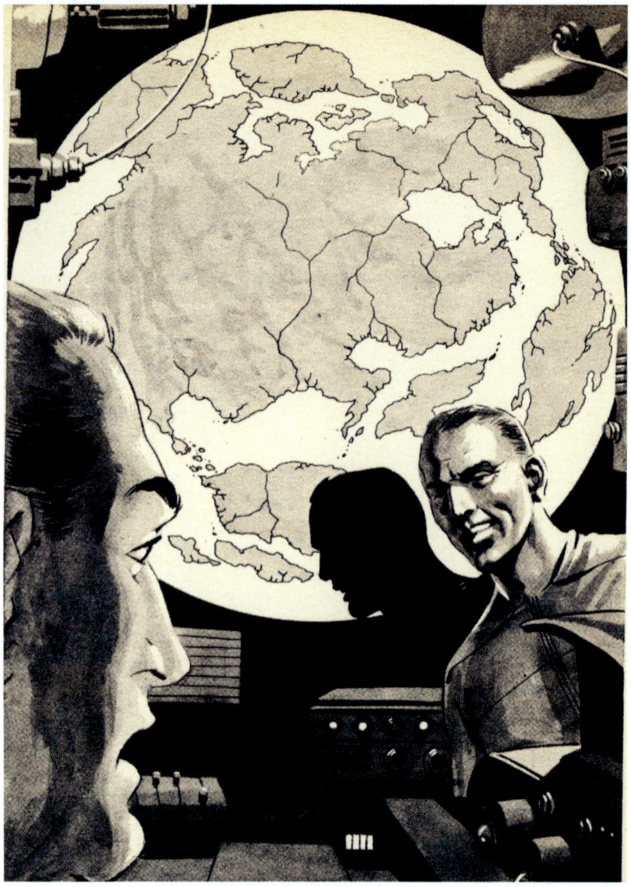

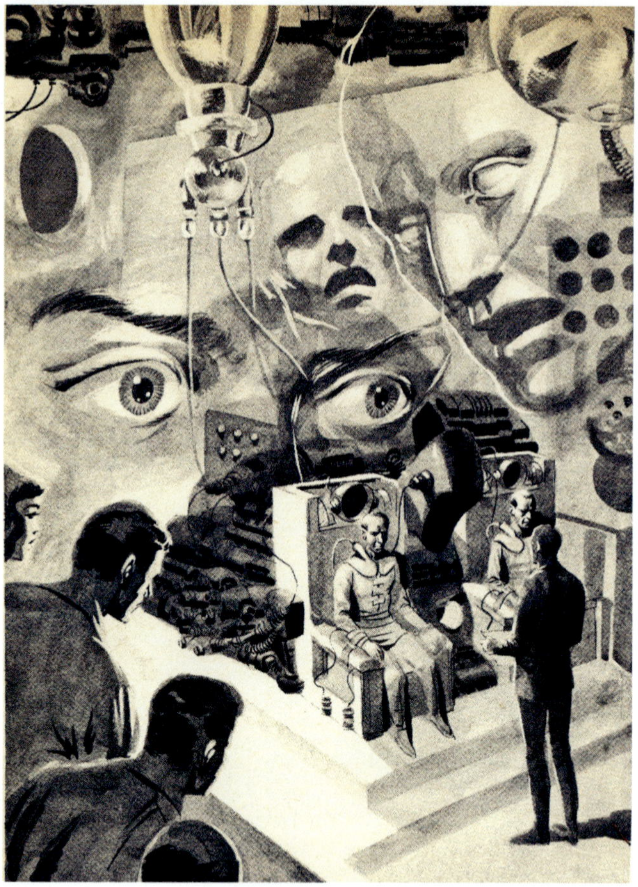

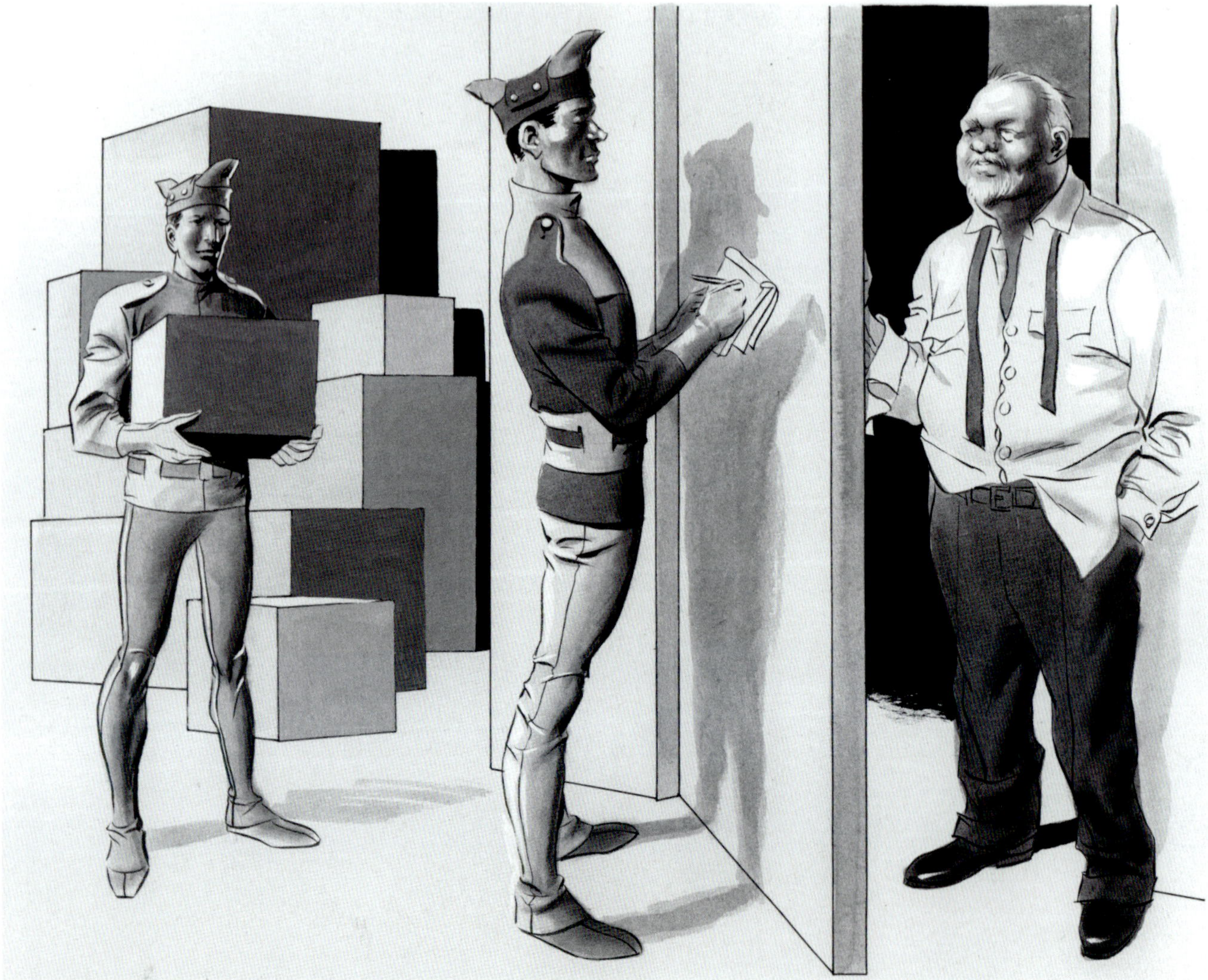

Top: *Galaxy Science Fiction* October 1958
Illustration for "Time Killer" part 1 by Robert Sheckley

Bottom: *Galaxy Science Fiction* July 1958
Illustration for "Innocent at Large" by Poul and Karen Anderson

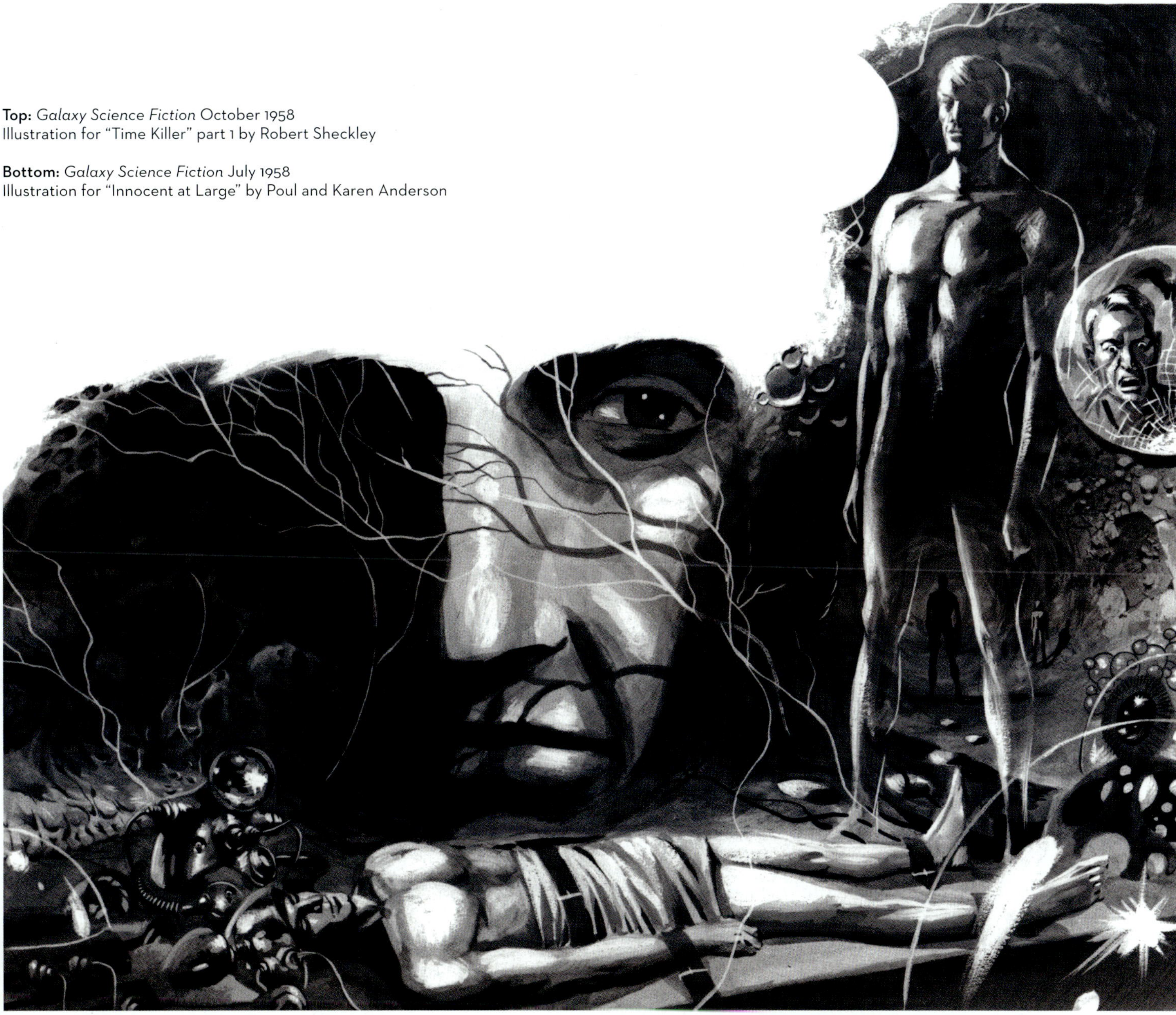

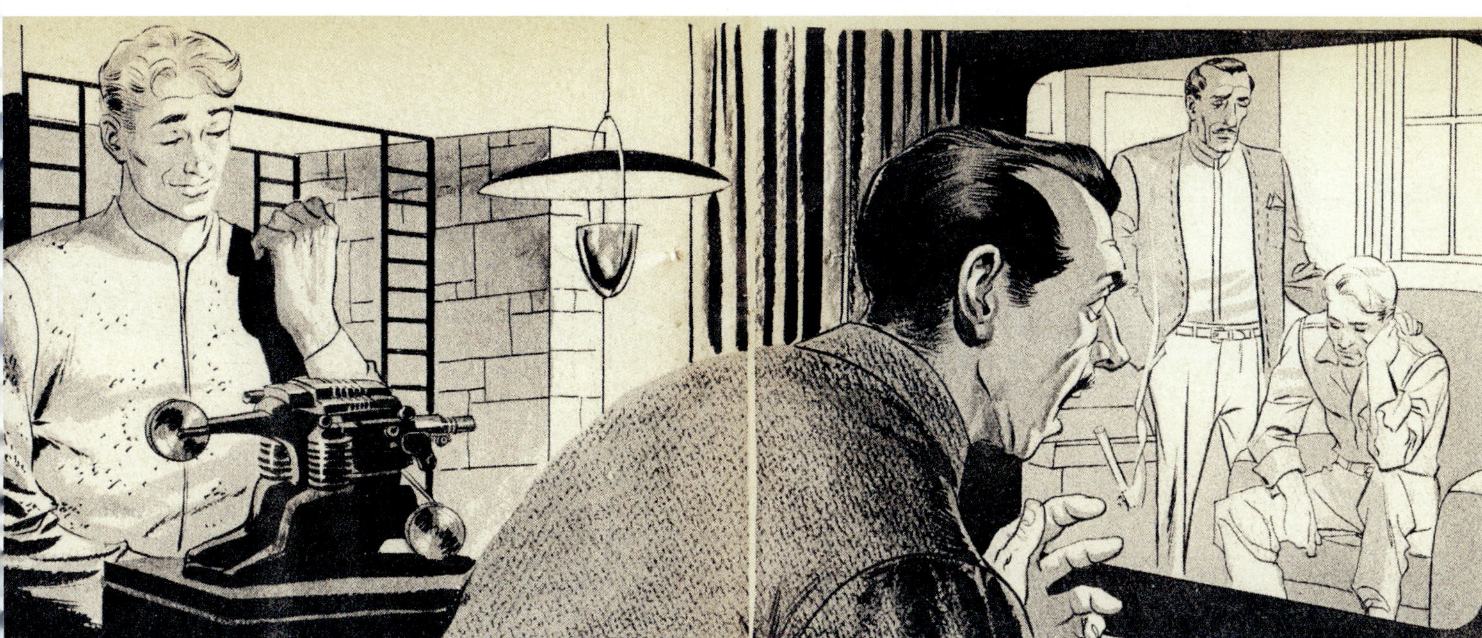

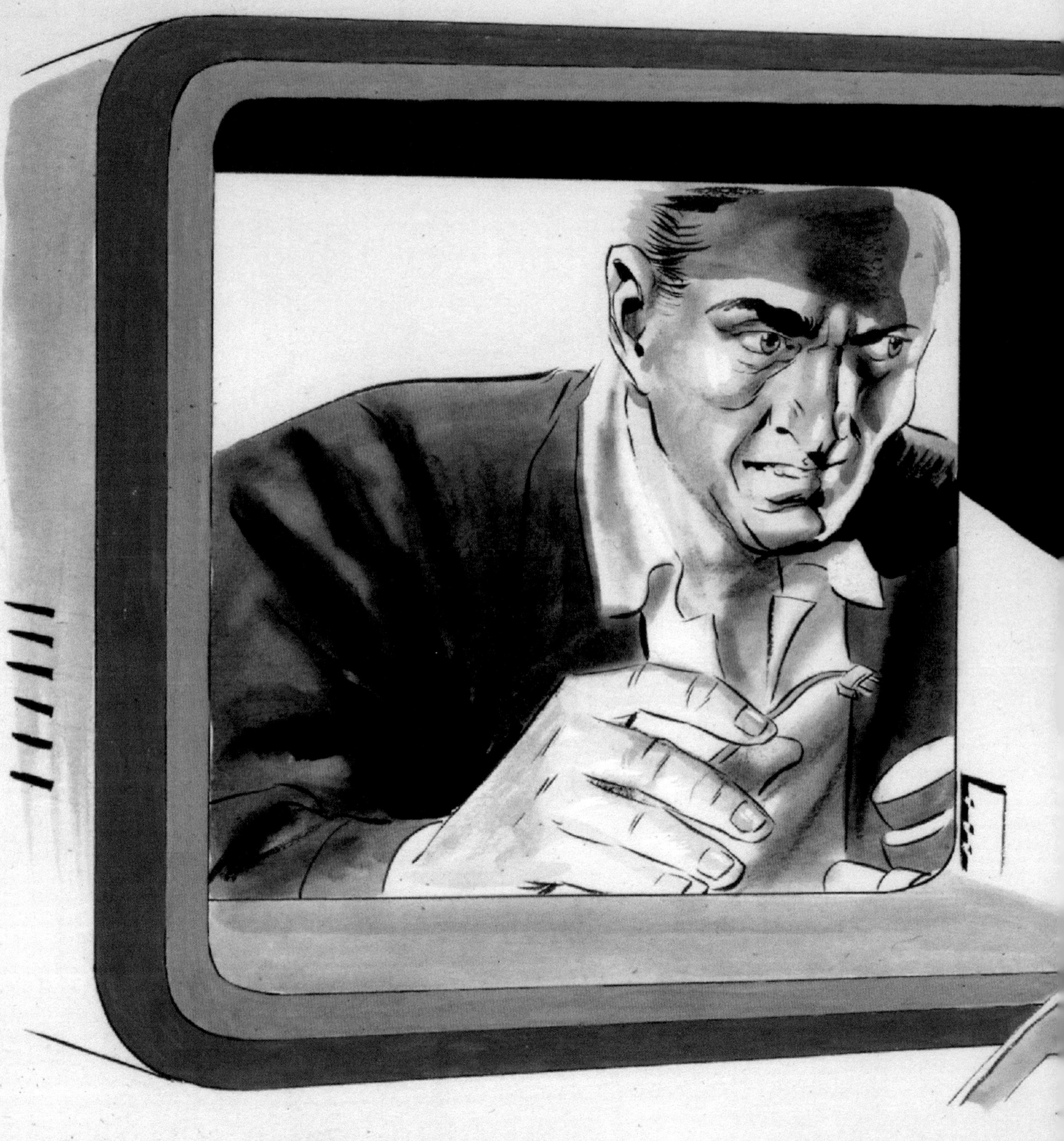

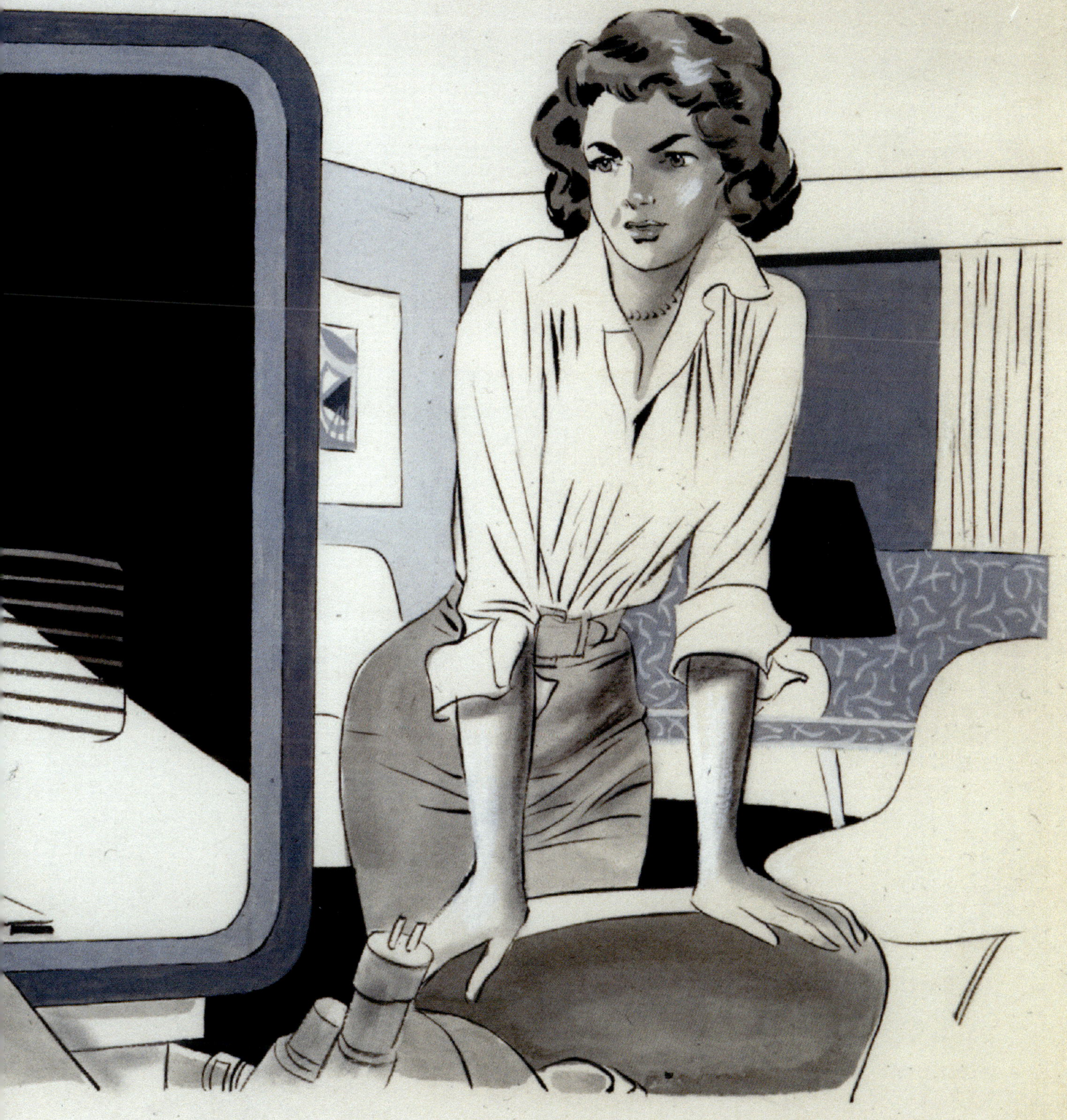

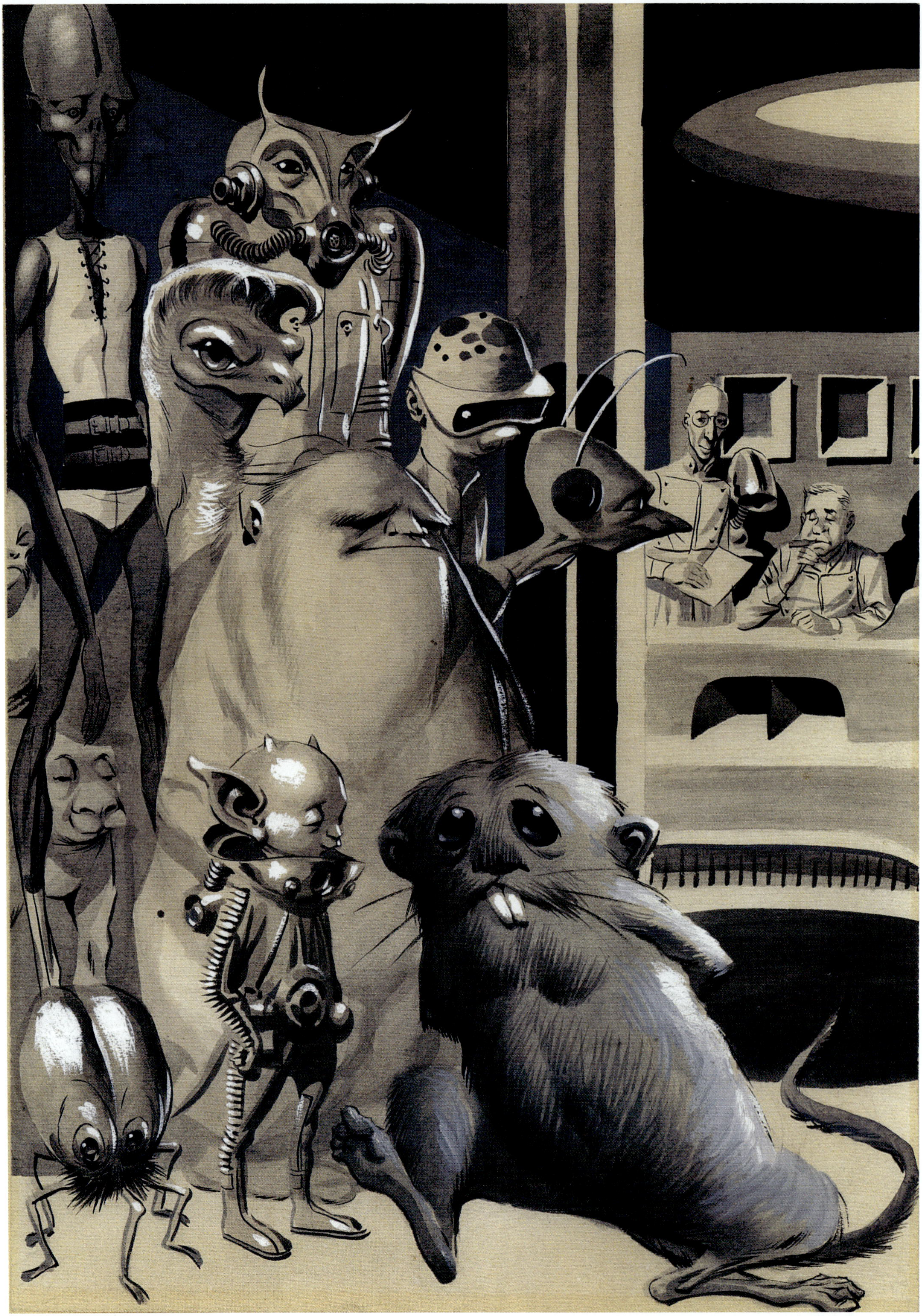

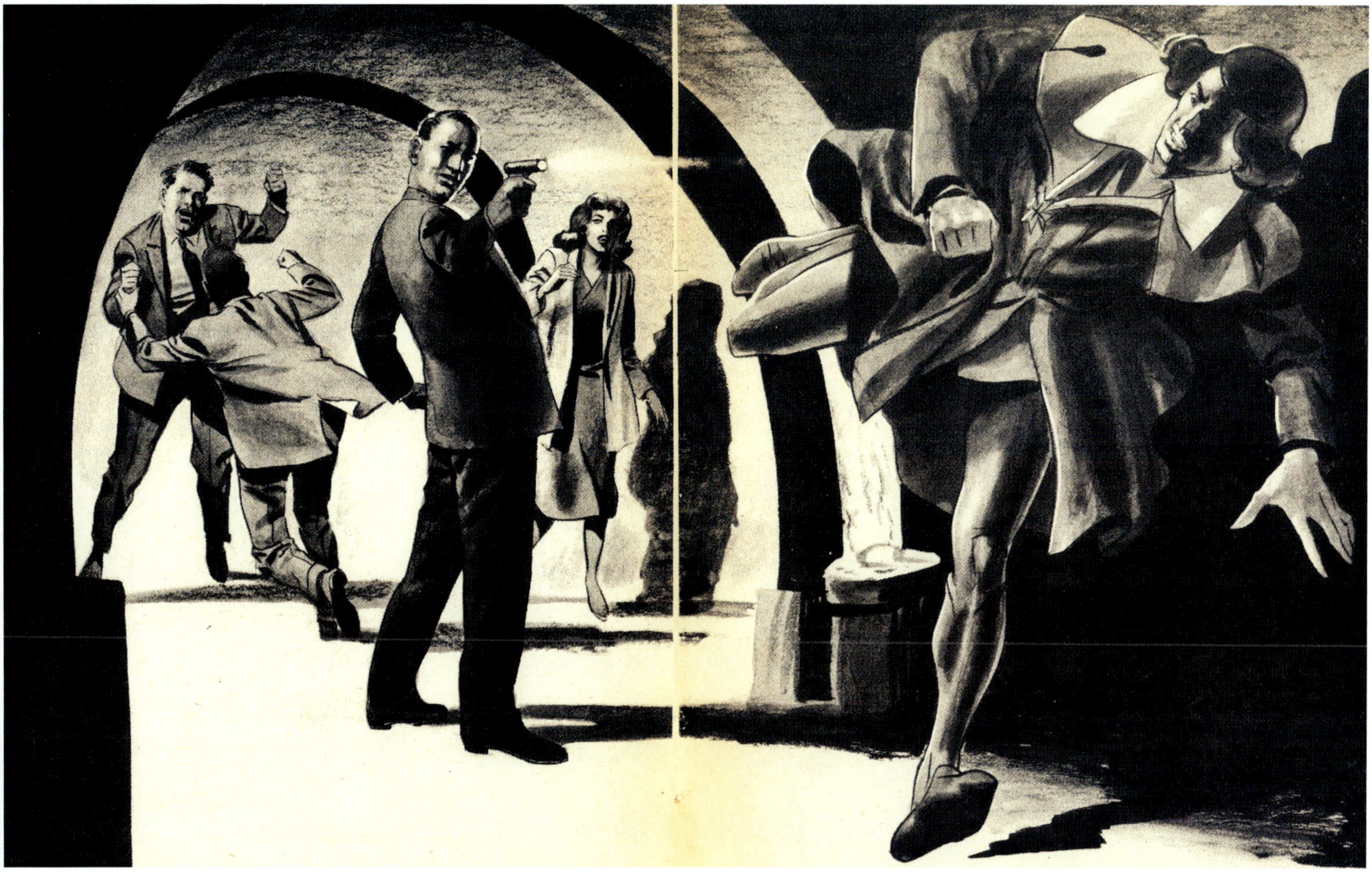

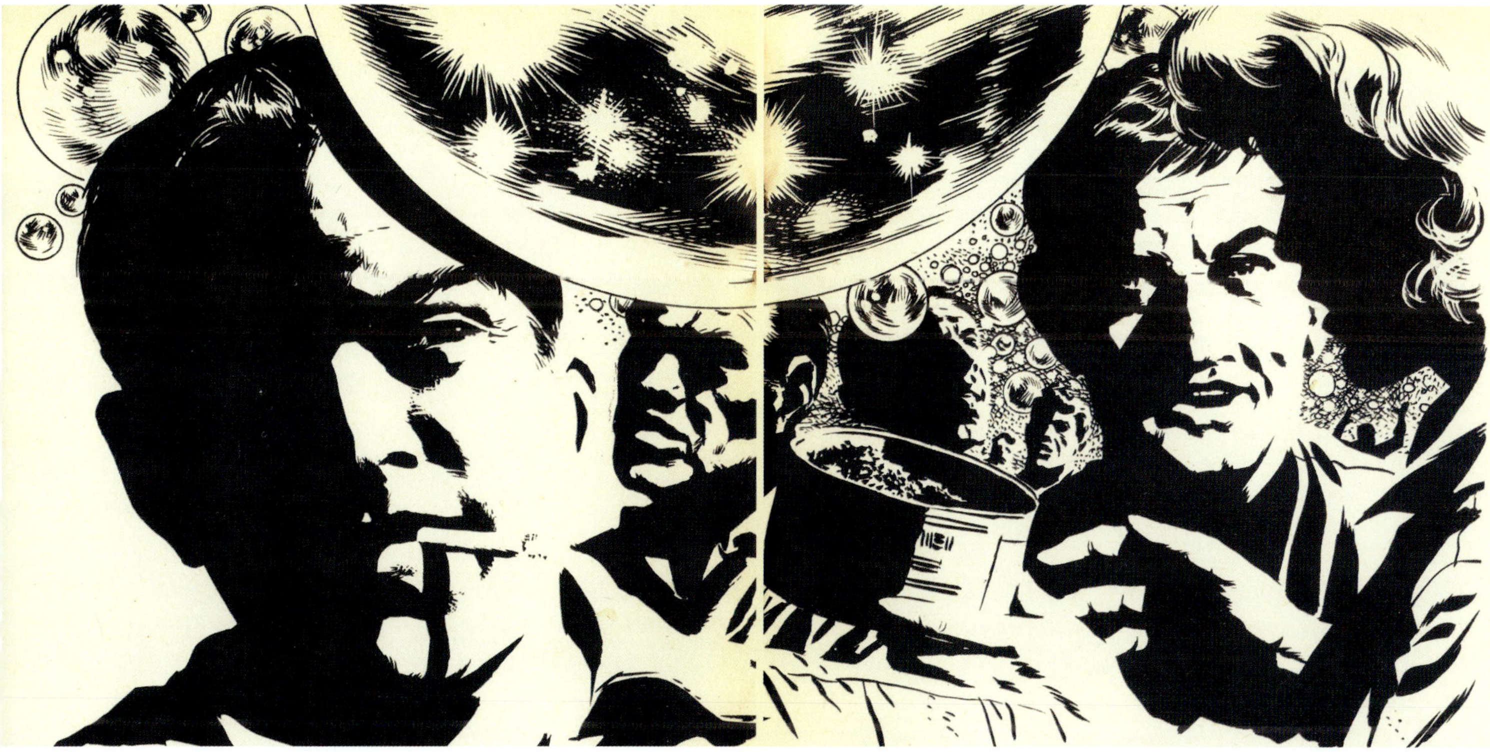

Top: *Galaxy Science Fiction* October 1958 Illustration for "Lisbon Cubed" by William Tenn

Bottom: *Galaxy Science Fiction* August 1959 Illustration for "The Waging of the Peace" by Frederik Pohl

Opposite: *Galaxy Science Fiction* November 1958 Illustration for "Birds of a Feather" by Robert Silverberg

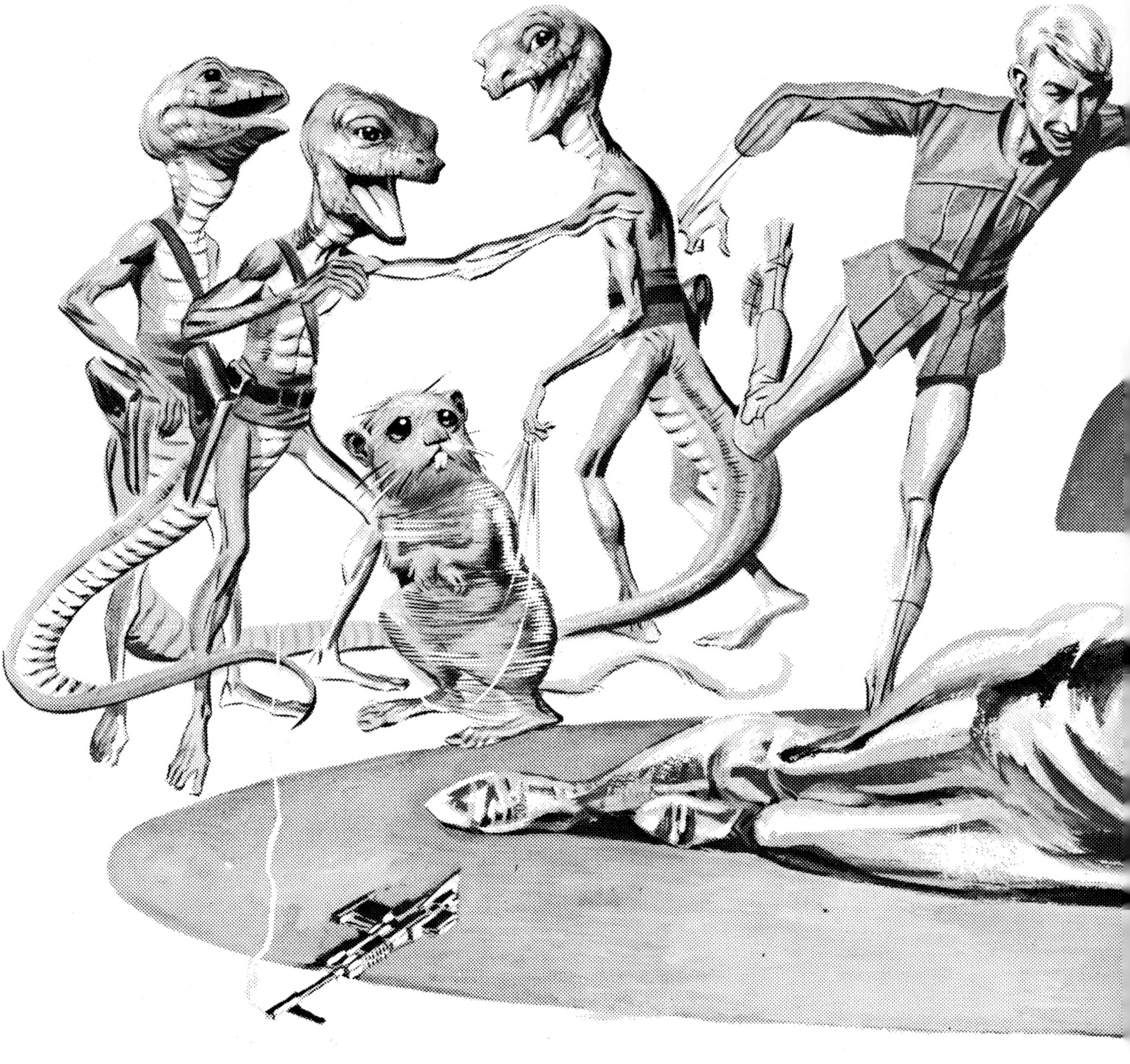

Galaxy Science Fiction November 1958 Illustration for "Birds of a Feather" by Robert Silverberg

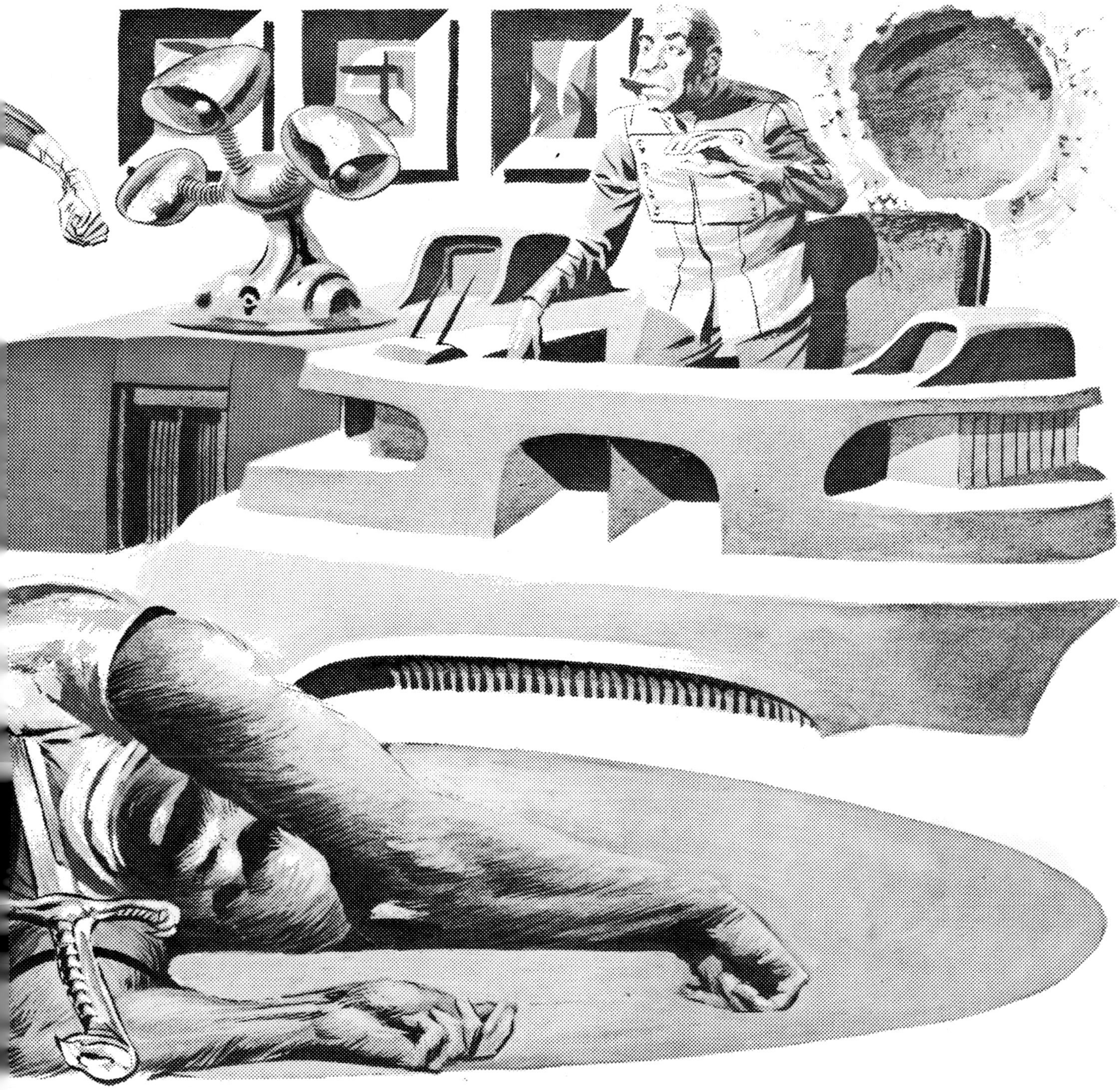

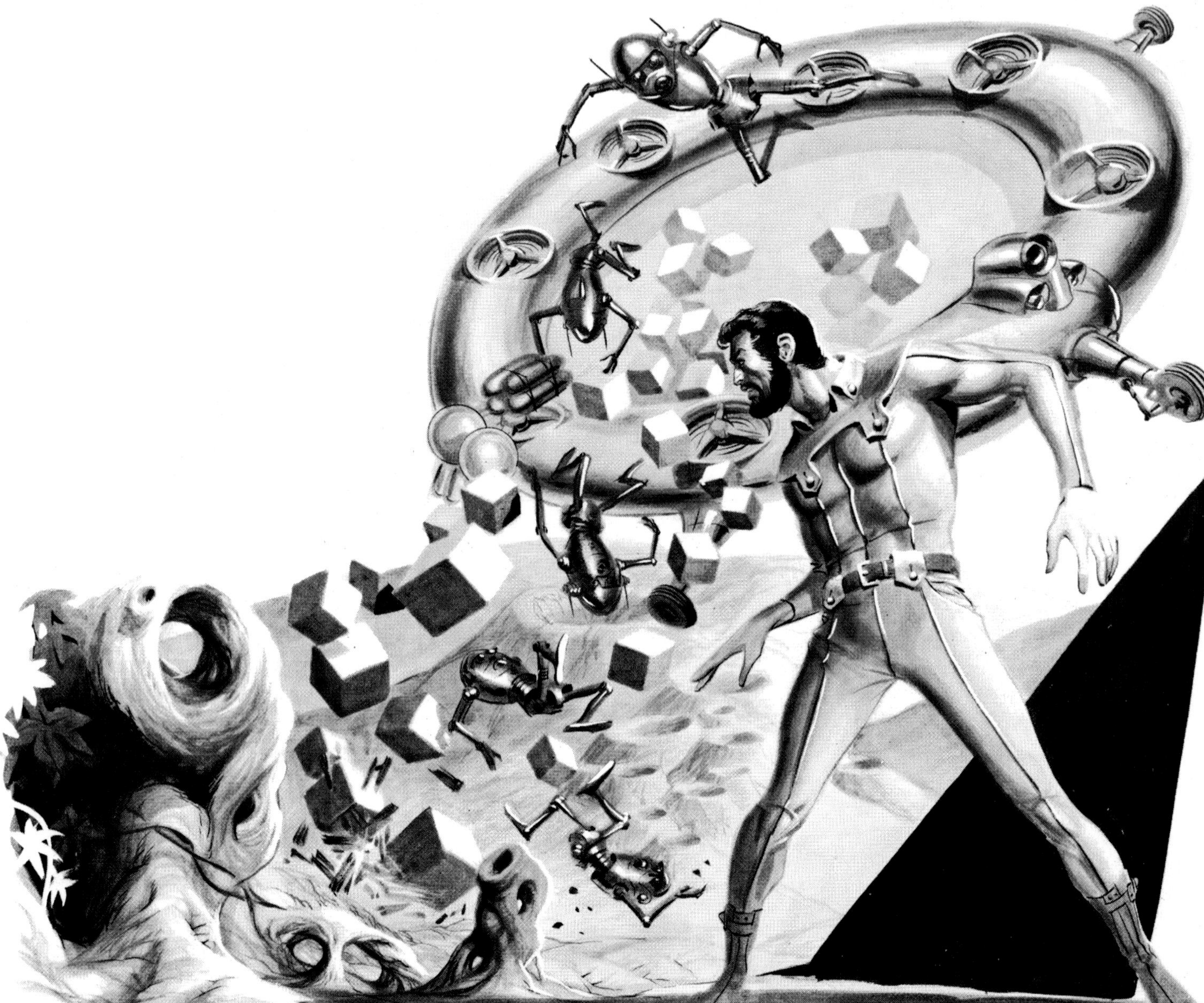

Galaxy Science Fiction February 1959 Illustration for "Installment Plan" by Clifford D. Simak

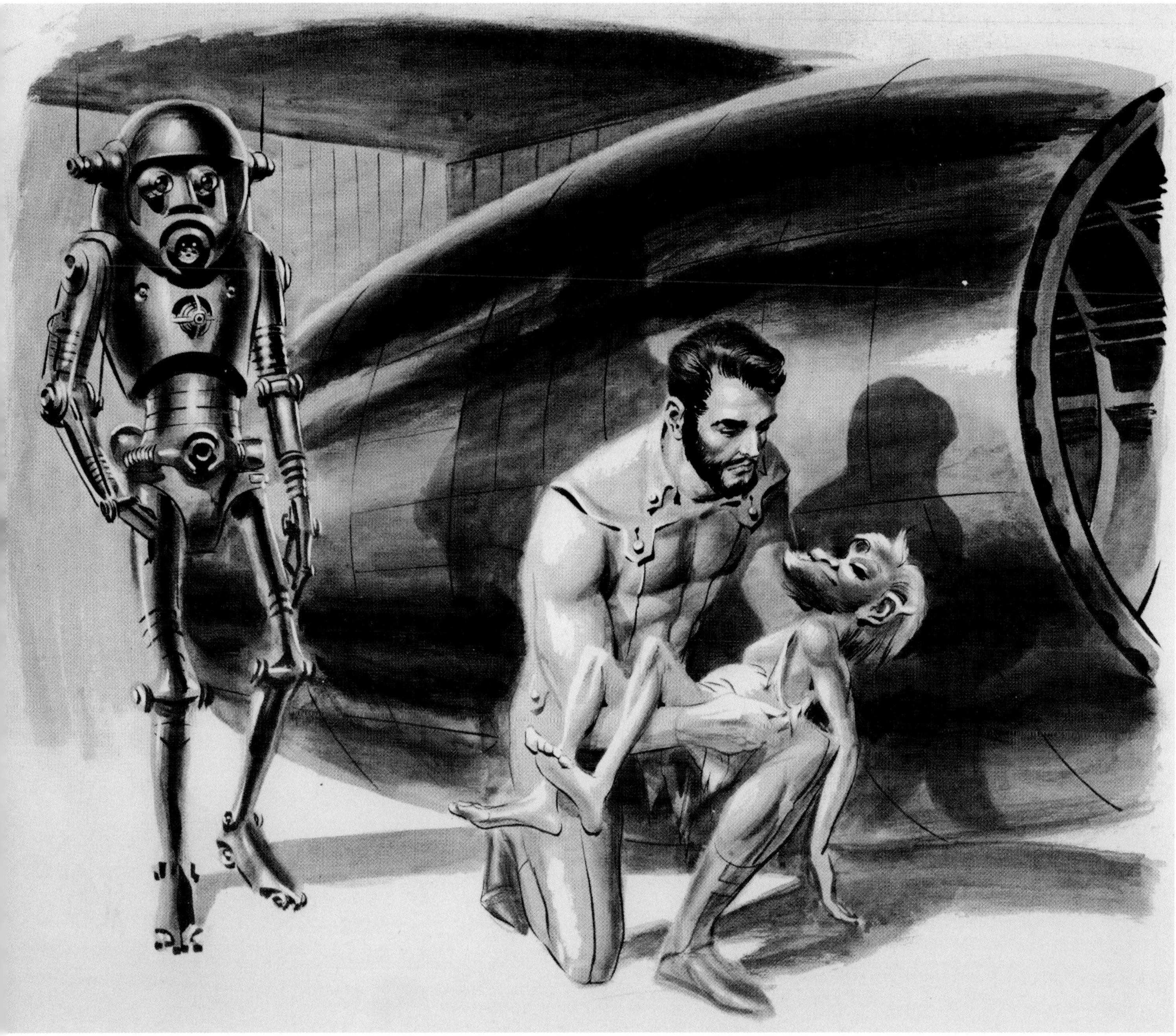

Galaxy Science Fiction February 1959 Illustration for "Installment Plan" by Clifford D. Simak

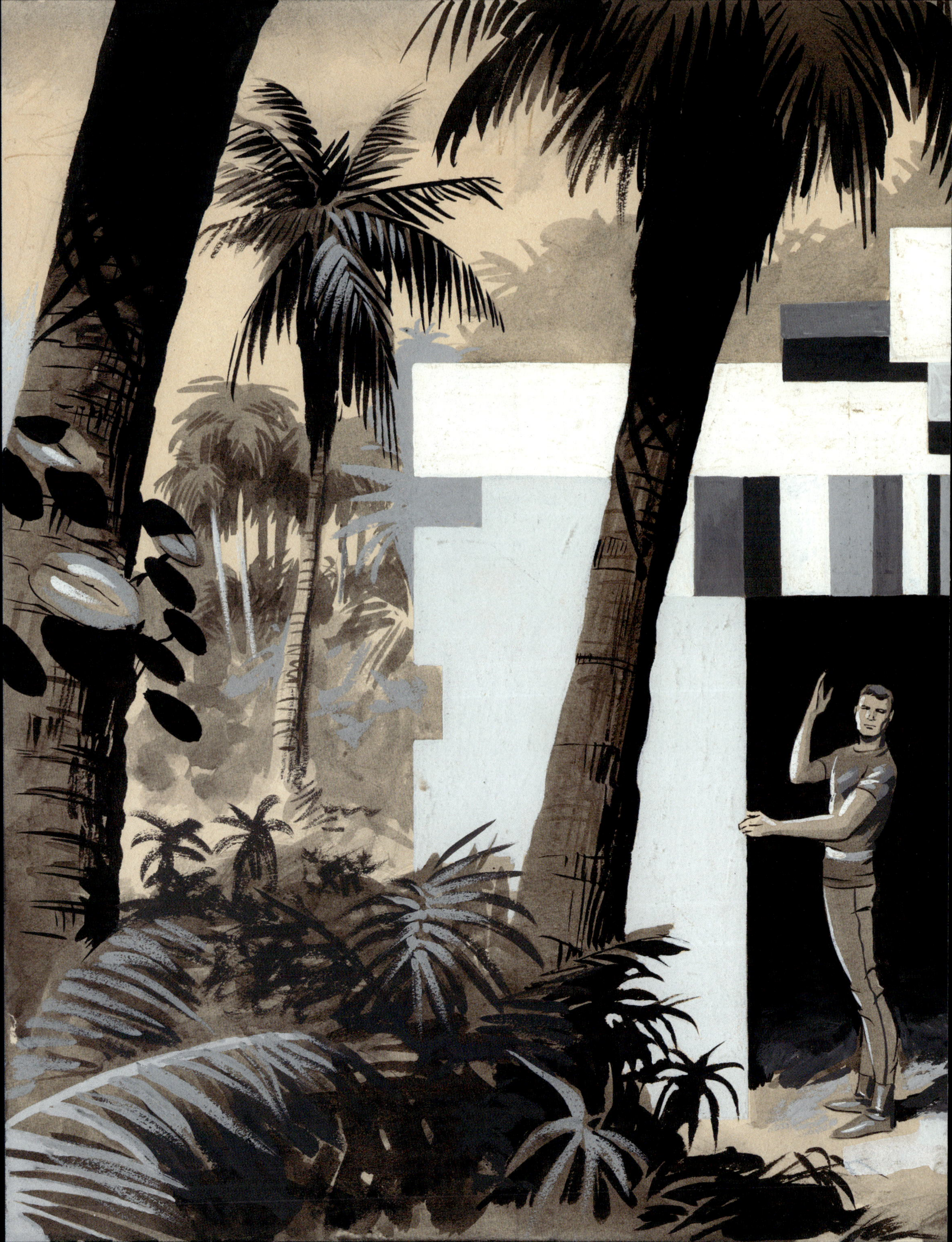

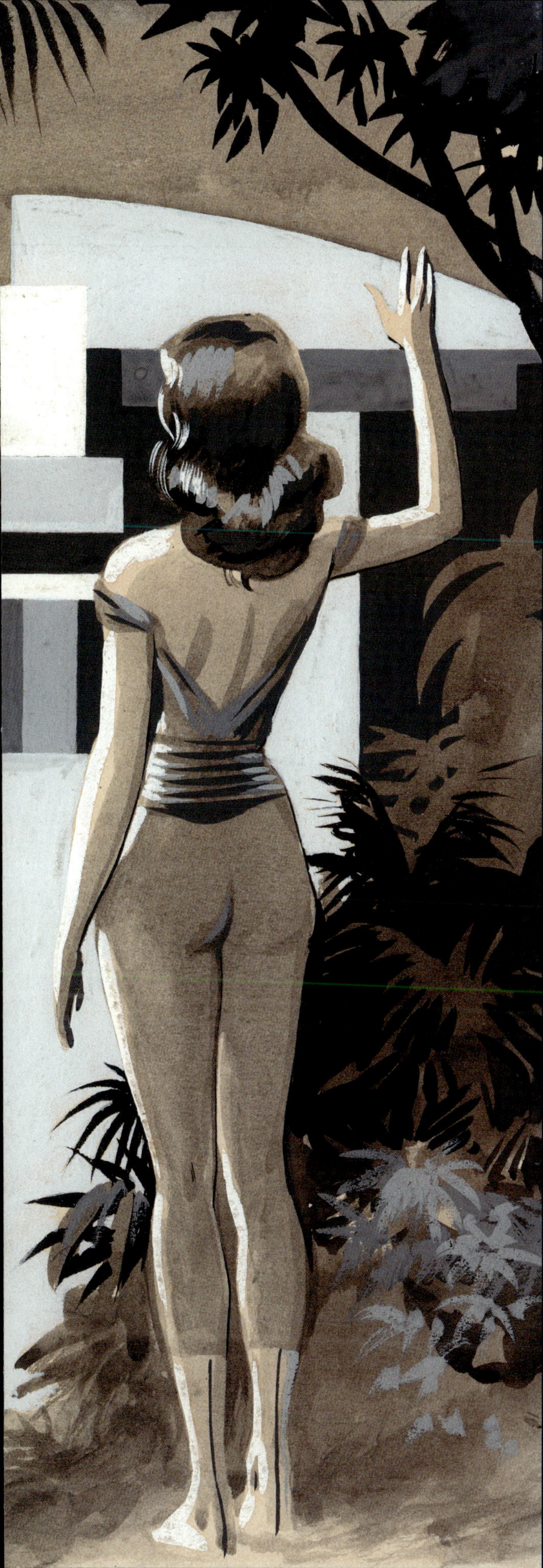

Galaxy Science Fiction February 1959
Illustration for "Time Killer" part 4 by Robert Sh

Next page:
Top: *Galaxy Science Fiction* December 1958
Illustration for "Ullward's Retreat" by Jack Vanc

Bottom: *Galaxy Science Fiction* December 1958
Illustration for "Time Killer" part 3 by Robert Sh

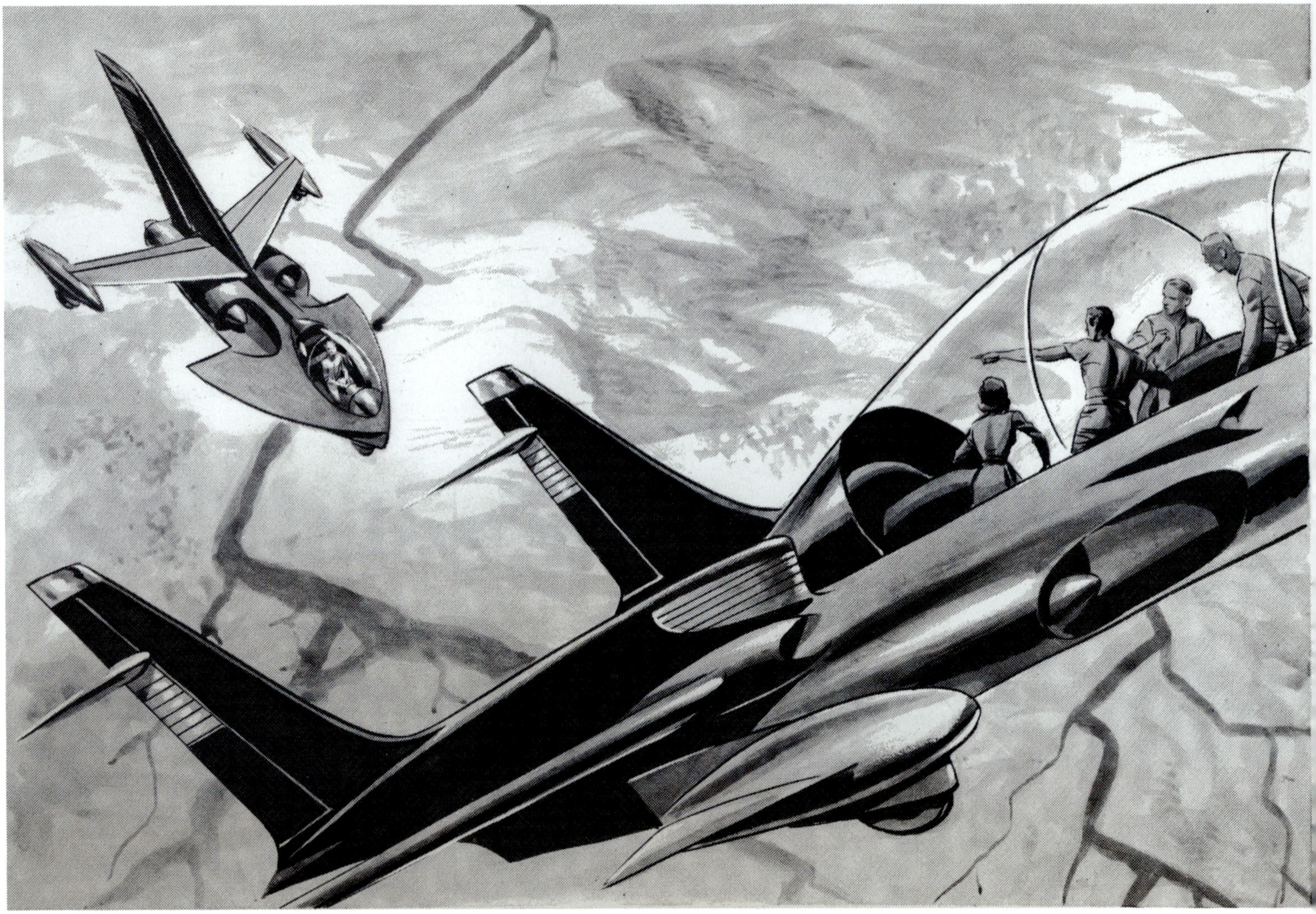

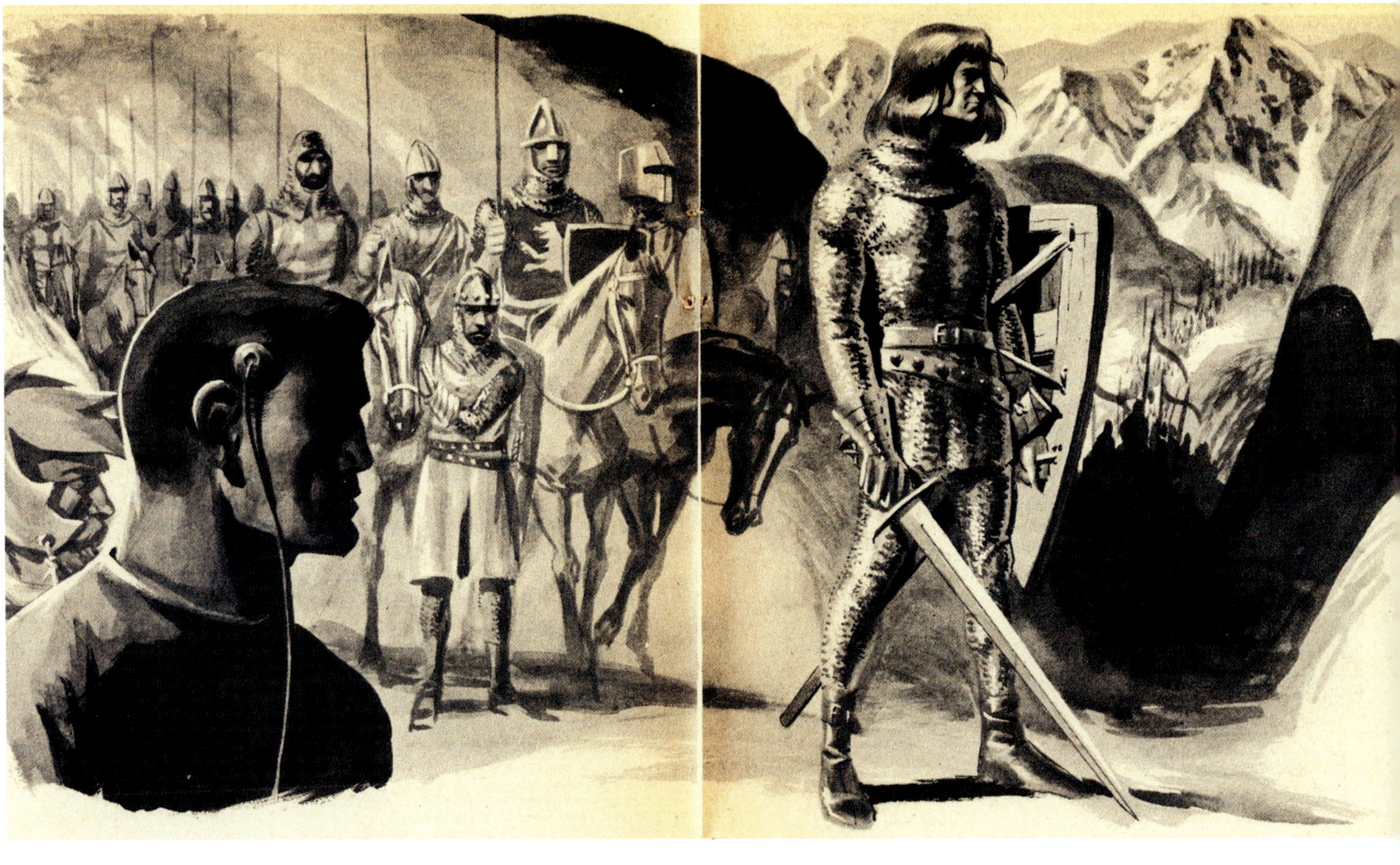

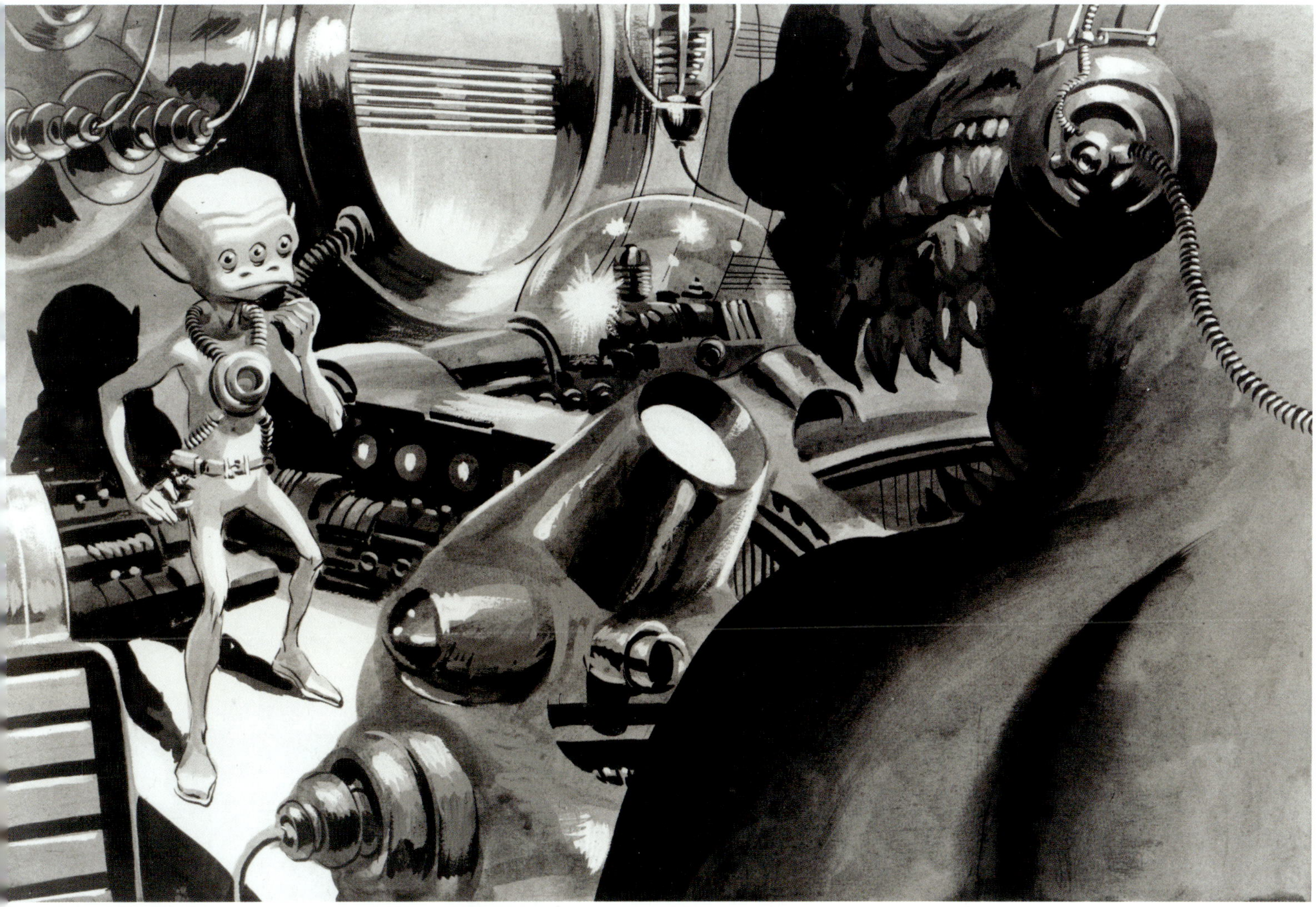

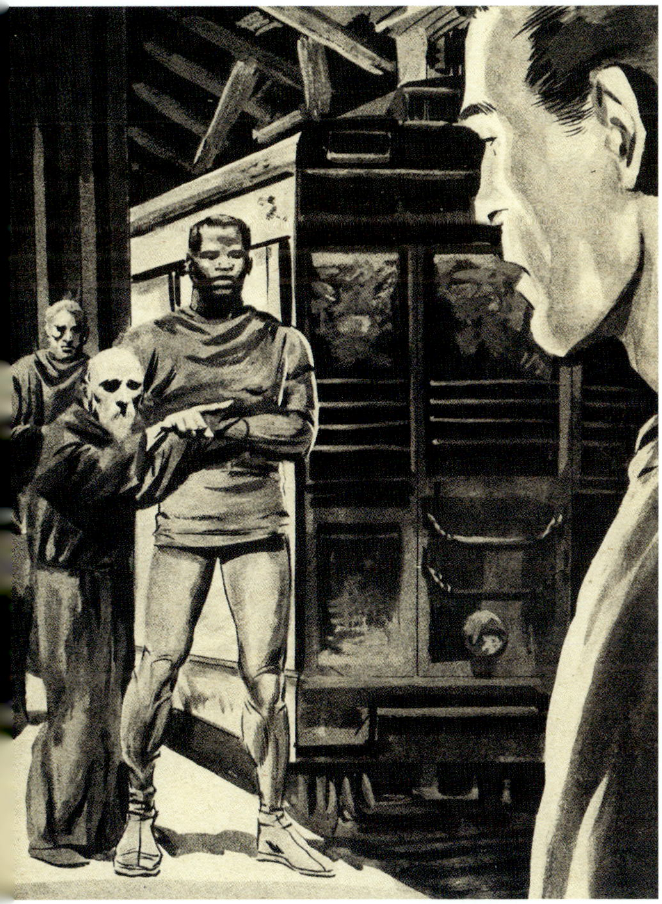

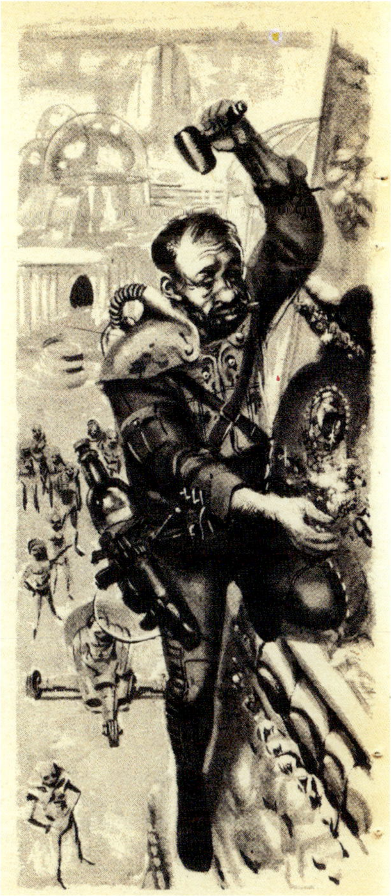

Clockwise from top:
Galaxy Science Fiction February 1959
Illustration for "I Plinglot—Who You?"
by Frederik Pohl

Galaxy Science Fiction June 1958
Illustration for "The Gentlest Unpeople"
by Frederik Pohl

Galaxy Science Fiction February 1959
Illustration for "Time Killer" part 4
by Robert Sheckley

Next spread:
Galaxy Science Fiction February 1959
Illustration for "Time Killer" part 4
by Robert Sheckley

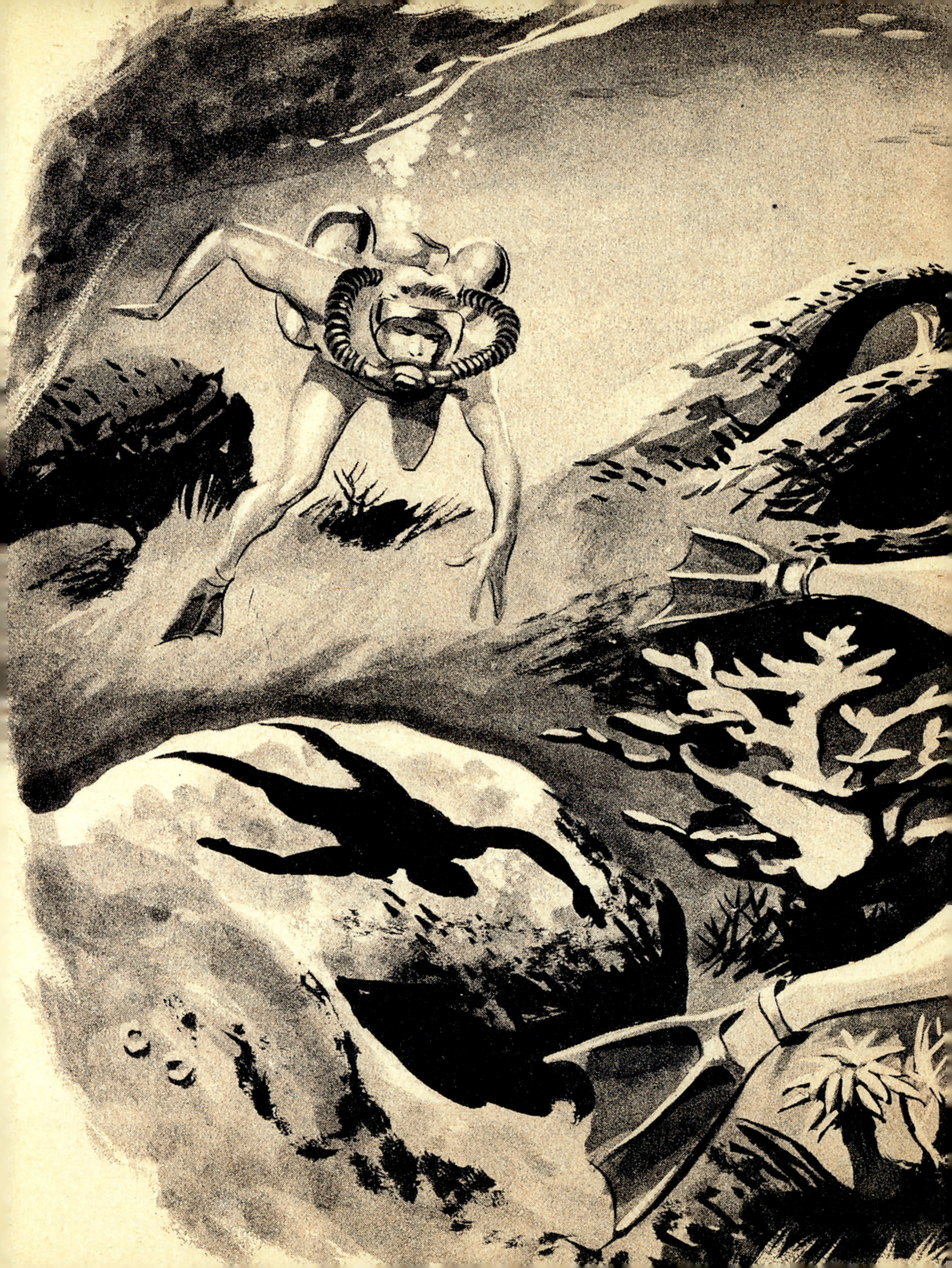

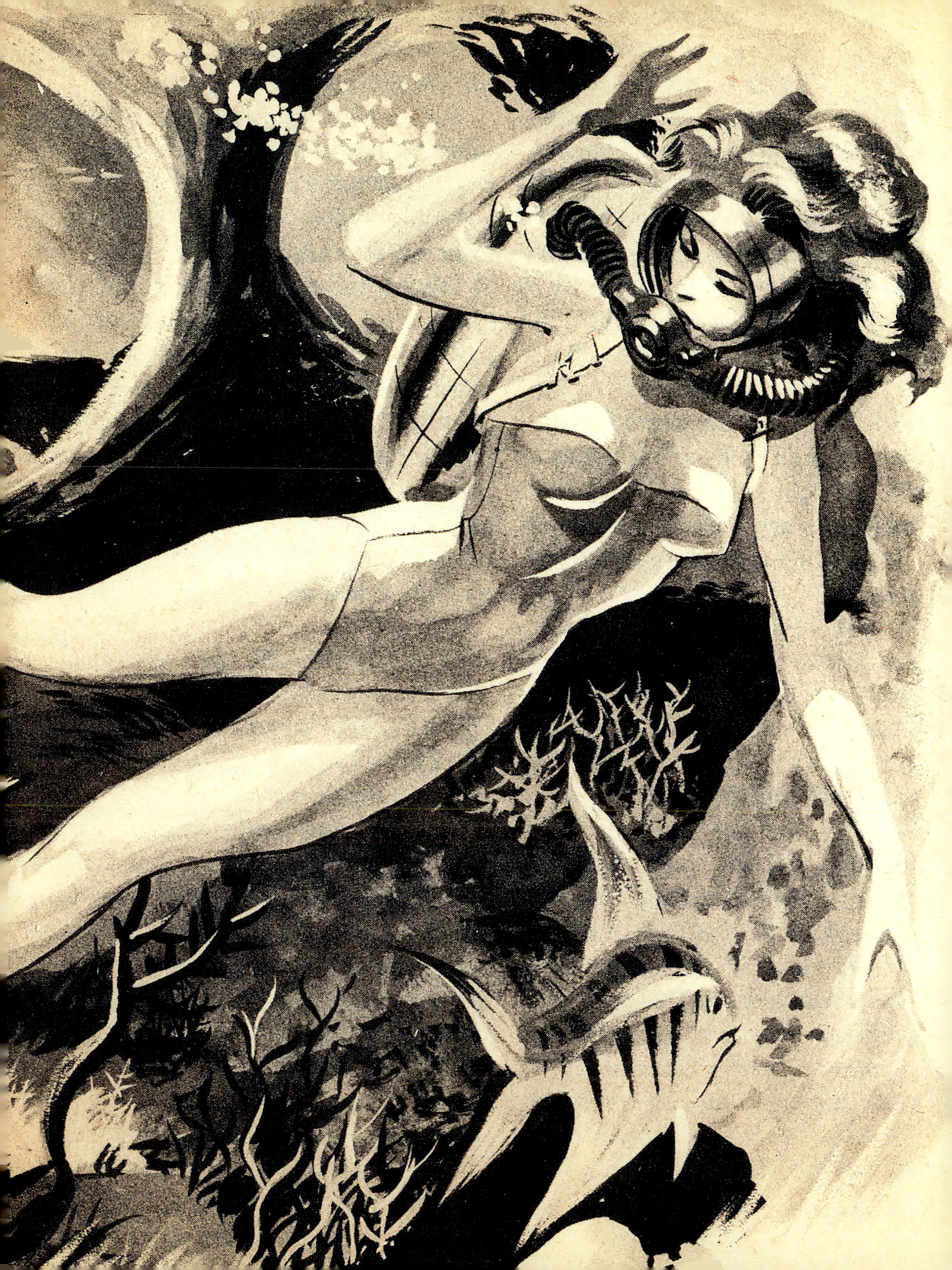

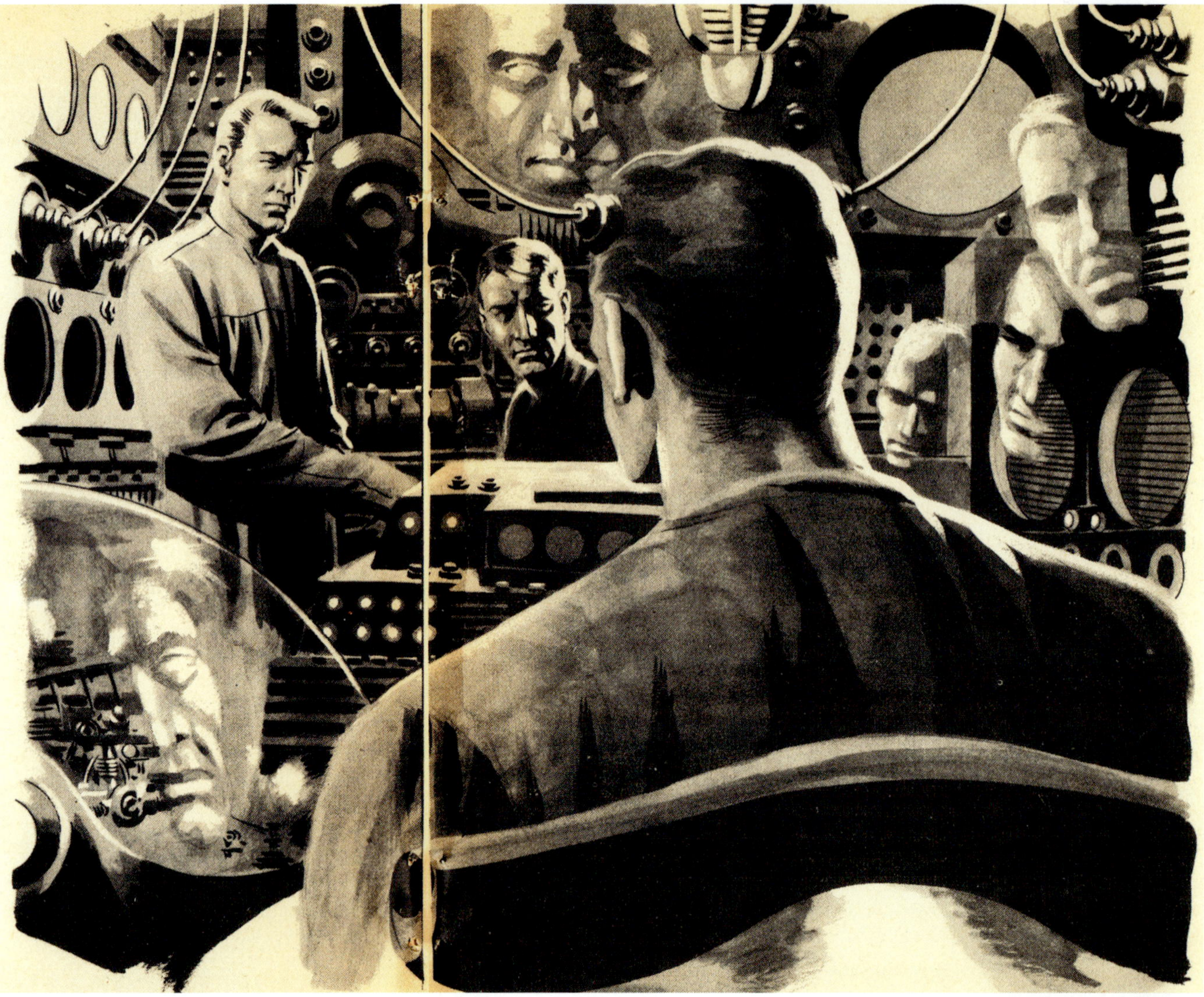

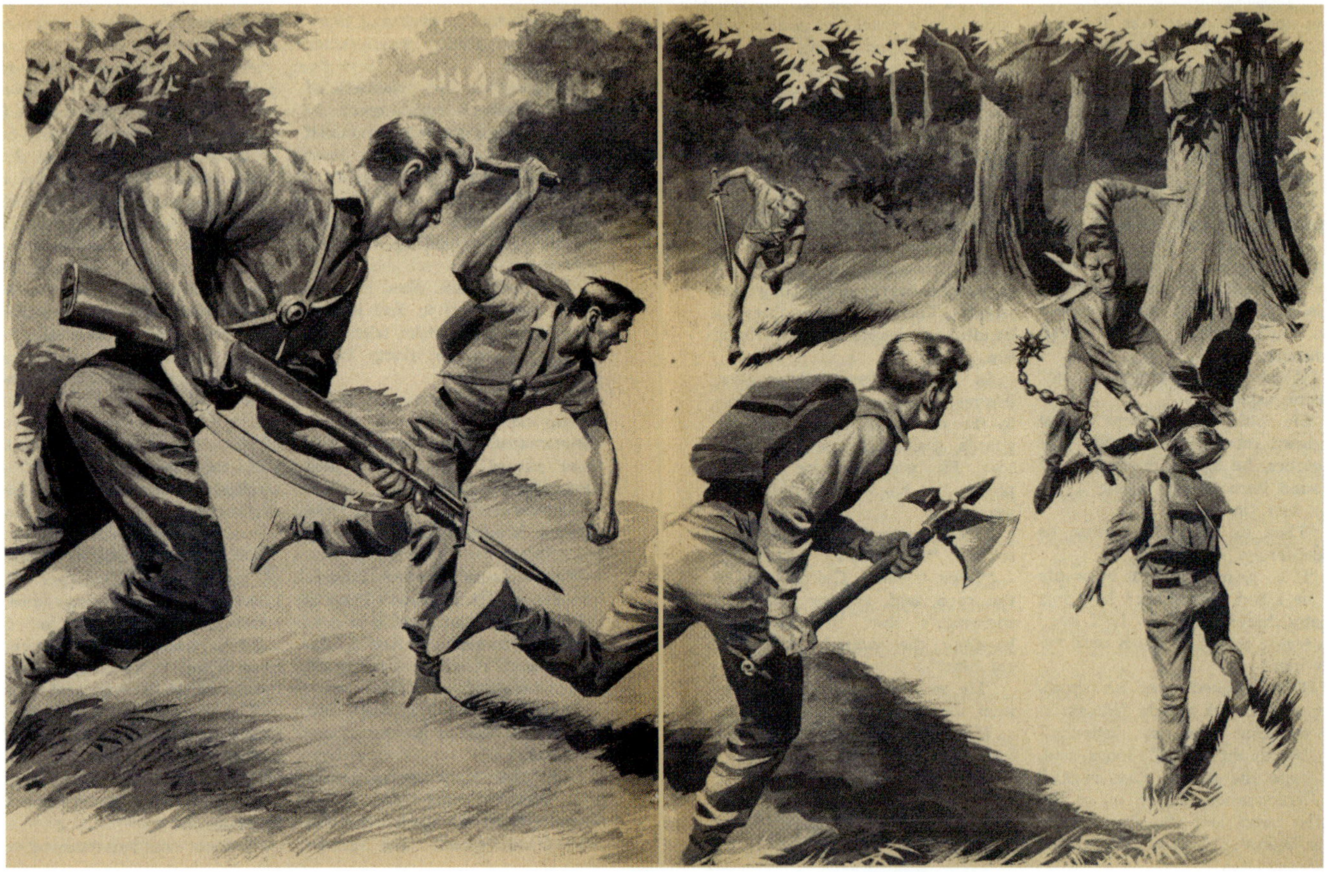

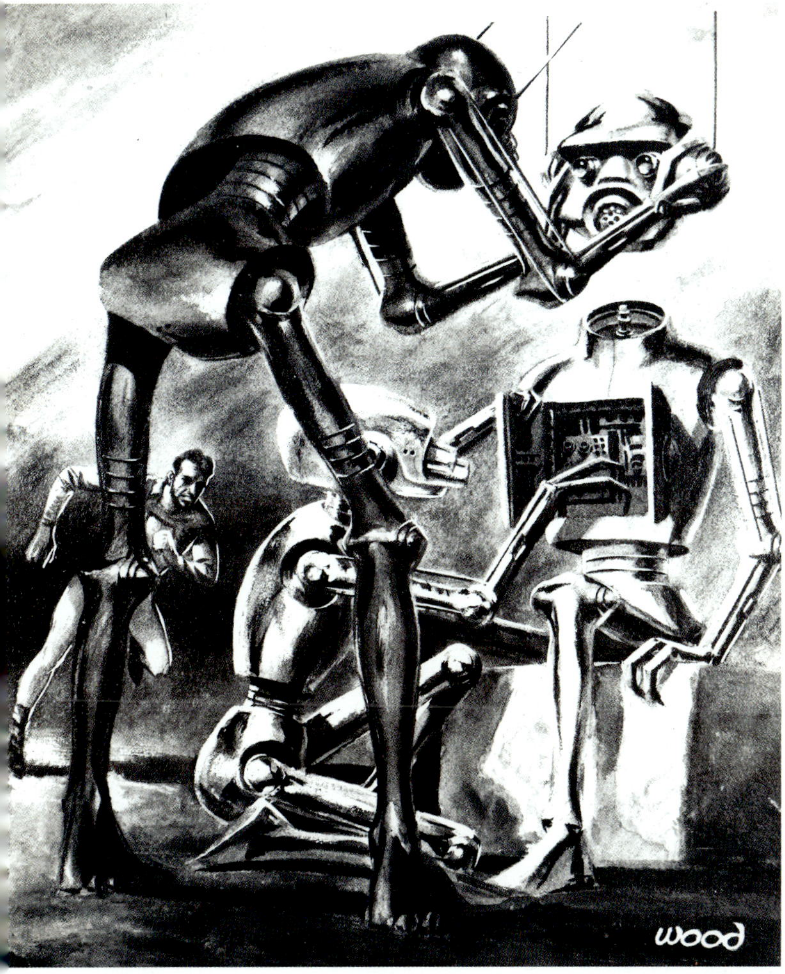
wood

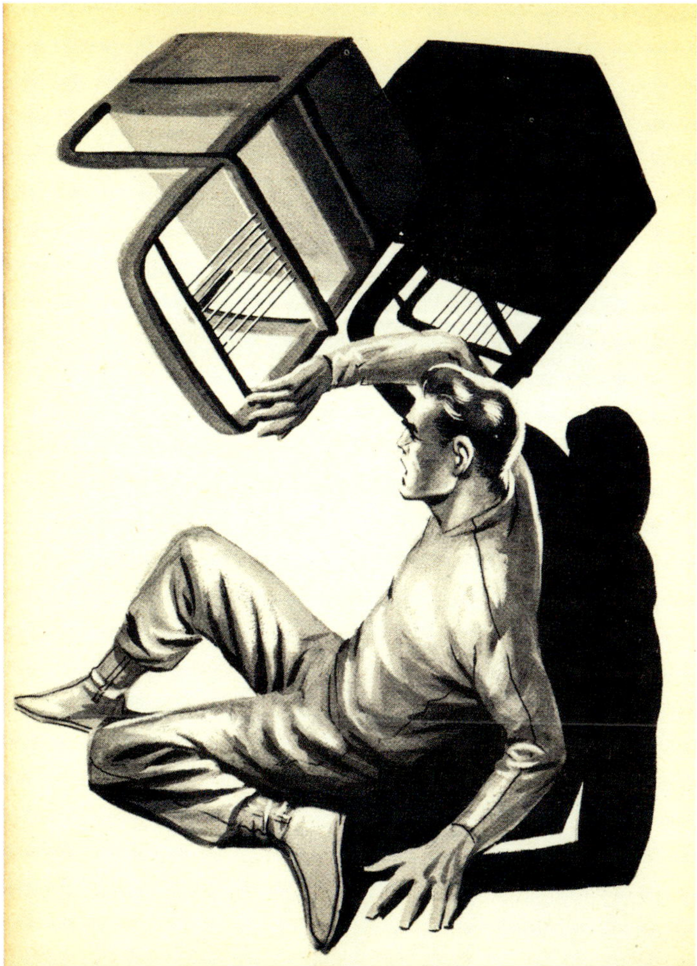

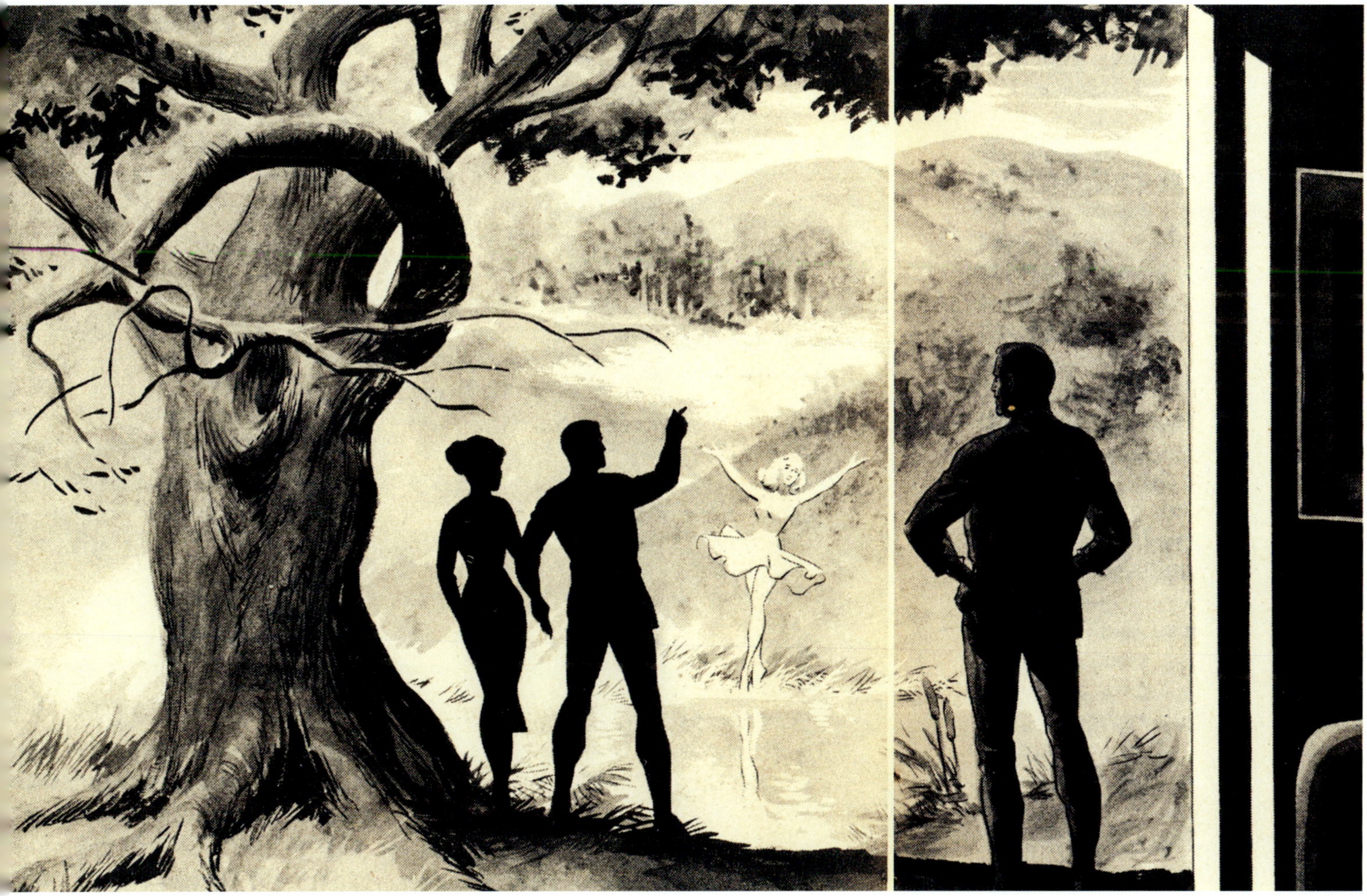

Previous spread:
Page 84, top: *Galaxy Science Fiction* December 1958
Illustration for "Time Killer" part 3 by Robert Sheckley

Page 84, bottom: *Galaxy Science Fiction* November 1958
Illustration for "Time Killer" part 2 by Robert Sheckley

Page 85, clockwise from upper left:
Galaxy Science Fiction February 1959 (Cover)
Illustration for "Installment Plan" by Clifford D. Simak

Galaxy Science Fiction December 1958
Illustration for "Time Killer" part 3 by Robert Sheckley

Galaxy Science Fiction December 1958
Illustration for "Ullward's Retreat" by Jack Vance

This page, clockwise from upper right:
Galaxy Science Fiction February 1959
Illustration for "Insidekick" by J. F. Bone

Galaxy Science Fiction February 1959
Illustration for "I Plinglot—Who You?" by Frederik Pohl

Galaxy Science Fiction February 1959
Illustration for "Installment Plan" by Clifford D. Simak

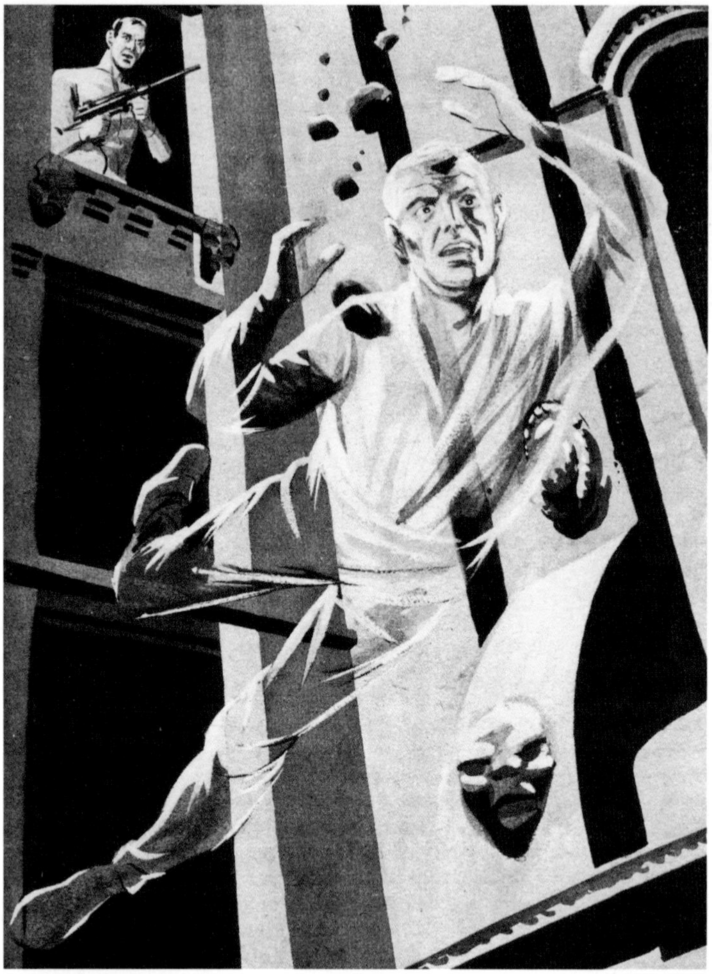

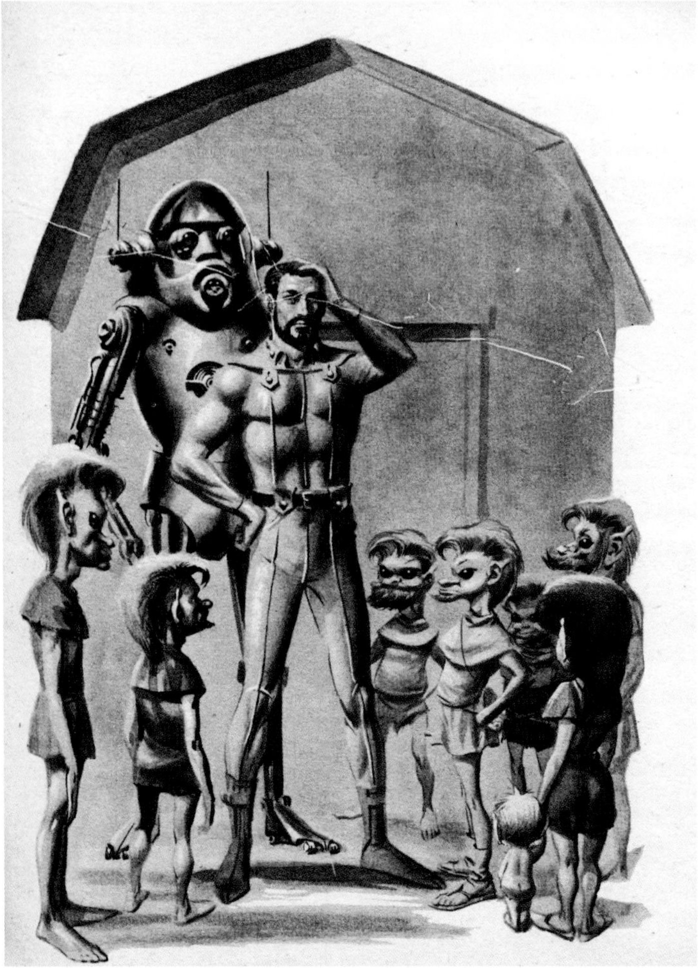

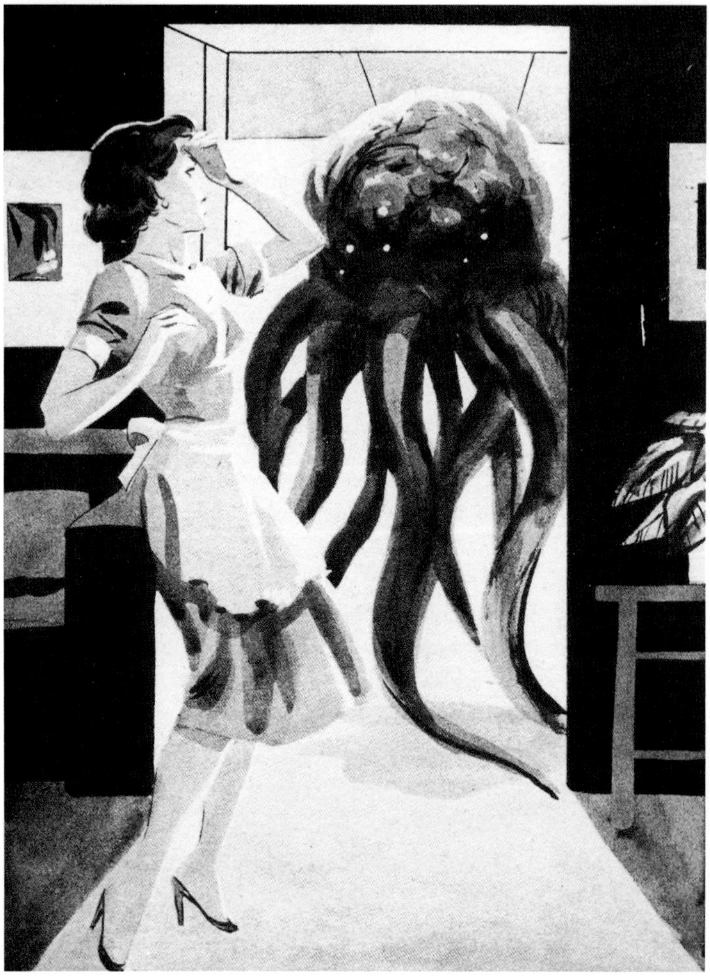

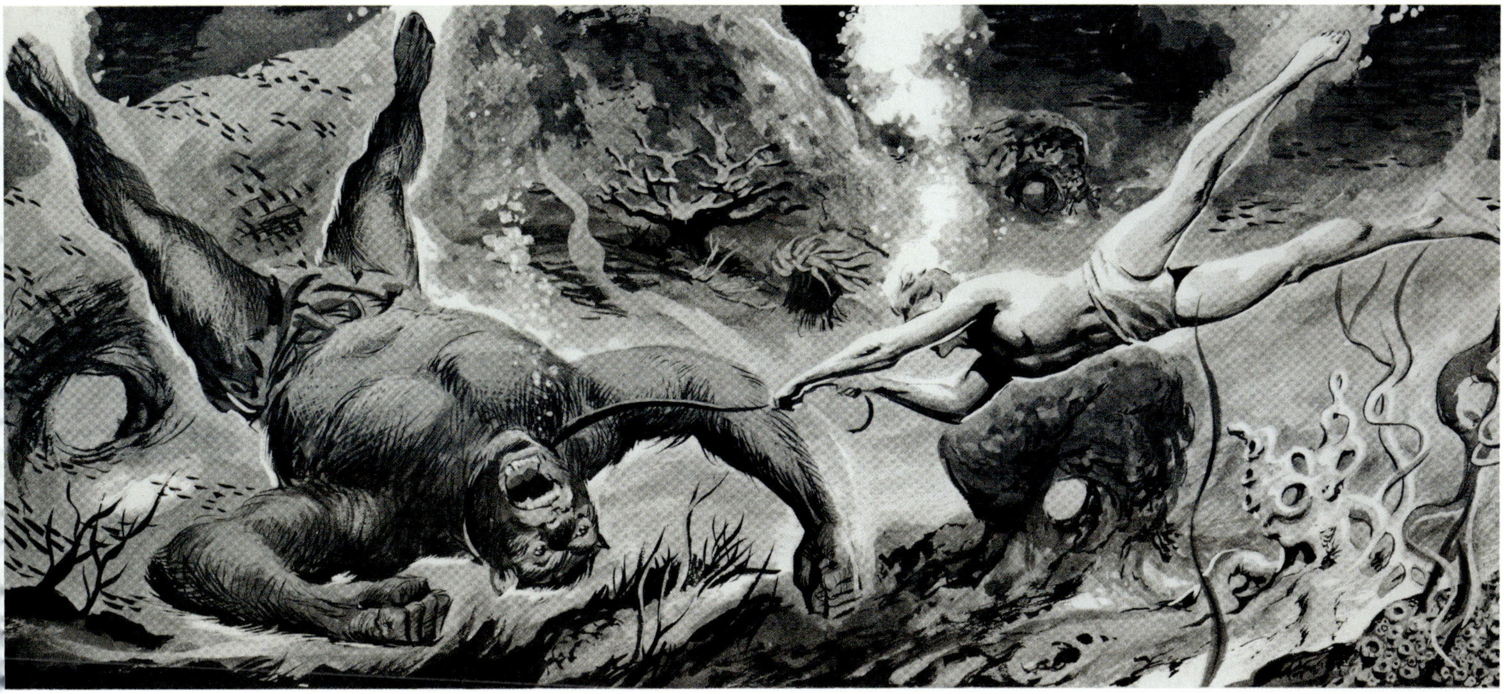

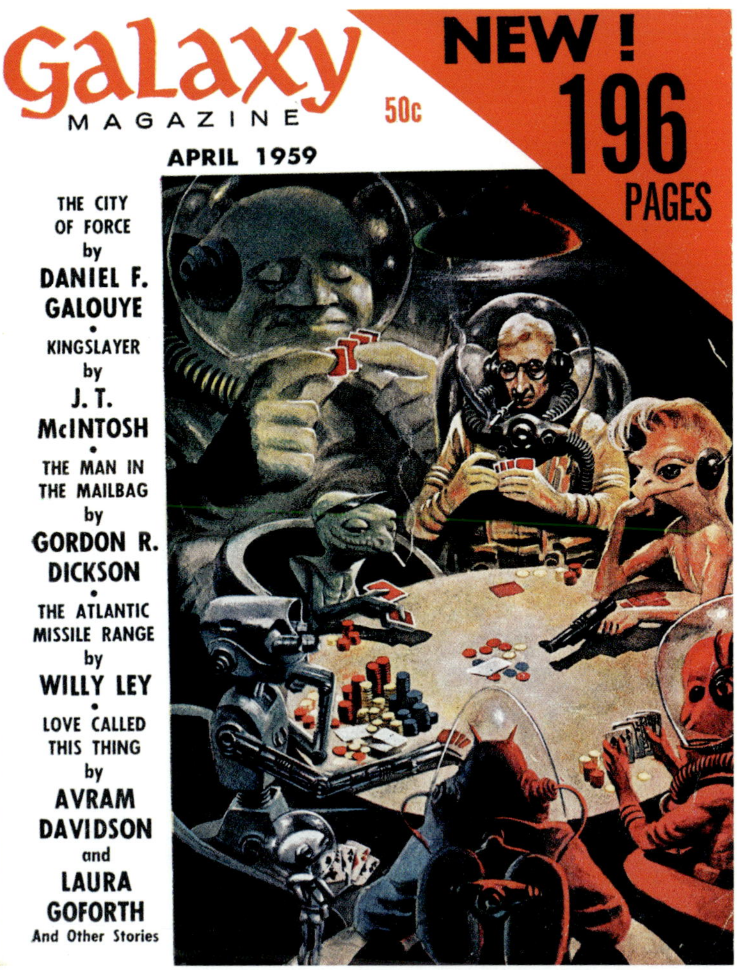

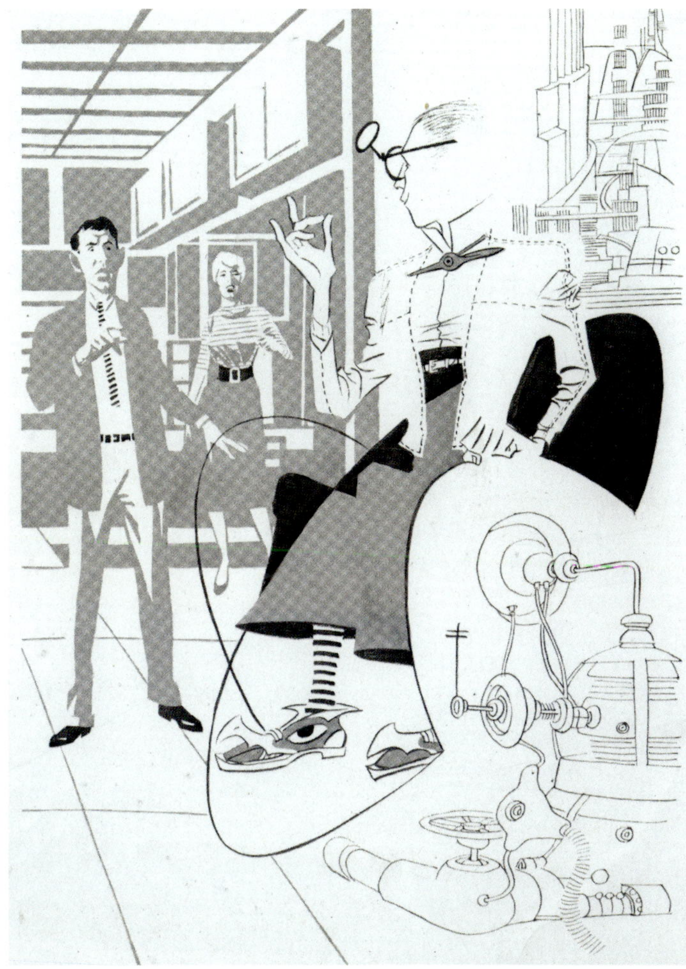

Clockwise from top:
Galaxy Science Fiction April 1959
Illustration for "The Man in the Mailbag" by Gordon R. Dickson

Galaxy Science Fiction April 1959
Illustration for "Security Plan" by Joseph Farrell

Galaxy Science Fiction April 1959 (Cover)
Painting titled "A Full House Beats Togetherness"

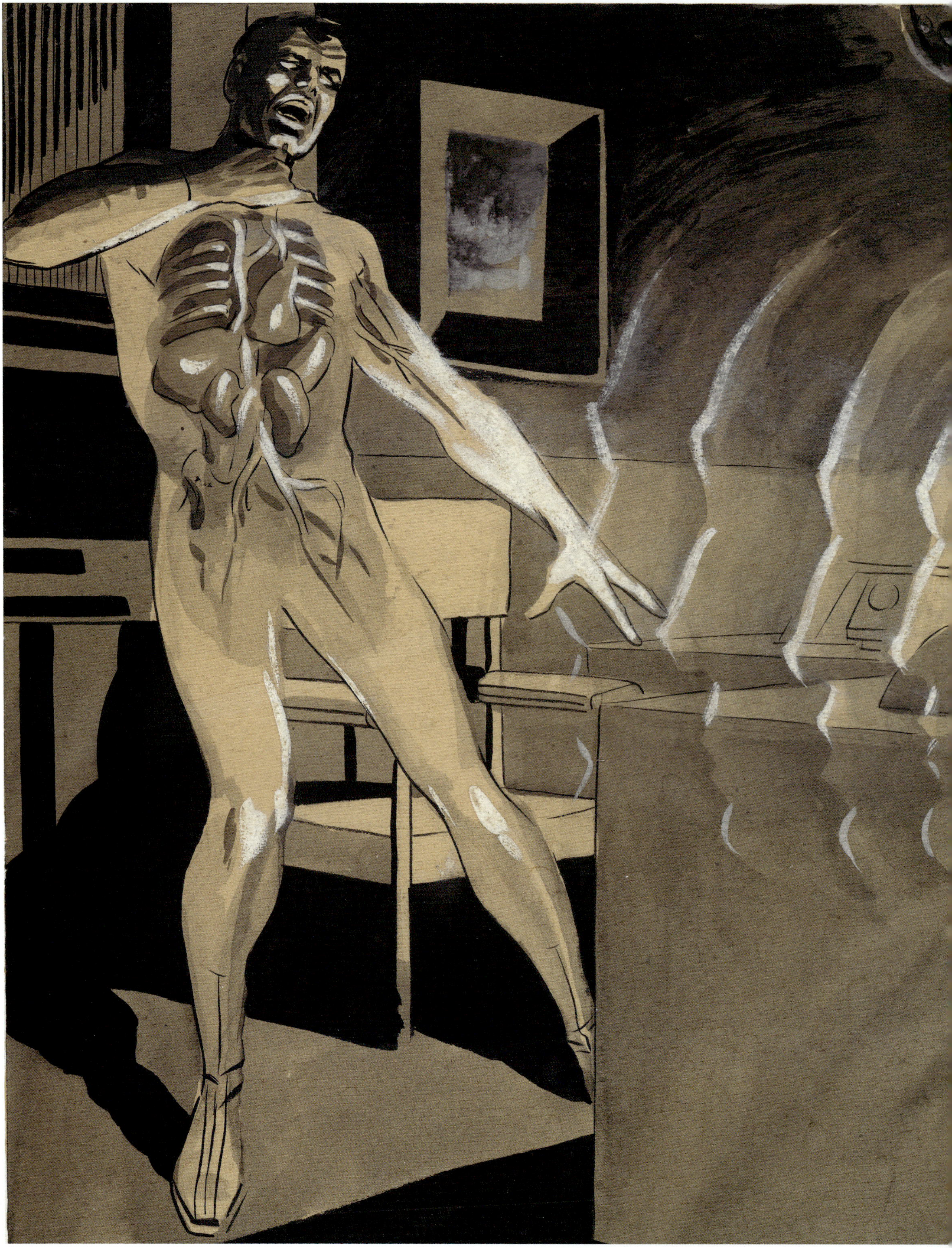

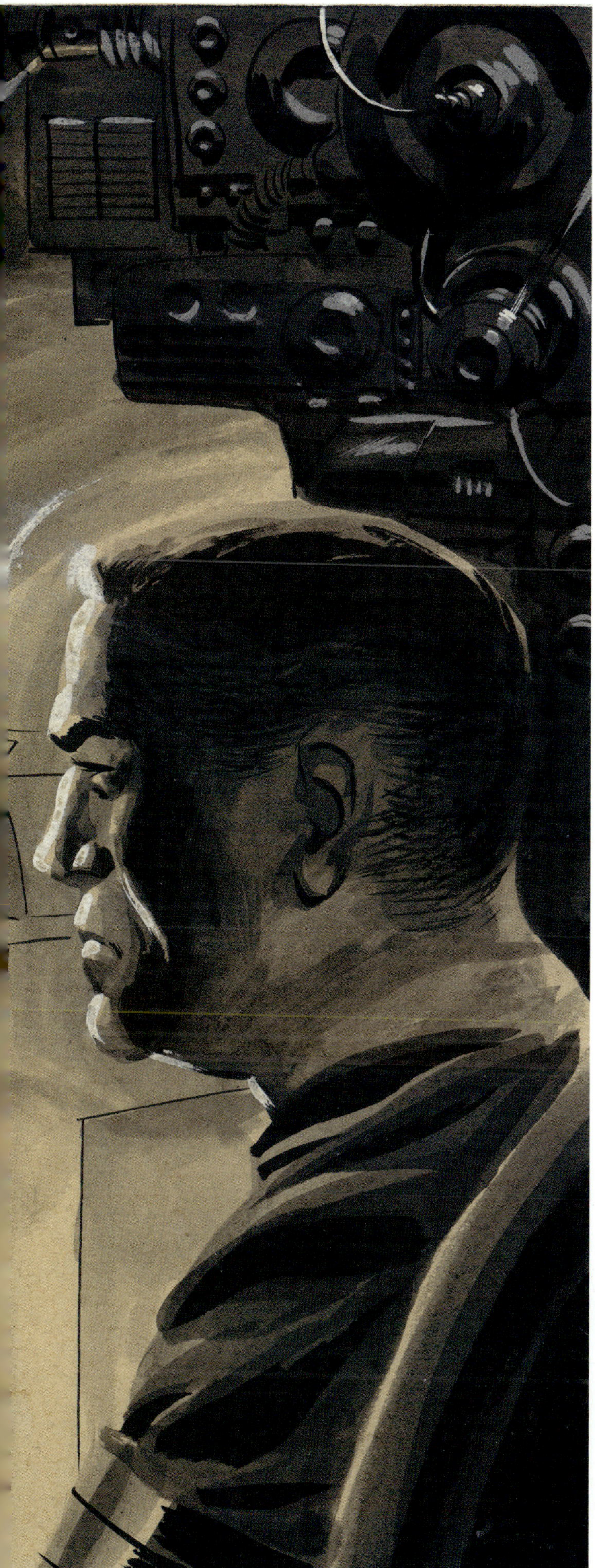

Galaxy Science Fiction February 1959
Illustration for "Insidekick" by J. F. Bone

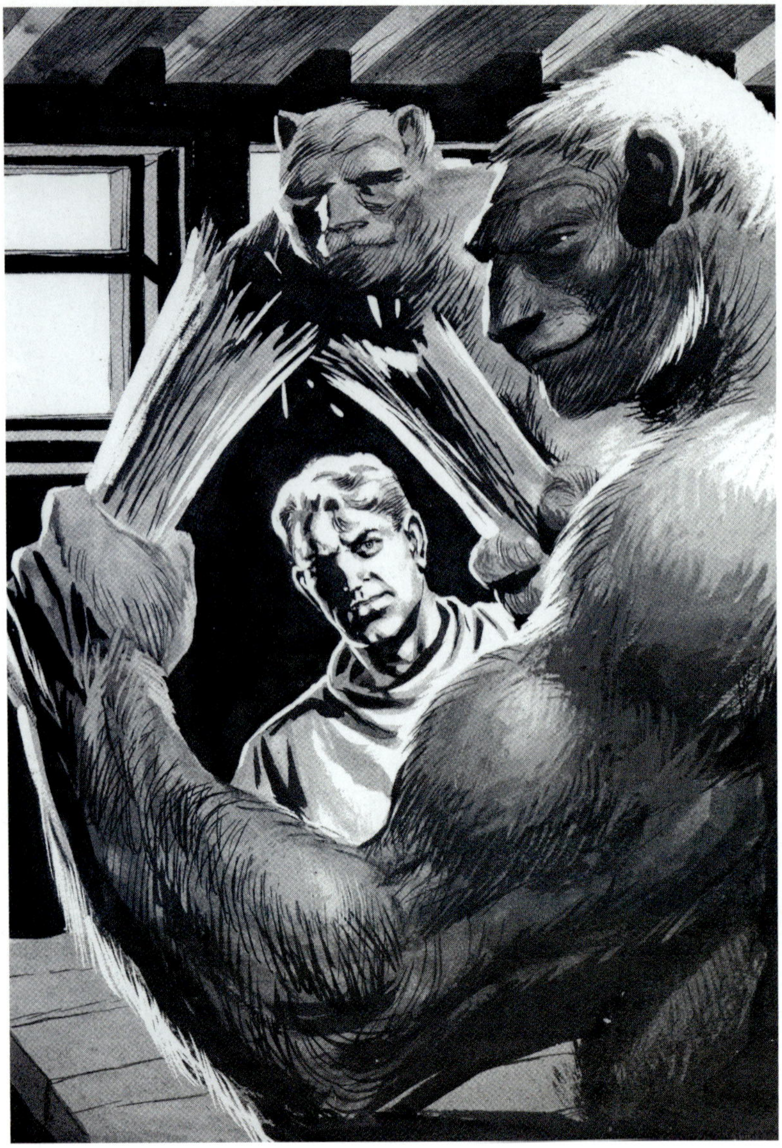

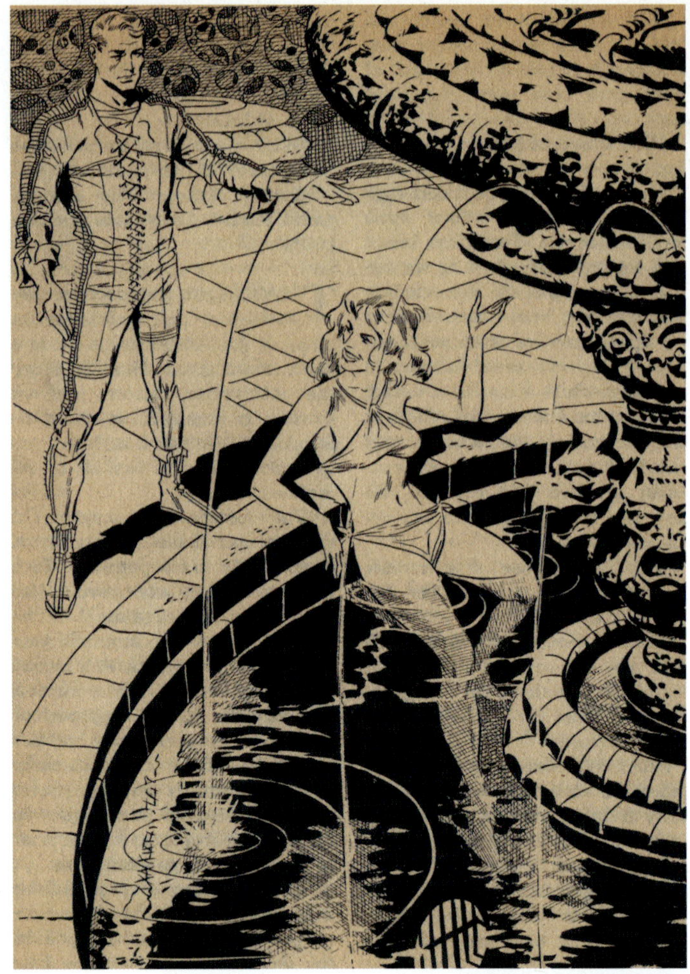

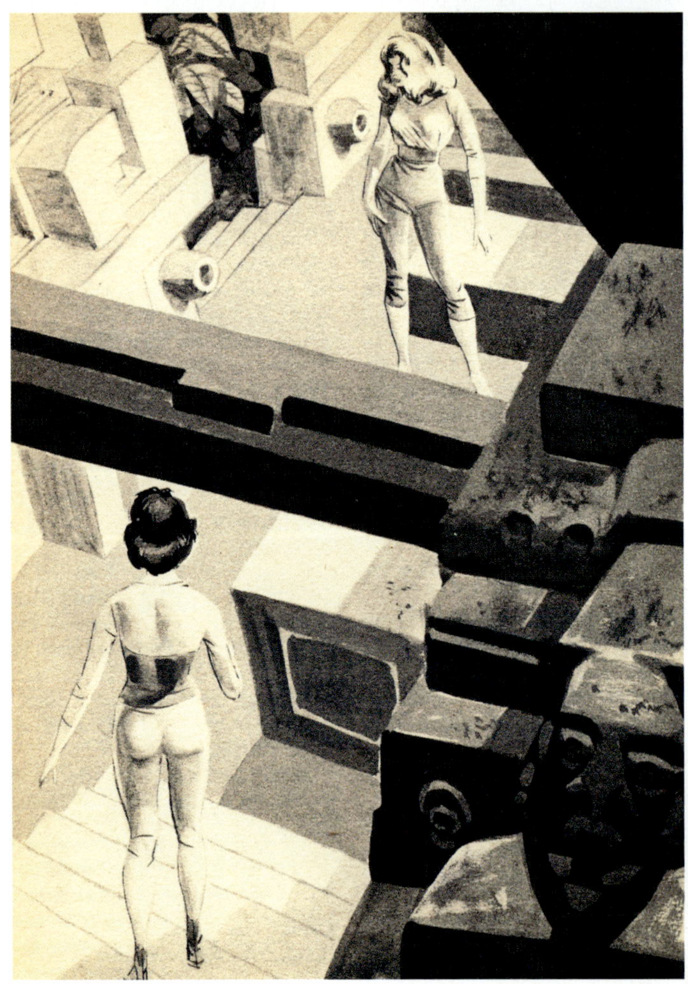

Clockwise from upper left:
Galaxy Science Fiction April 1959
Illustration for "The Man in the Mailbag" by Gordon R. Dickson

Galaxy Science Fiction April 1959
Illustration for "Kingslayer" by J. T. McIntosh

Galaxy Science Fiction June 1959
Illustration for "Whatever Counts" by Frederik Pohl

Opposite: *Galaxy Science Fiction* August 1959
Illustration for "No Life of Their Own" by Clifford D. Simak

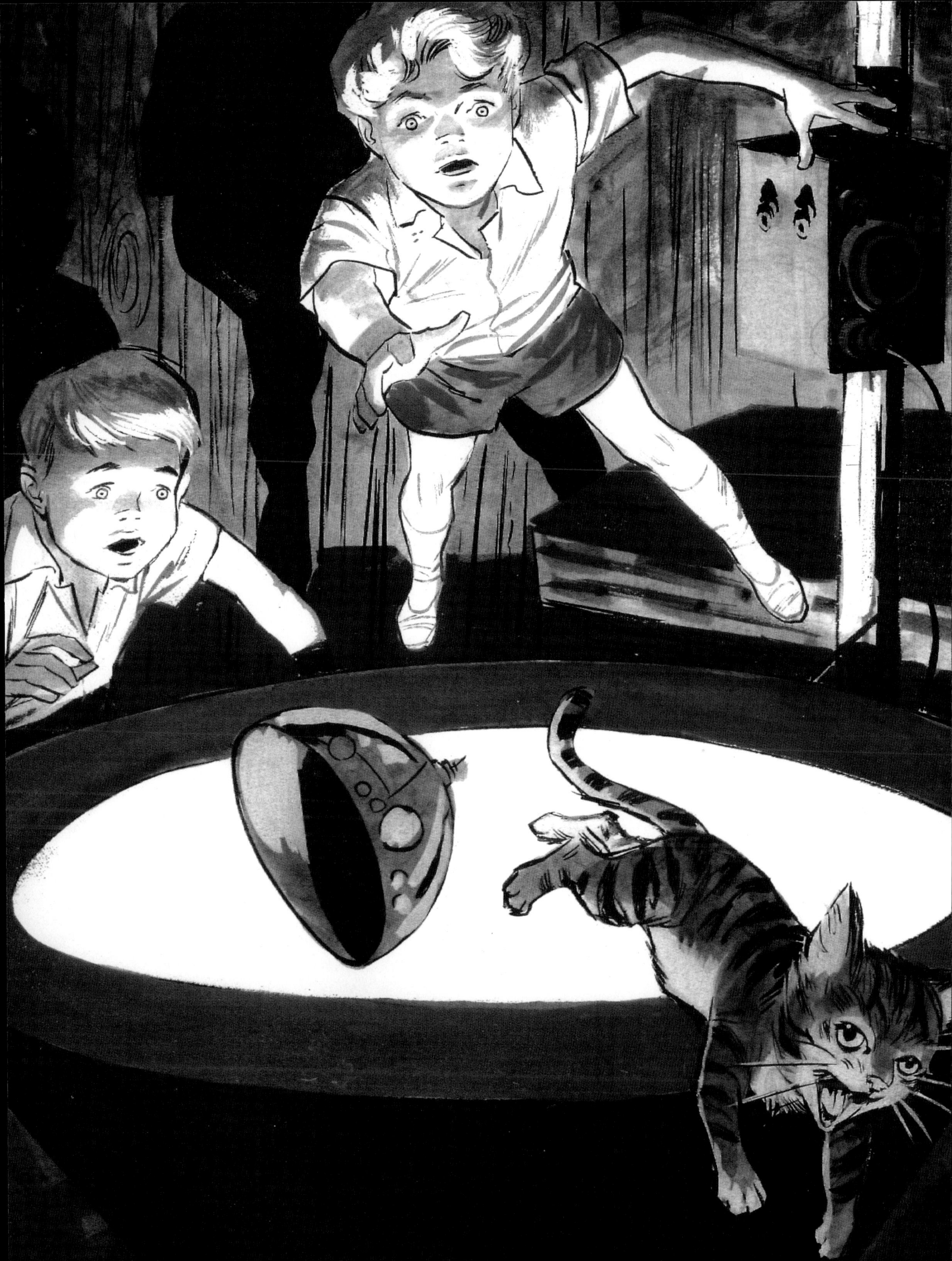

Top: *Galaxy Science Fiction* February 1959
Illustration for "I Plinglot—Who You?" by Frederik Pohl

Bottom: *Galaxy Science Fiction* February 1959
Illustration for "Insidekick" by J. F. Bone

Opposite: *Galaxy Science Fiction* August 1959 (Cover)
Painting titled "1 Down, 1 Up"

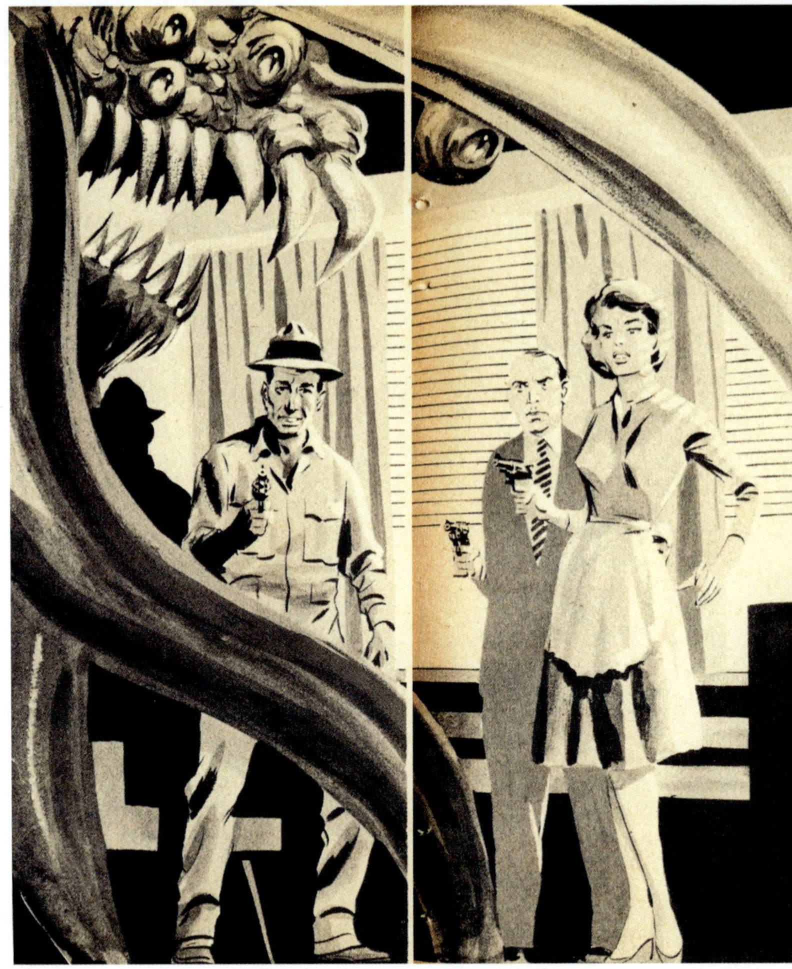

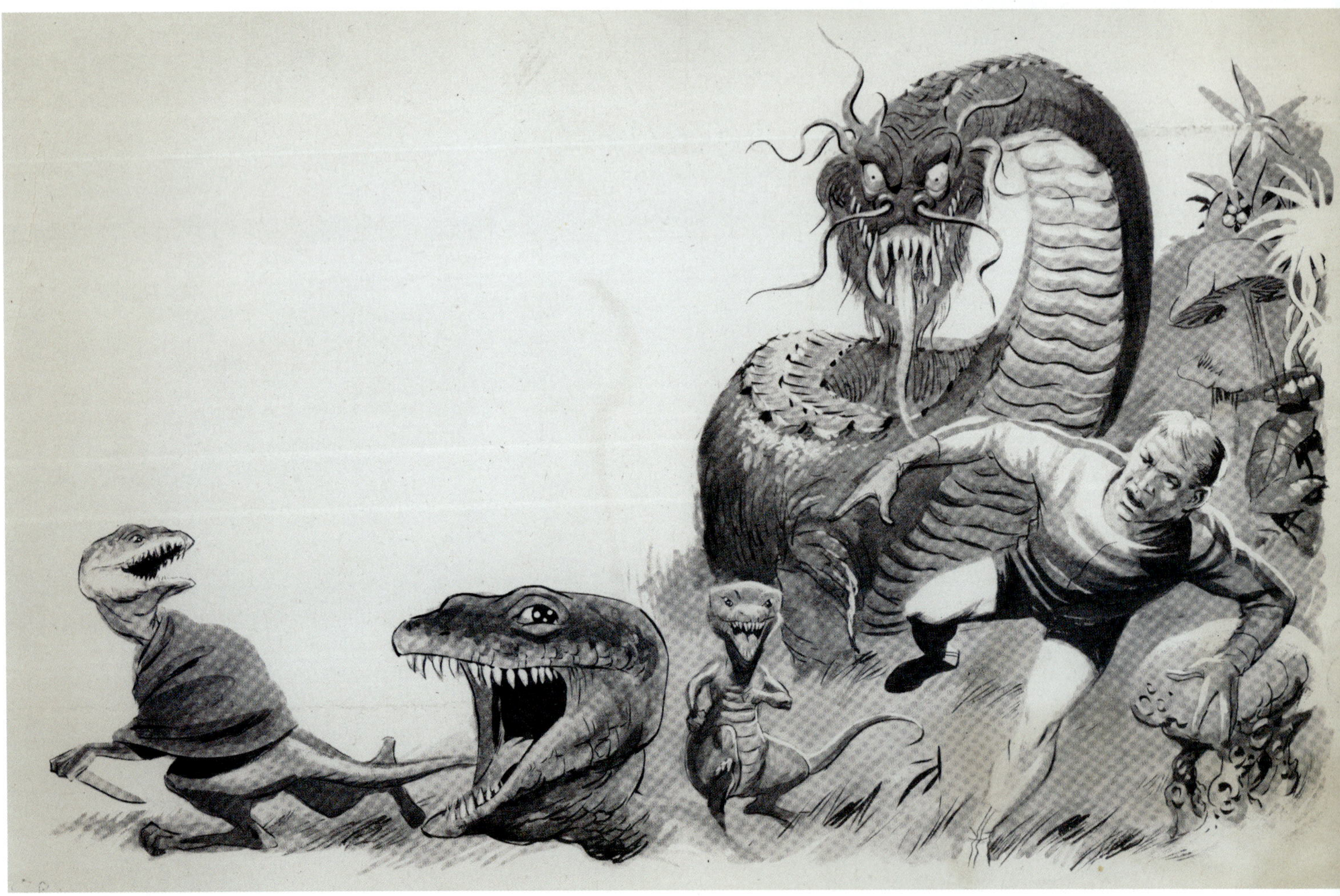

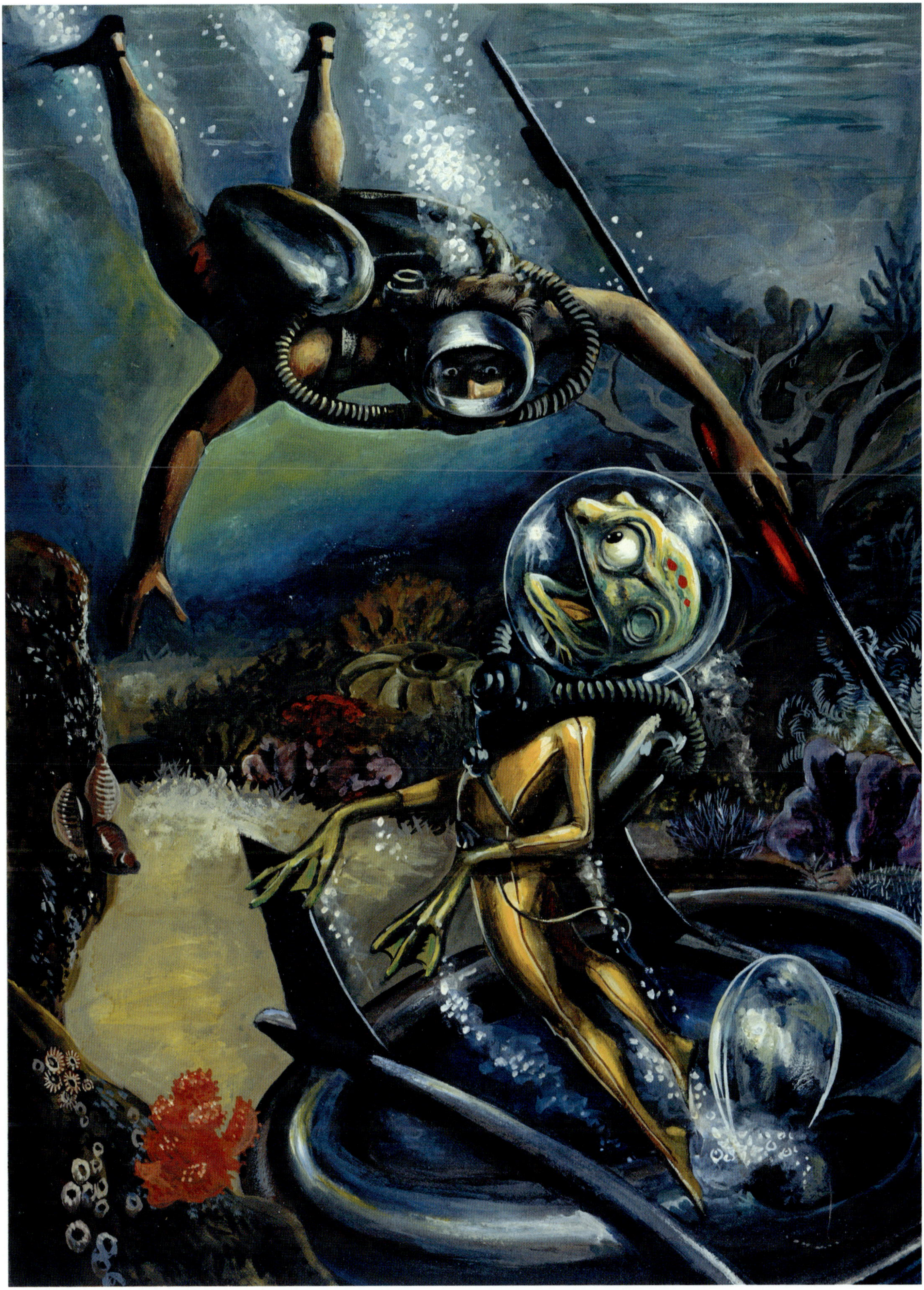

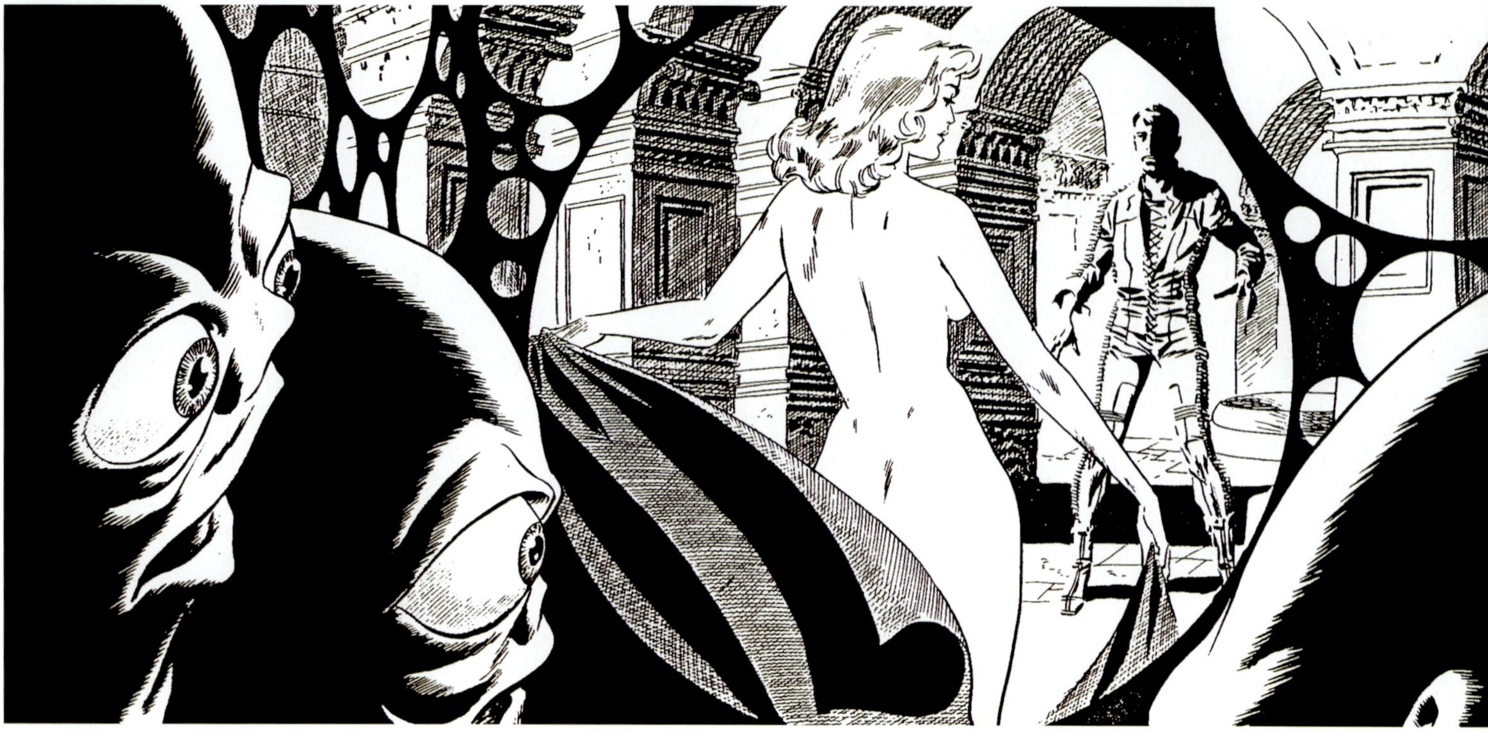

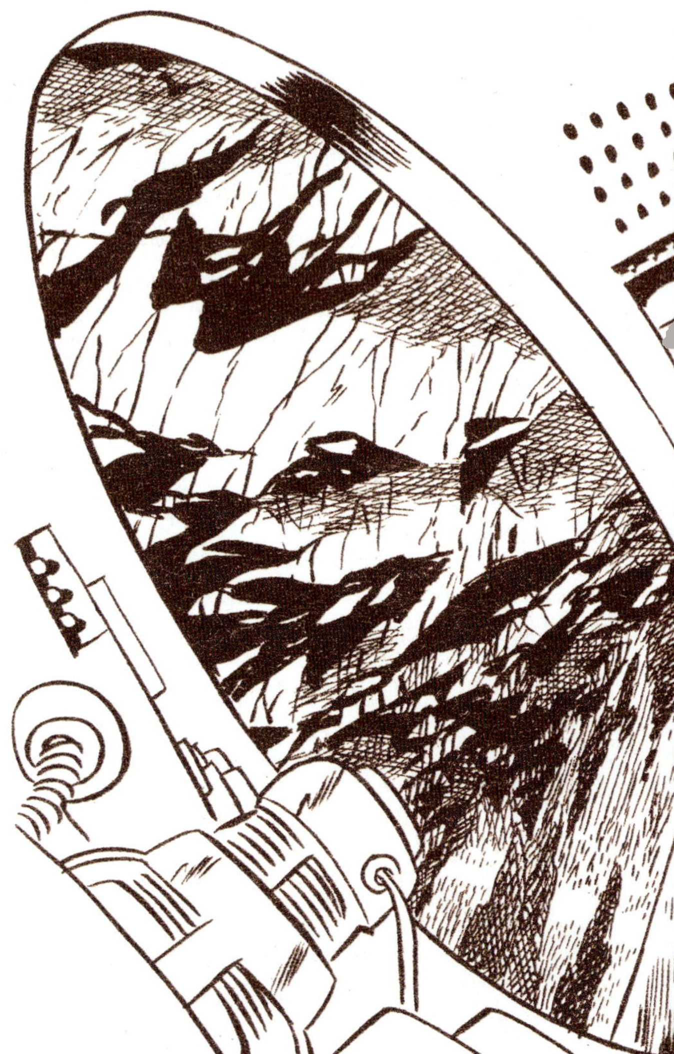

Top and opposite:
Galaxy Science Fiction April 1959
Illustration for "Kingslayer" by J. T. McIntosh

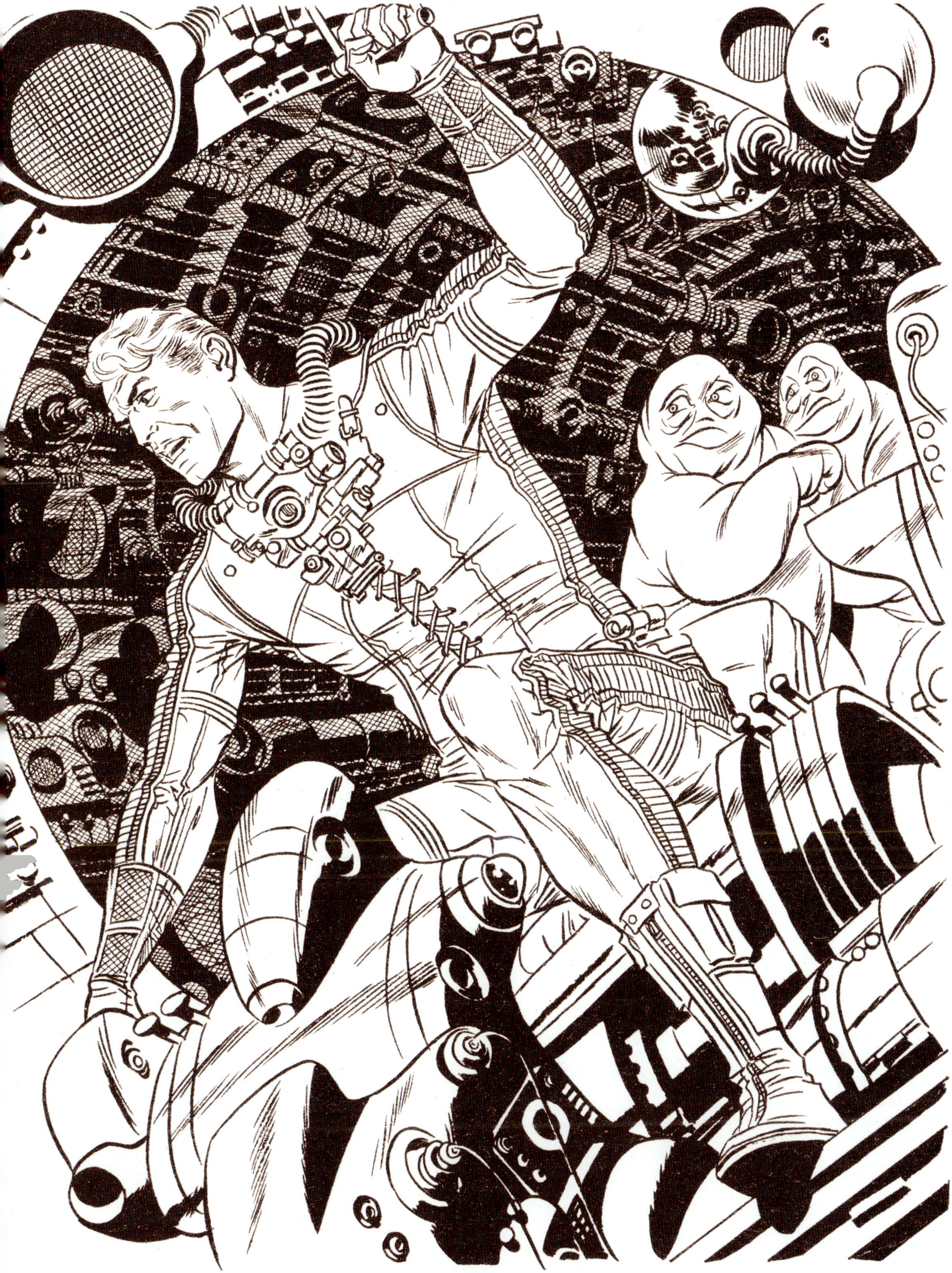

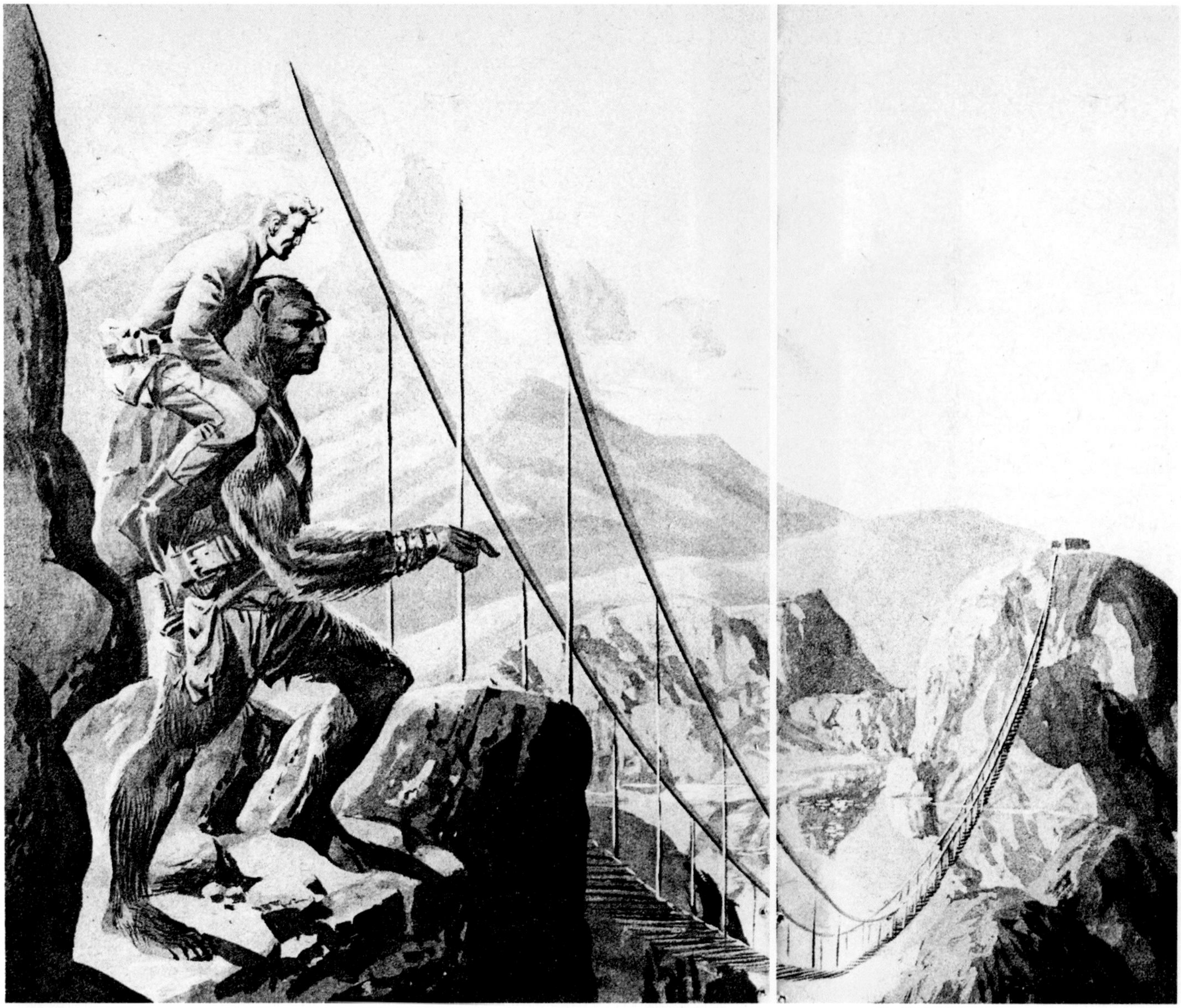

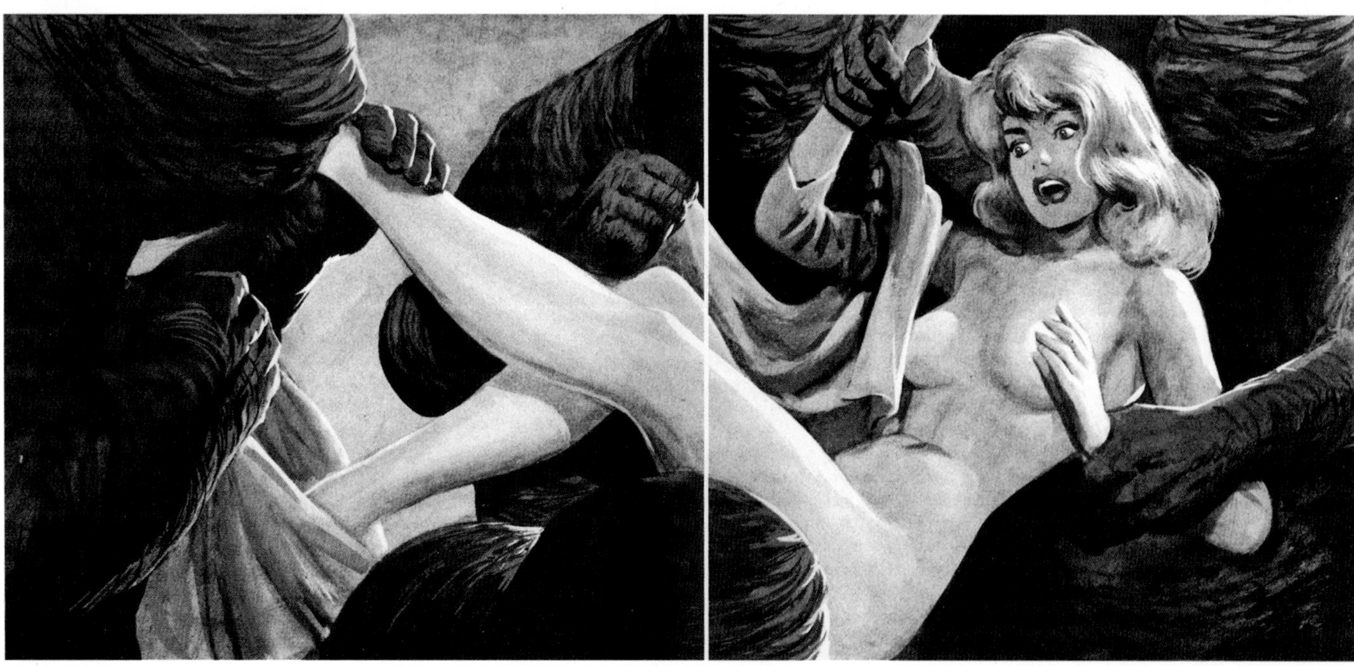

Top: *Galaxy Science Fiction* April 1959 Illustration for "The Man in the Mailbag" by Gordon R. Dickson

Bottom: *Galaxy Science Fiction* June 1959 Illustration for "Whatever Counts" by Frederik Pohl

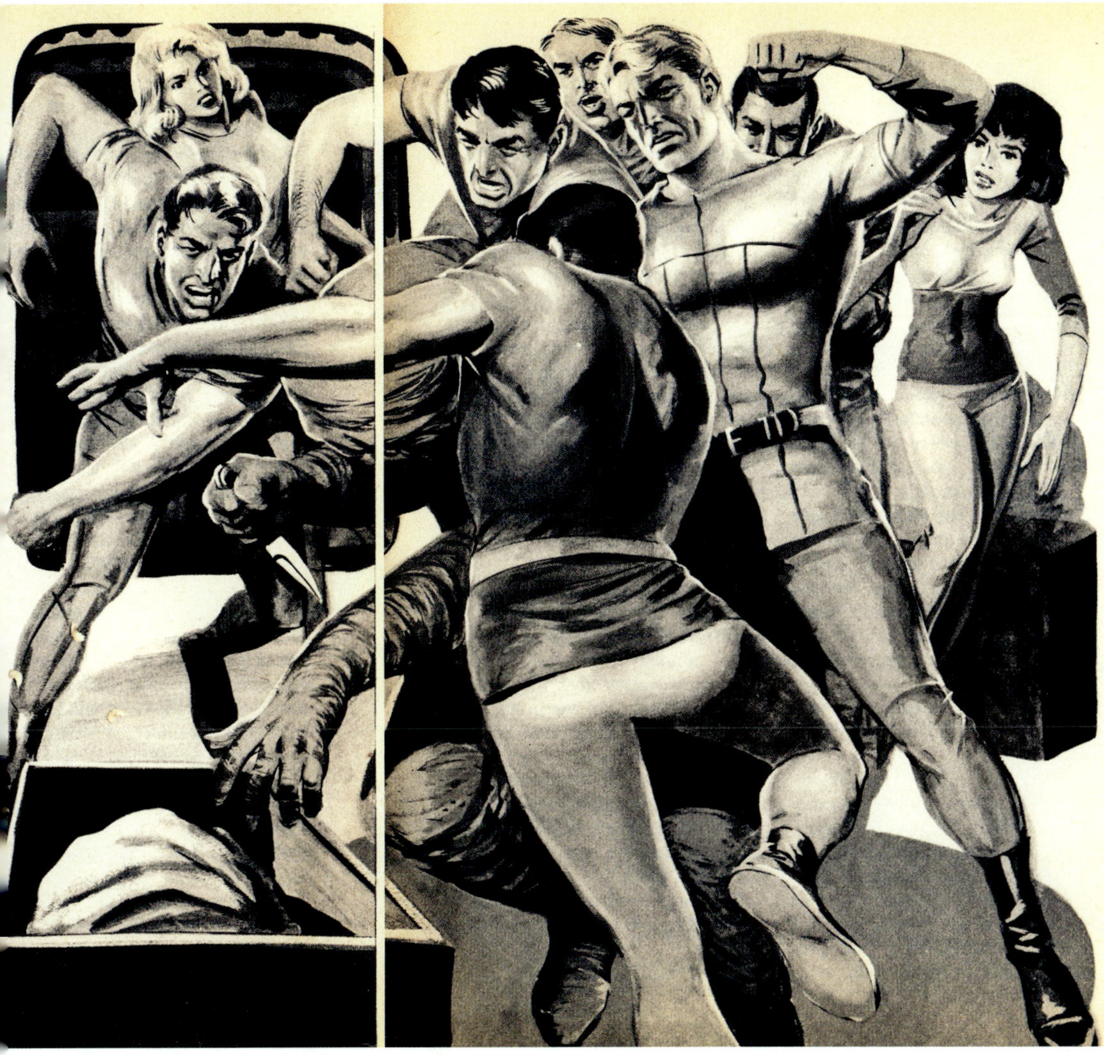

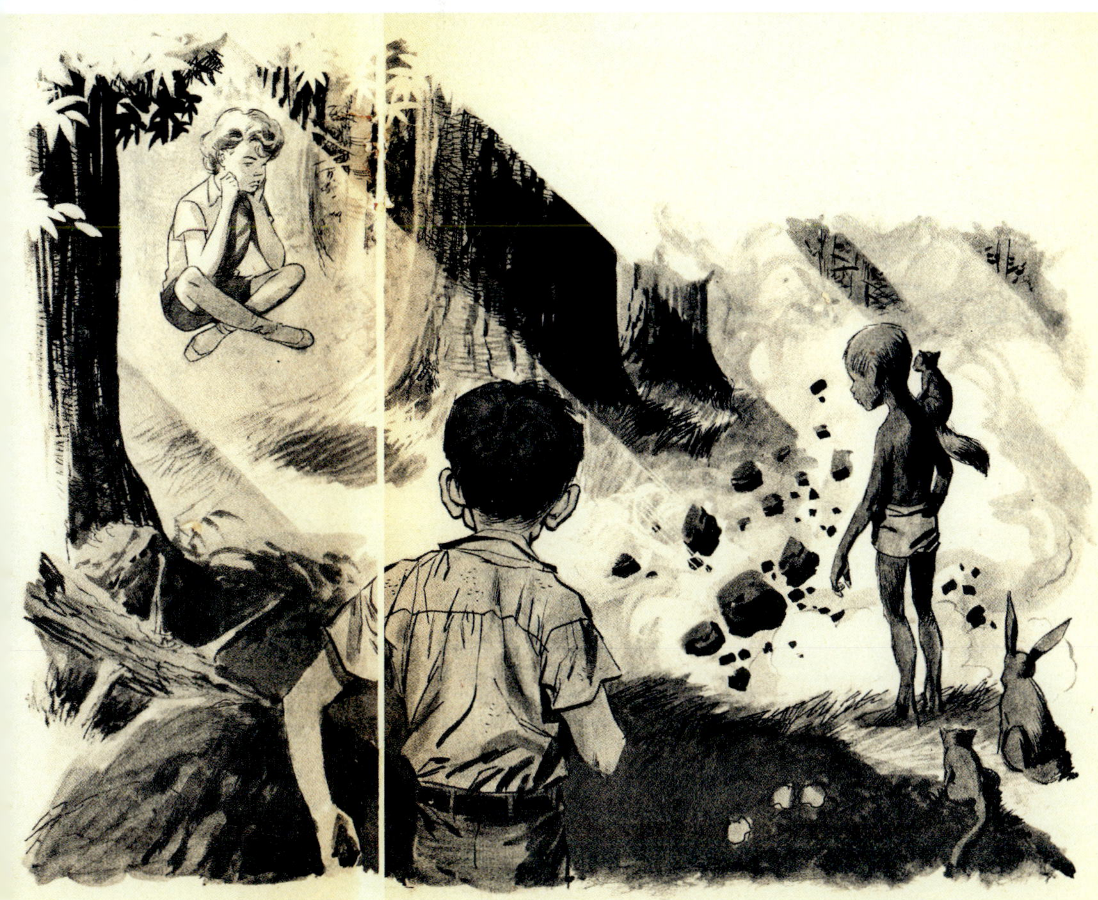

Top: *Galaxy Science Fiction*
August 1959
Illustration for "Whatever Counts"
by Frederik Pohl

Bottom: *Galaxy Science Fiction*
August 1959
Illustration for "No Life of Their Own"
by Clifford D. Simak

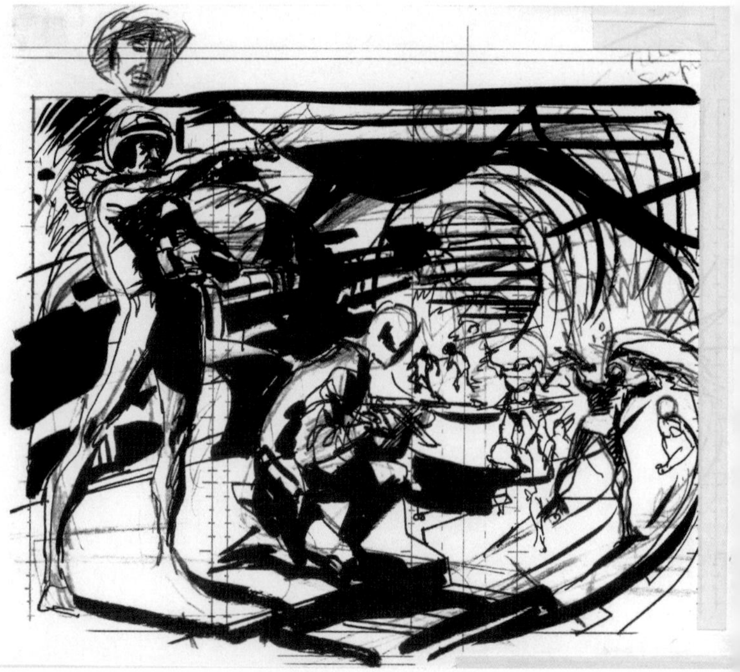

Galaxy Science Fiction August 1959
Illustration for "The Waging of the Peace" by Frederik Pohl

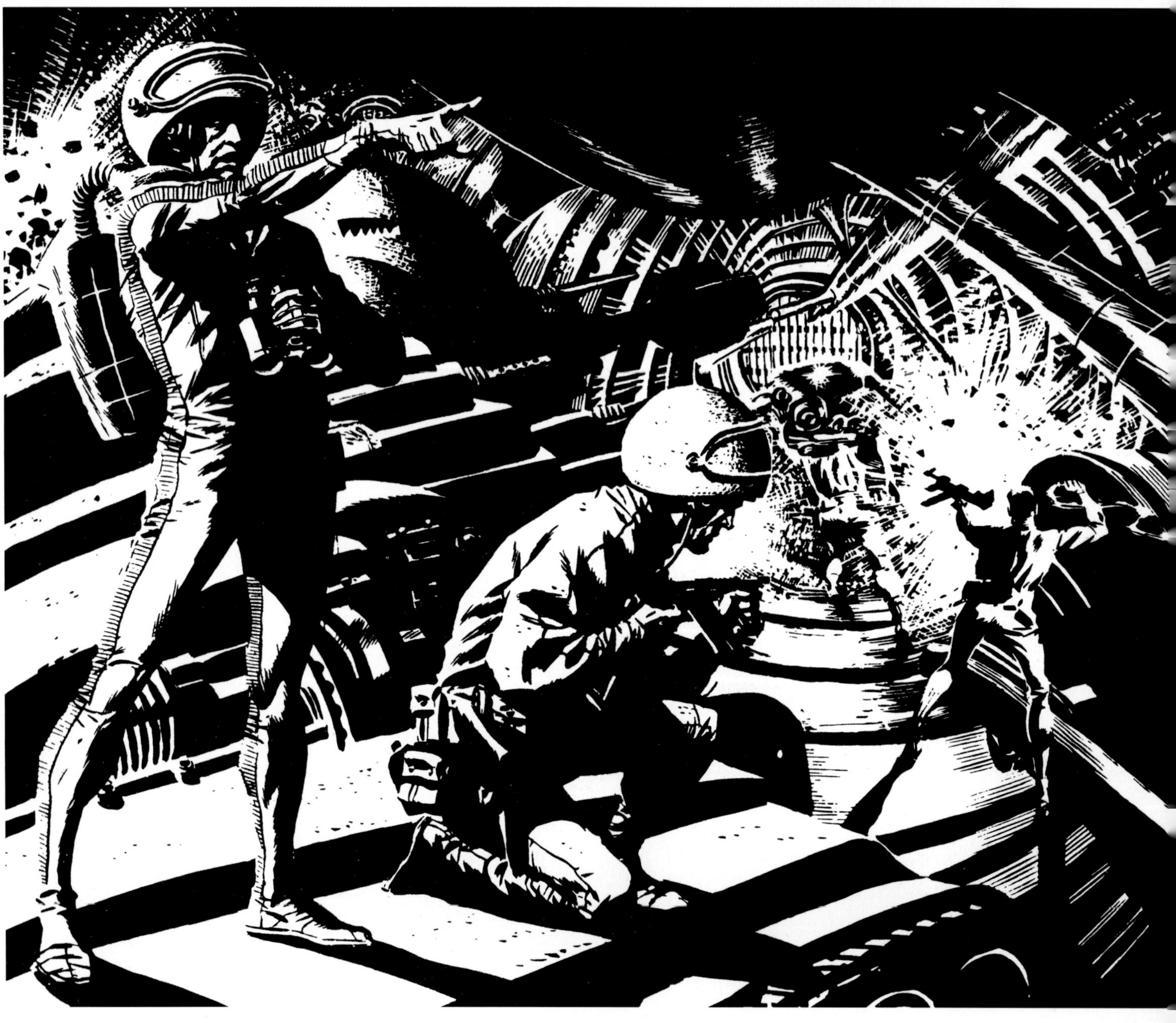

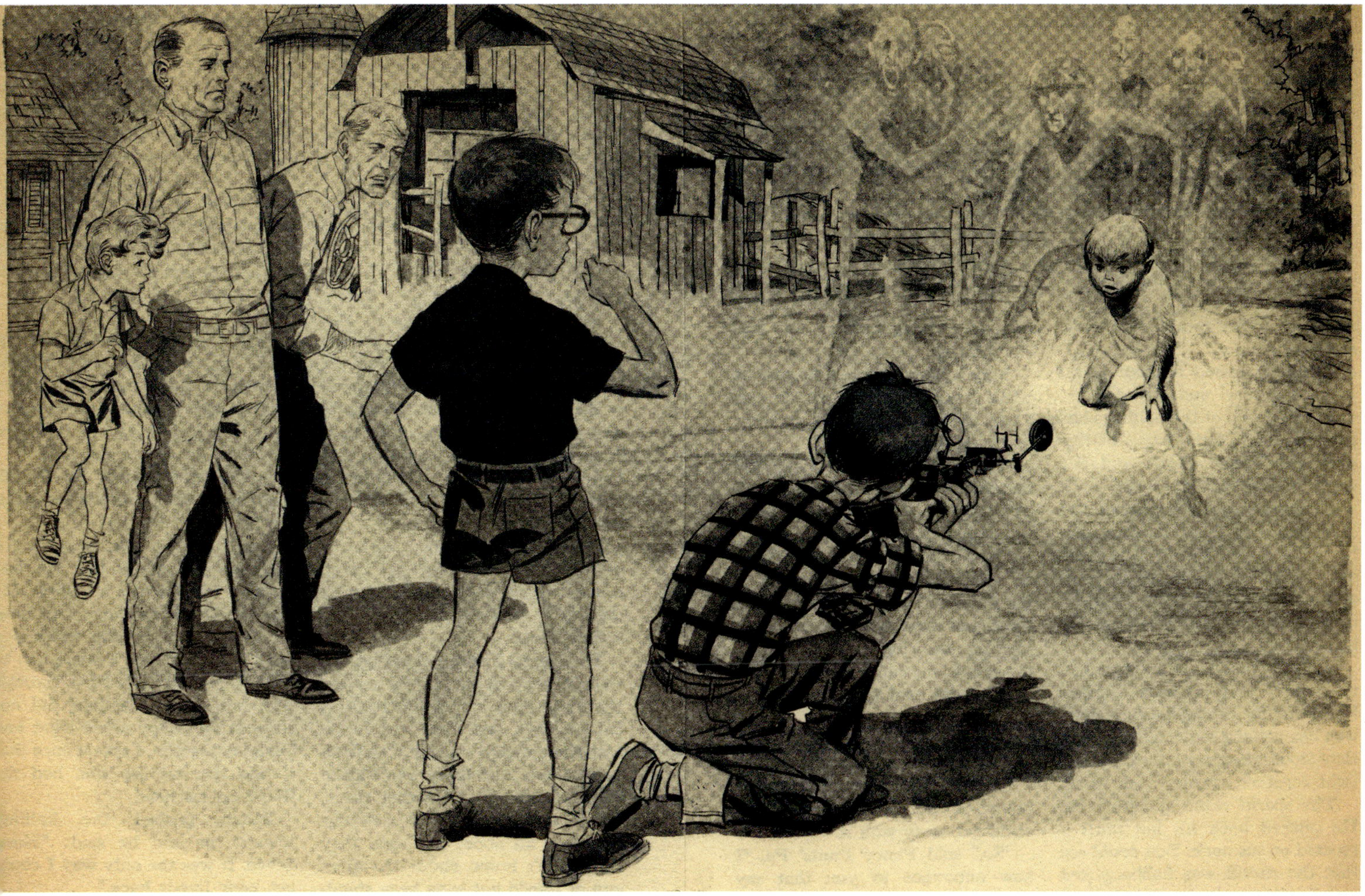

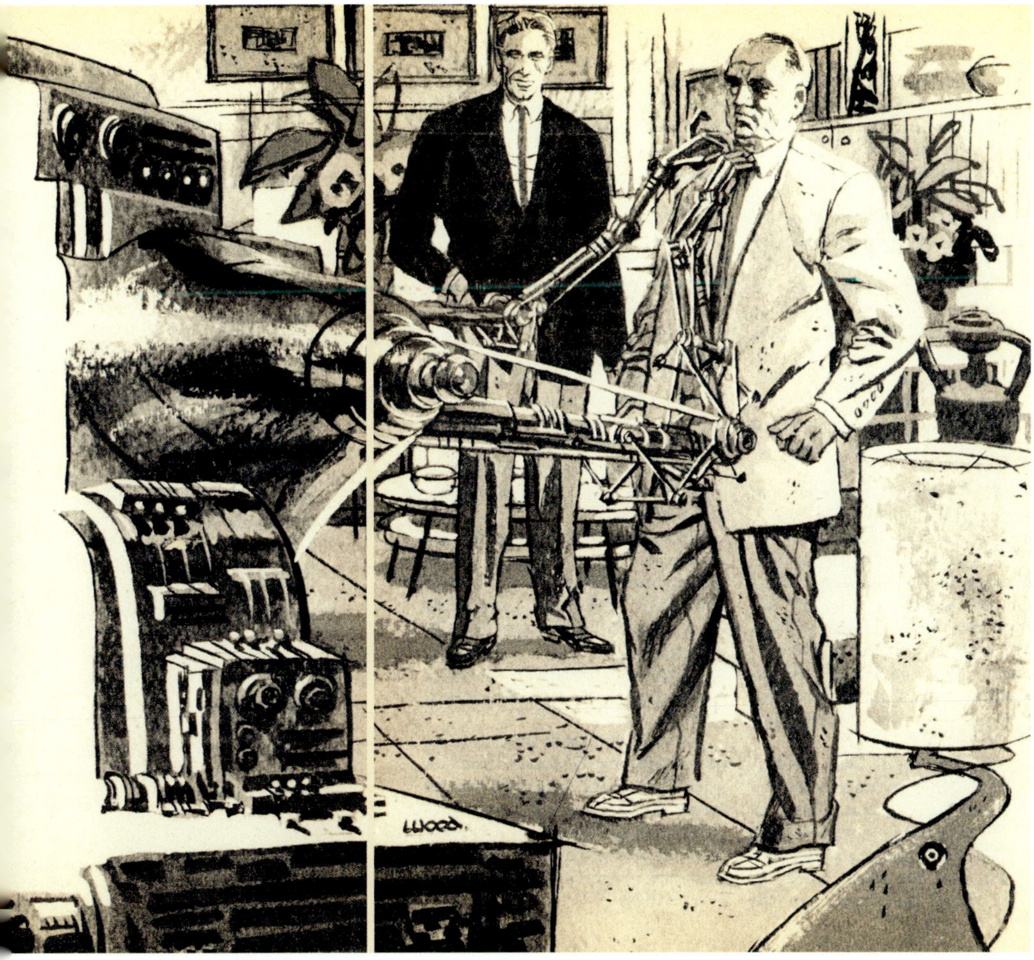

Top: *Galaxy Science Fiction* August 1959
Illustration for "No Life of Their Own"
by Clifford D. Simak

Bottom: *Galaxy Science Fiction* August 1959
Illustration for "Lex" by W. T. Haggert

This page, both:
Galaxy Science Fiction December 1959
Illustration for "Blacksword" by A. J. Offutt

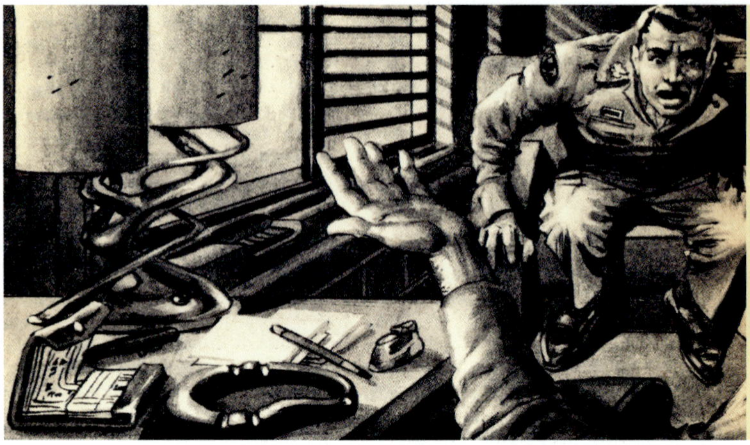

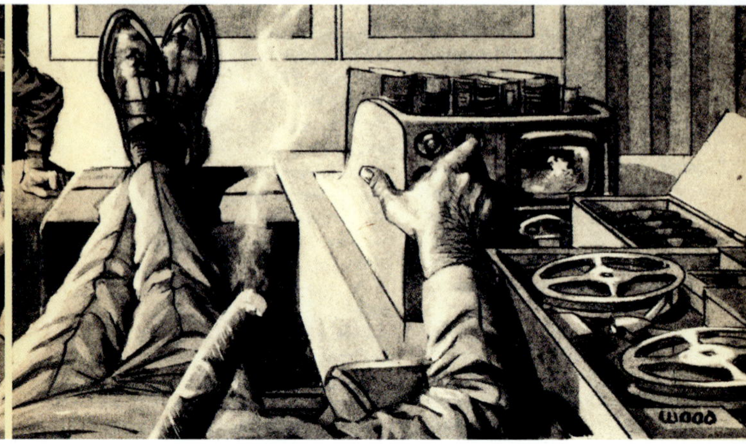

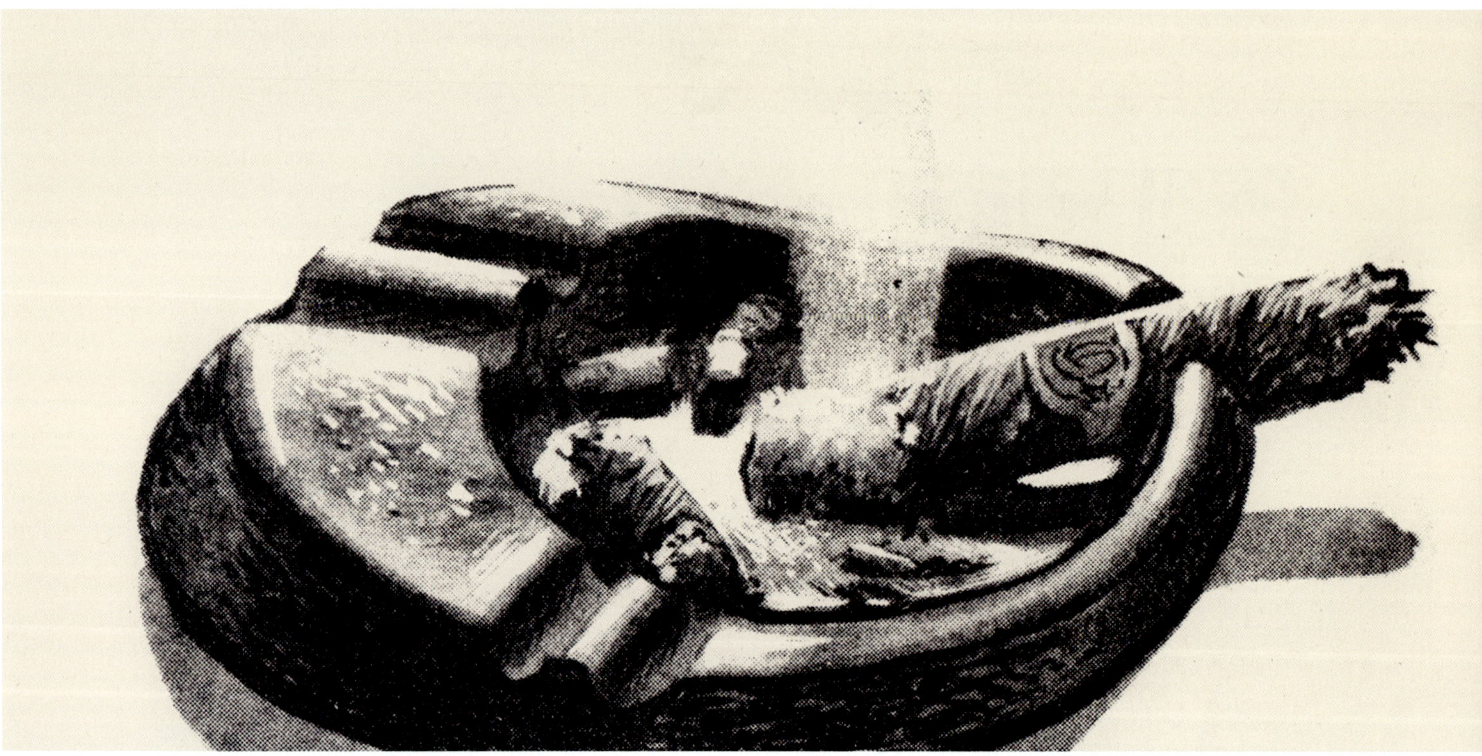

Opposite, both
Galaxy Science Fiction August 195
Illustration for "License to Steal" by Louis Newma

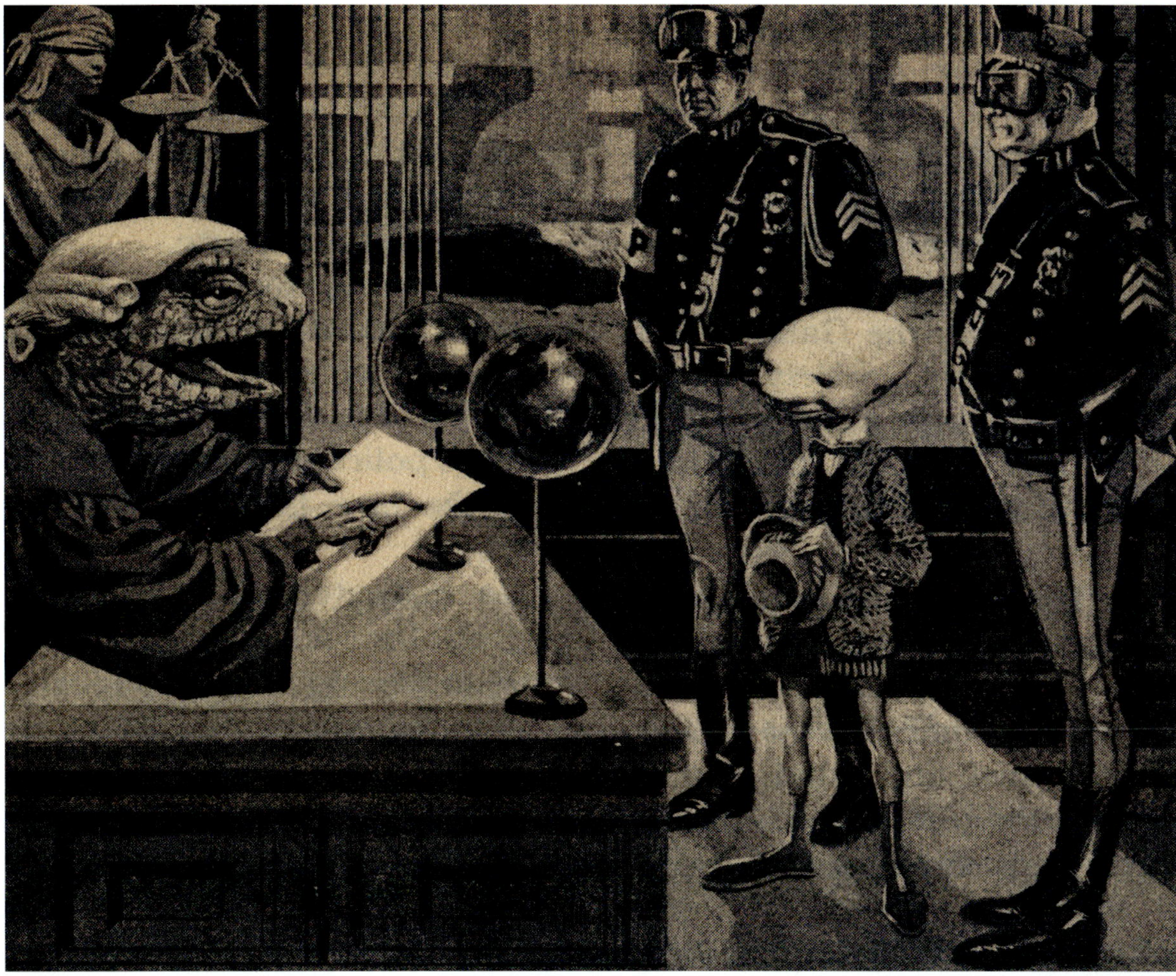

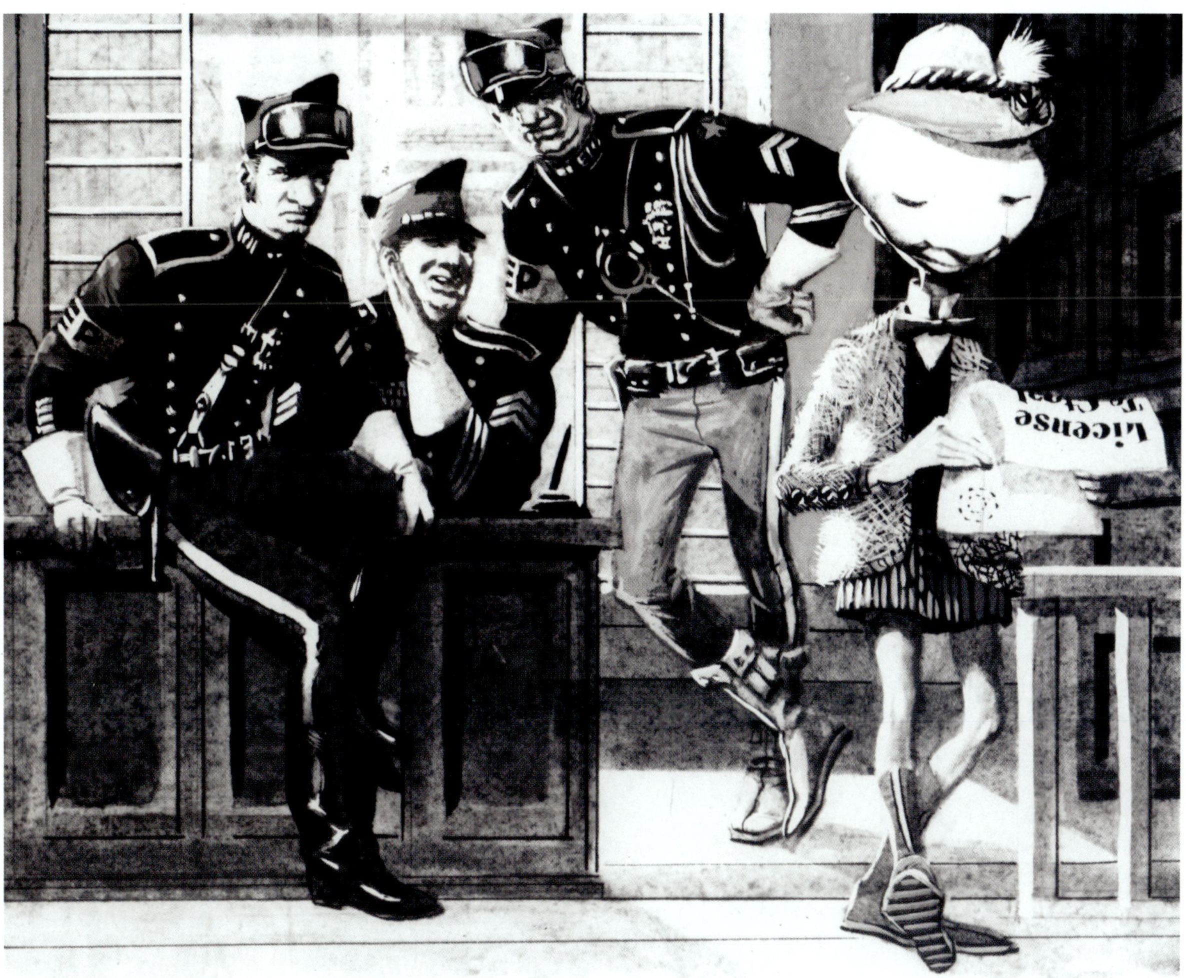
License

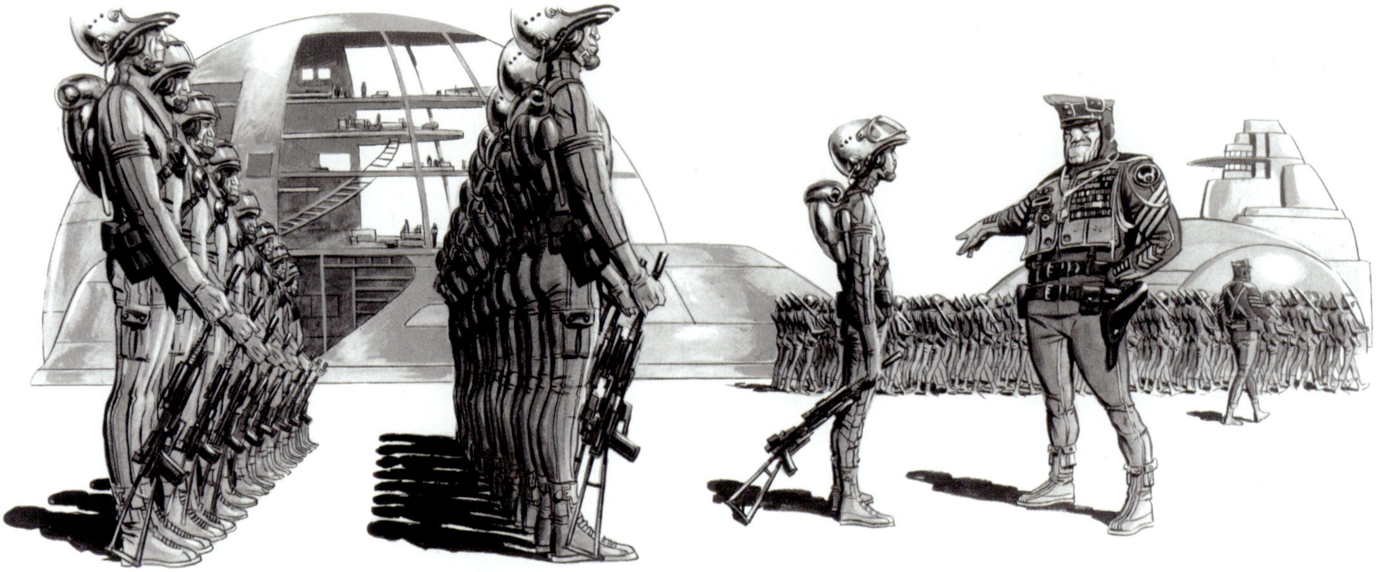

102

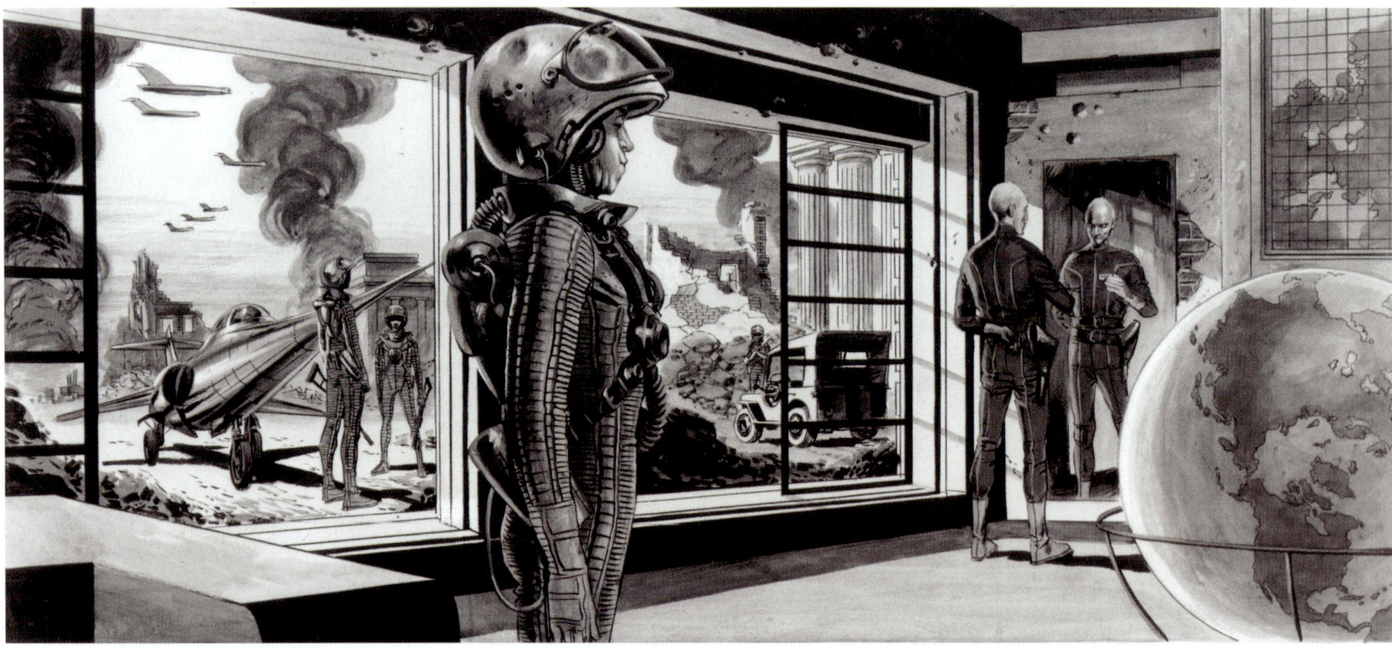

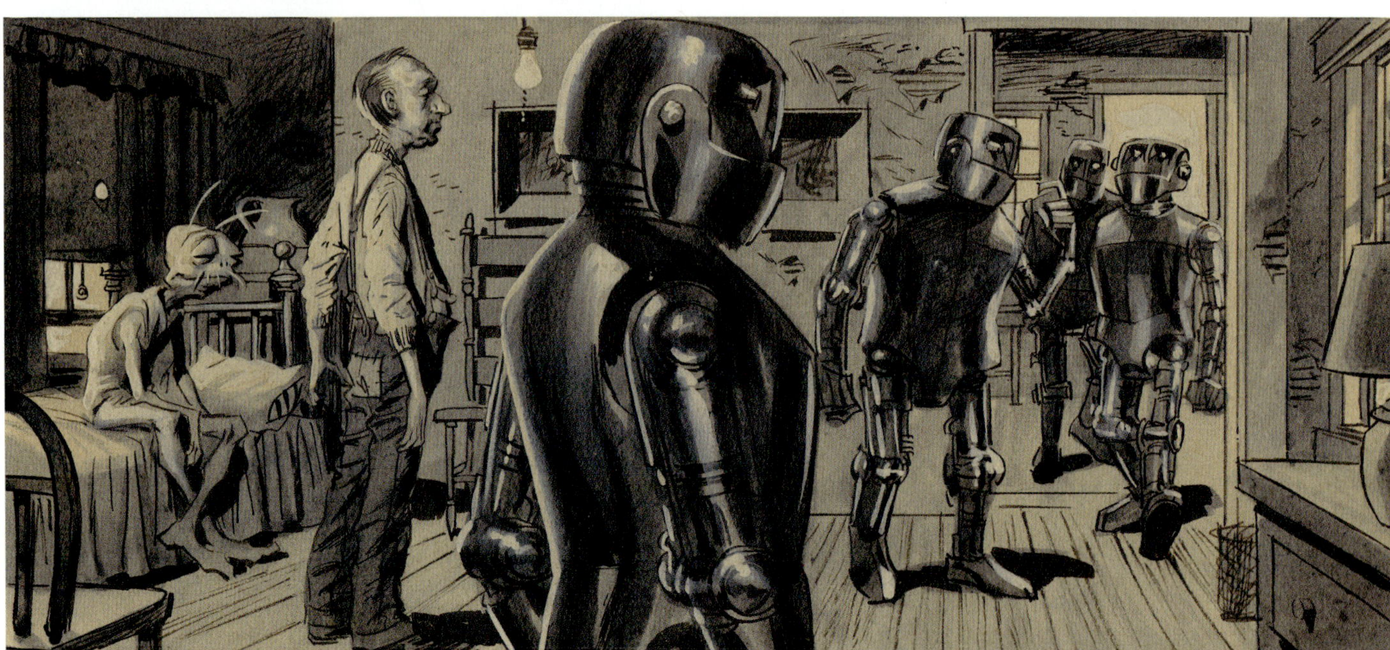

Galaxy Science Fiction June 1959
Illustration for "Whatever Counts" by Frederik Pohl

Opposite:
Top and middle: *Galaxy Science Fiction* April 1960
Illustration for "Success Story" by Earl Goodale

Bottom: *Galaxy Science Fiction* February 1960
Illustration for "Crying Jag" by Clifford D. Simak

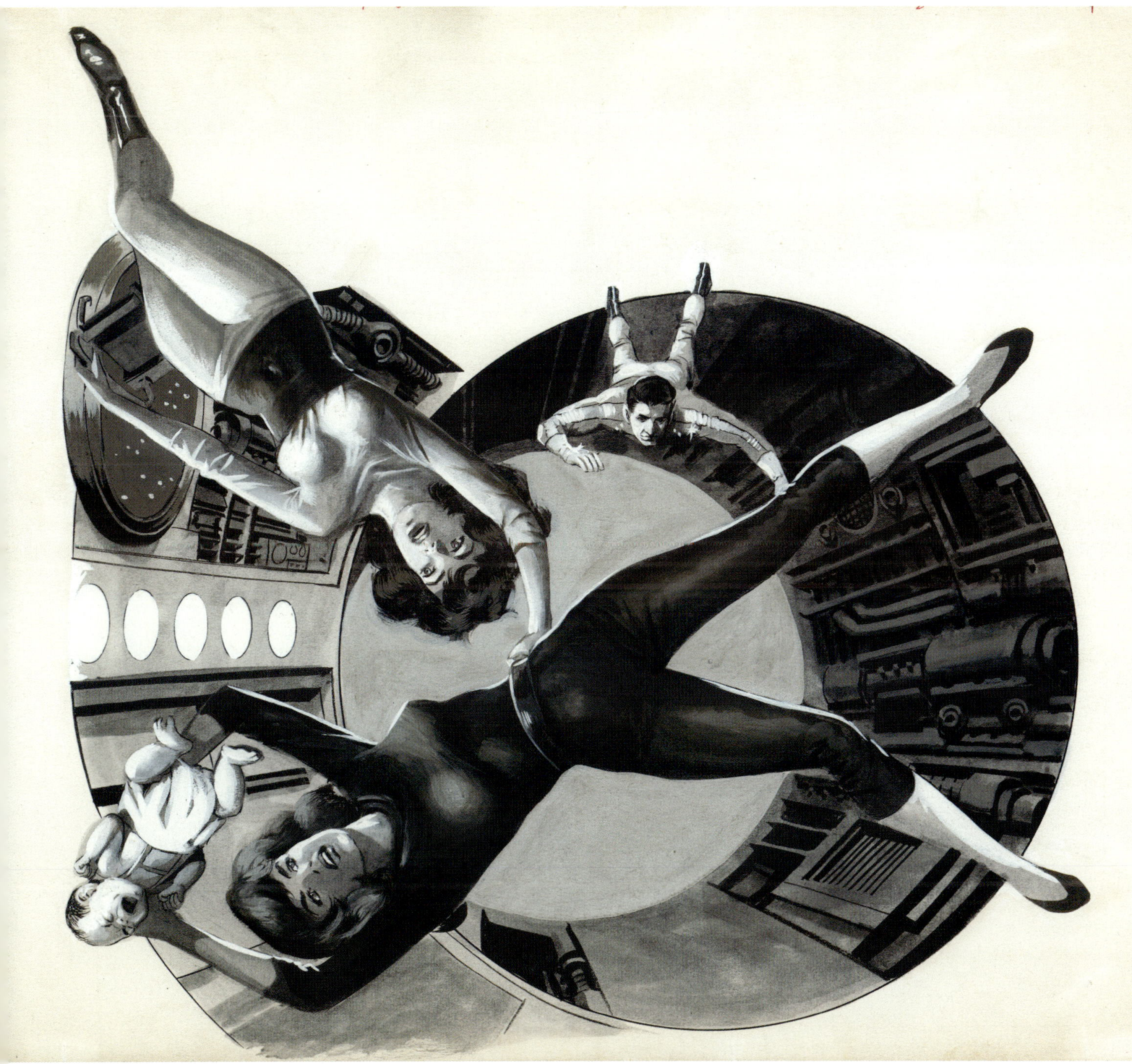

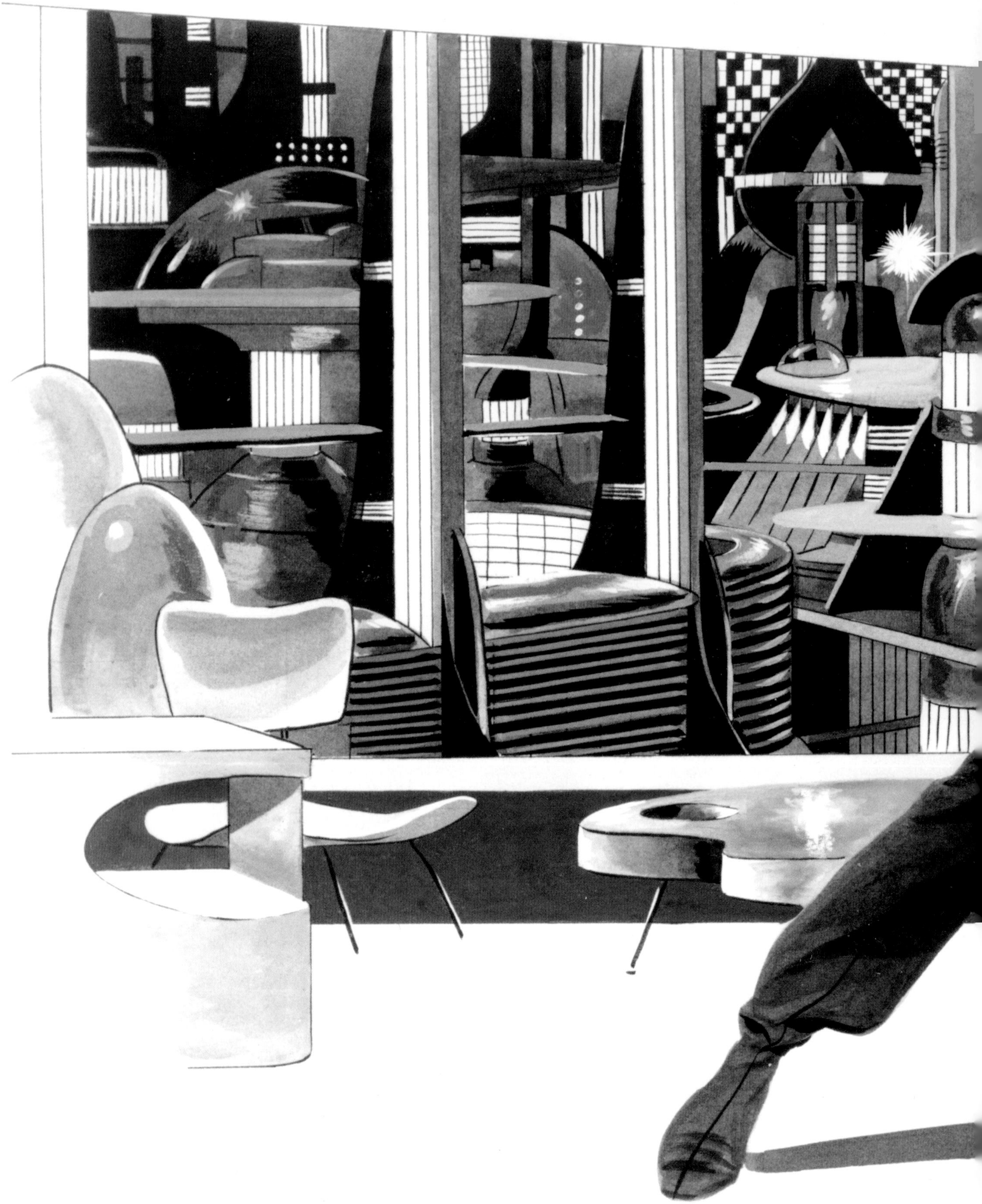

Galaxy Science Fiction November 1958
Illustration for "Time Killer" part 2 by Robert Sheckley

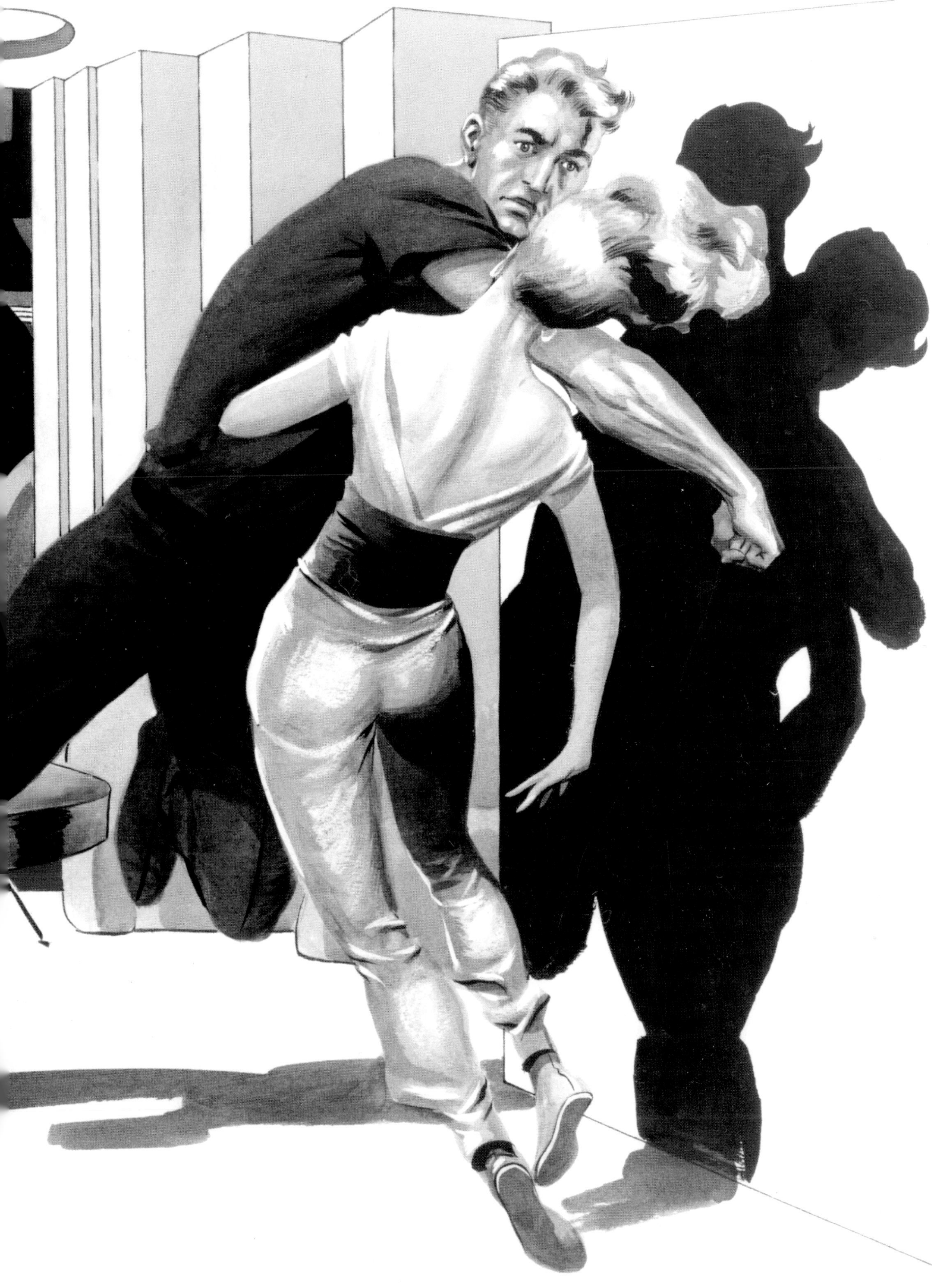

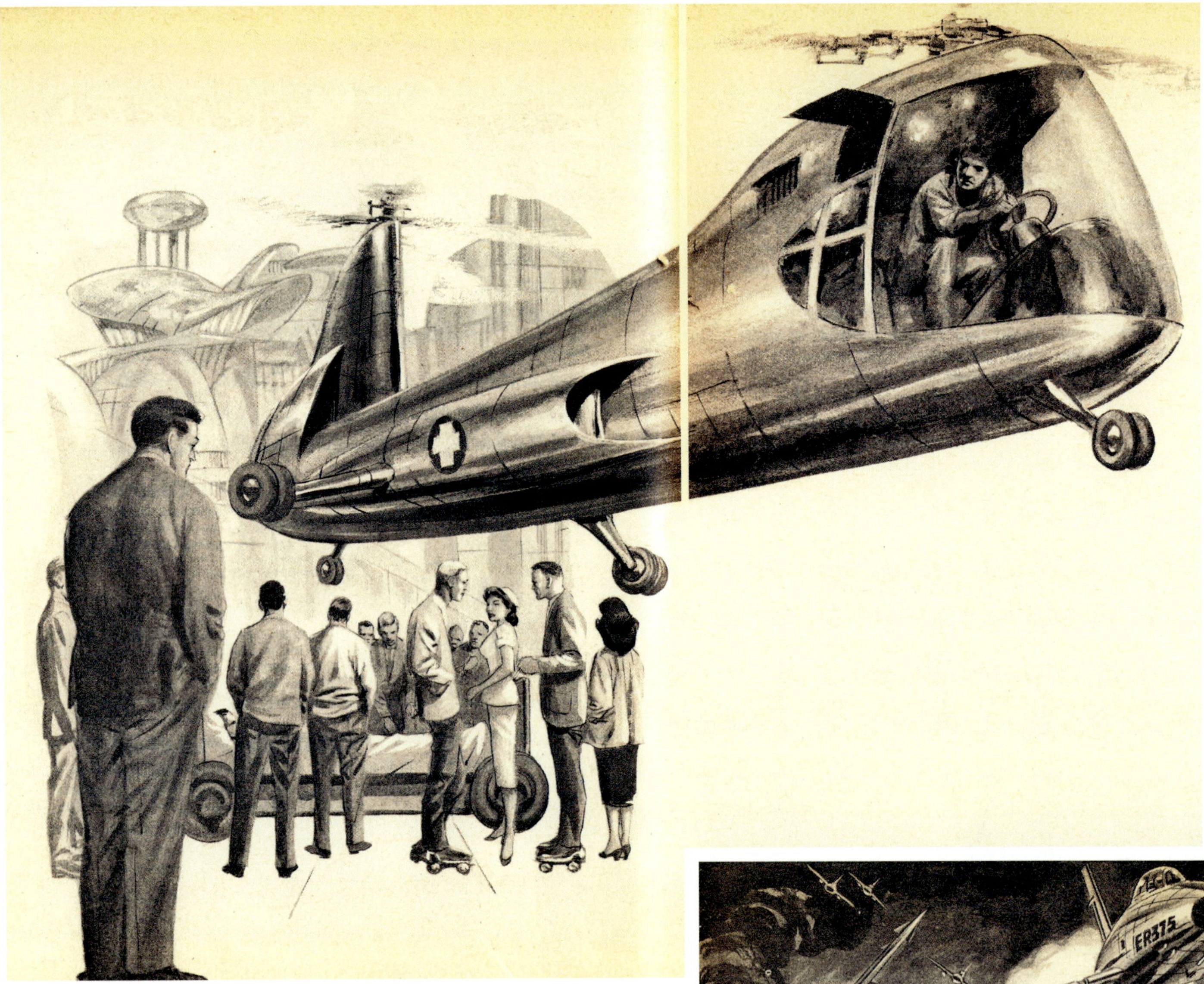

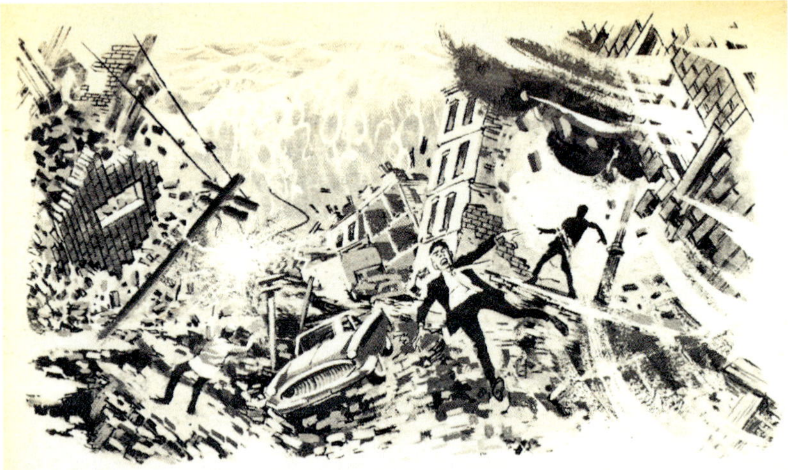

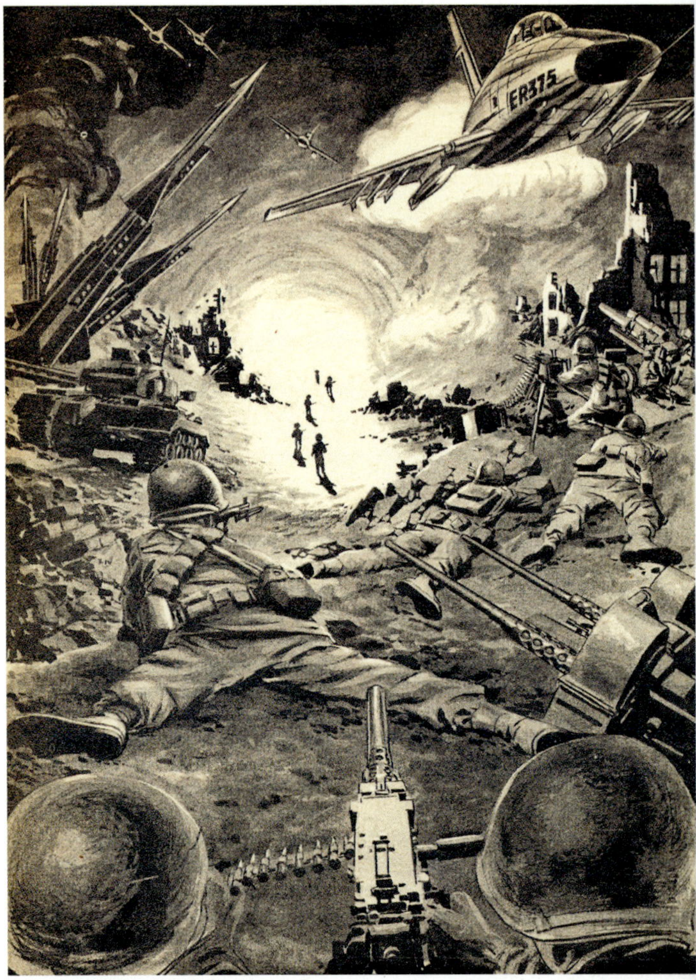

Clockwise from top: *Galaxy Science Fiction* April 1960
Illustration for "Don't Look Now" by Leonard Rubin

Galaxy Science Fiction April 1960
Illustration for "Success Story" by Earl Goodale

Galaxy Science Fictionn August 1963
Illustration for "The Great Nebraska Sea" by Allan Danzig

Opposite: *Galaxy Science Fiction* October 1959 Cover (Prelim)
Illustration for "Someone To Watch Over Me" by Christopher Grimm

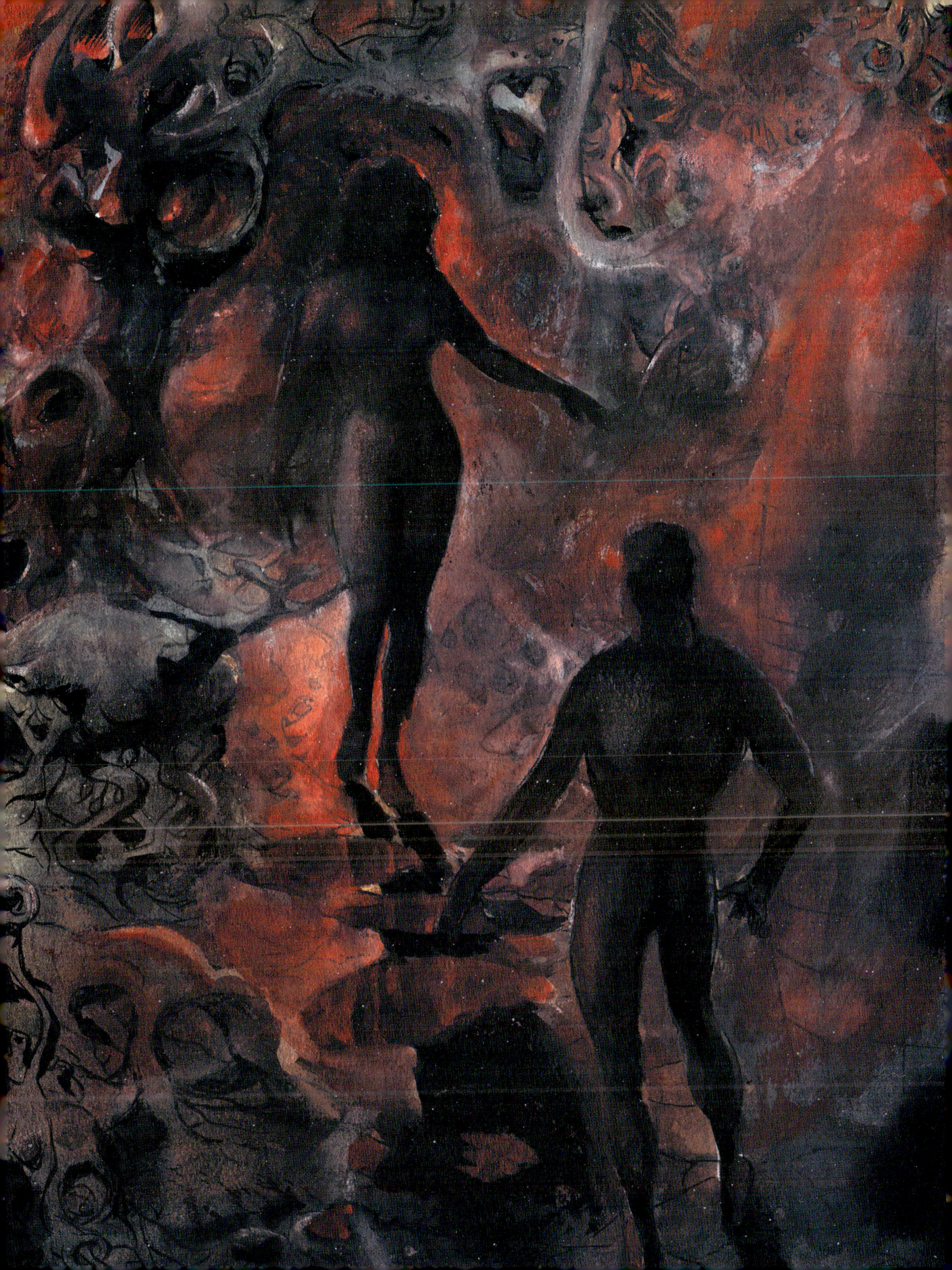

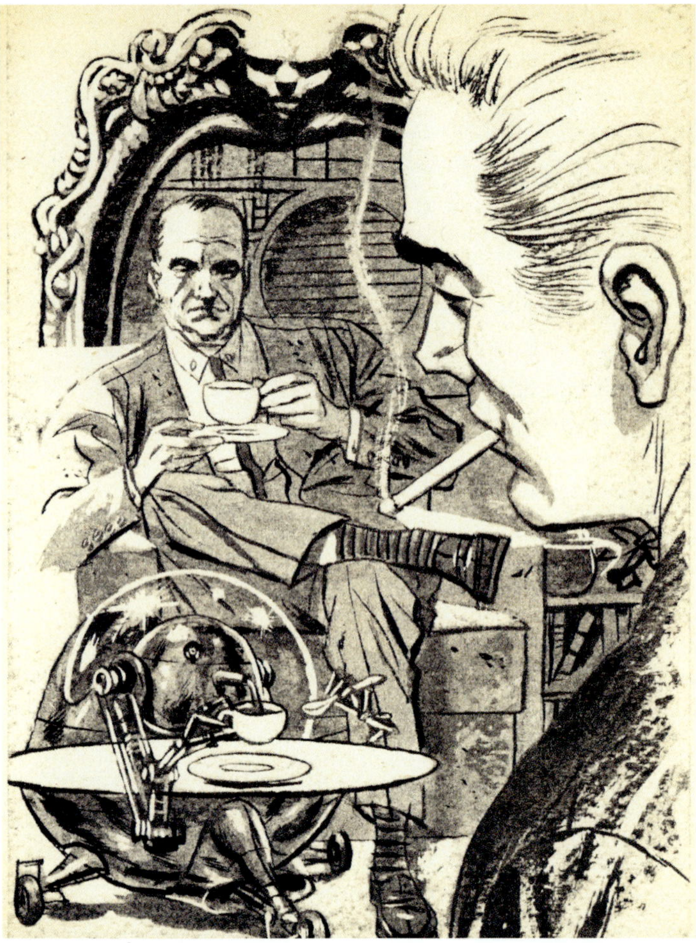
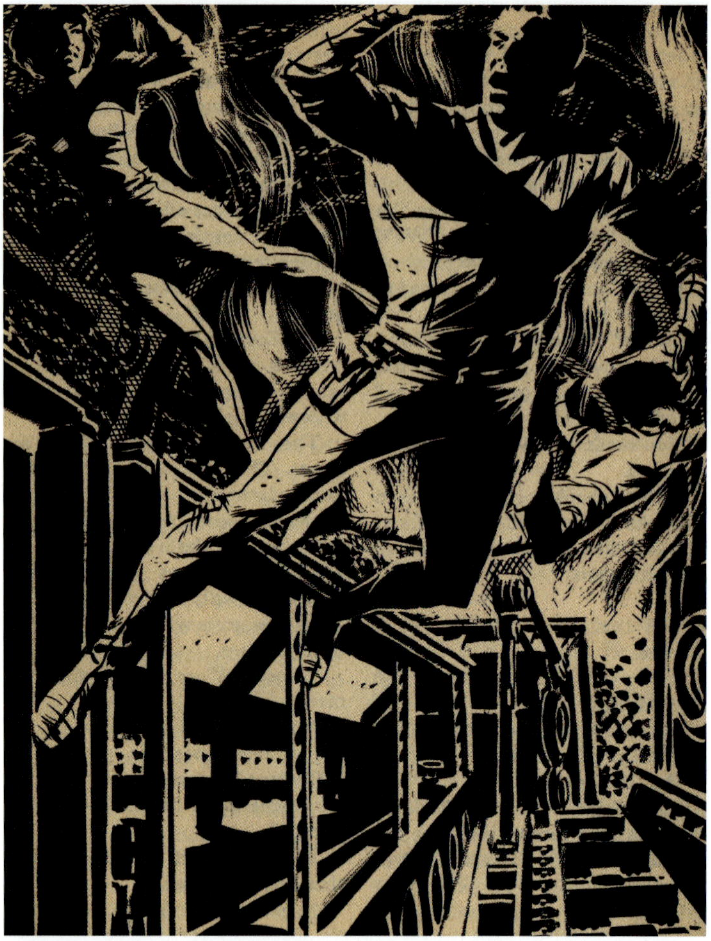
108
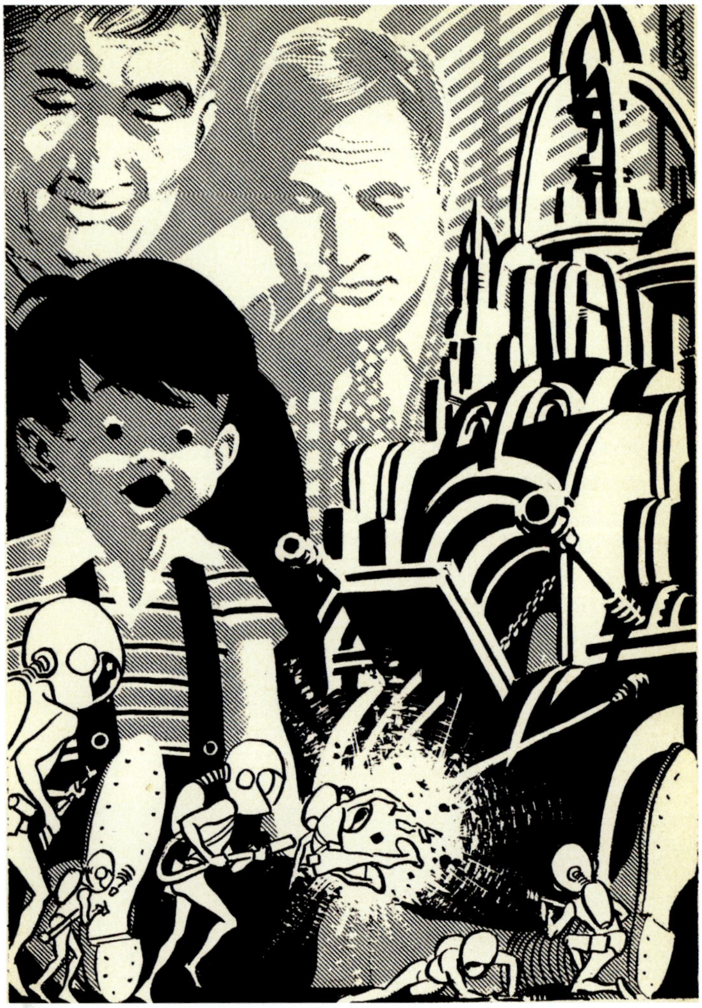
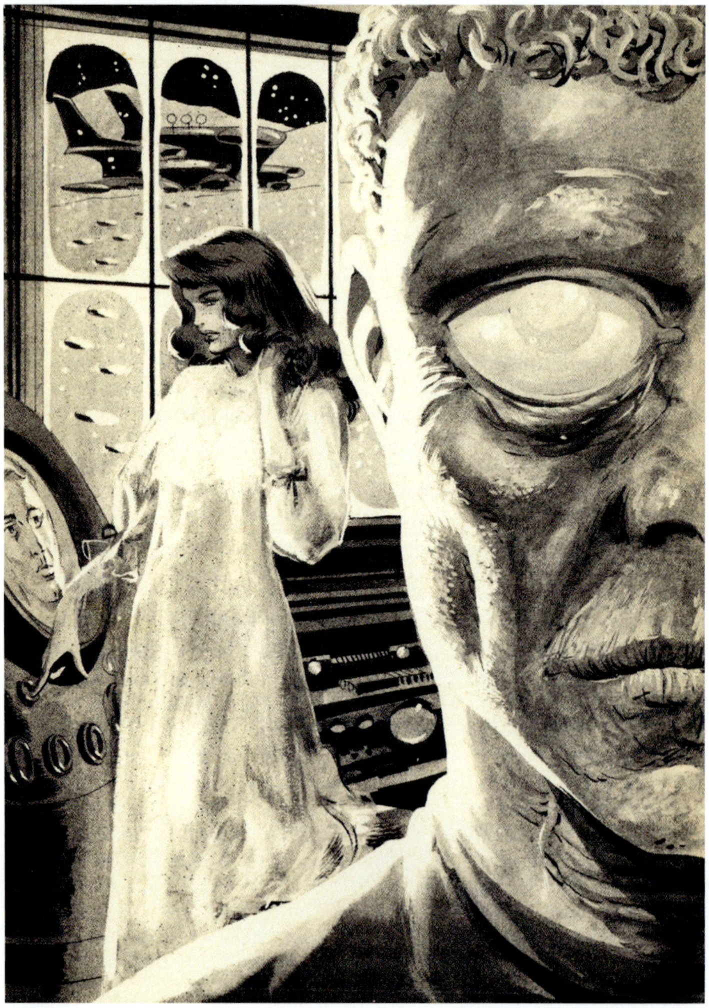

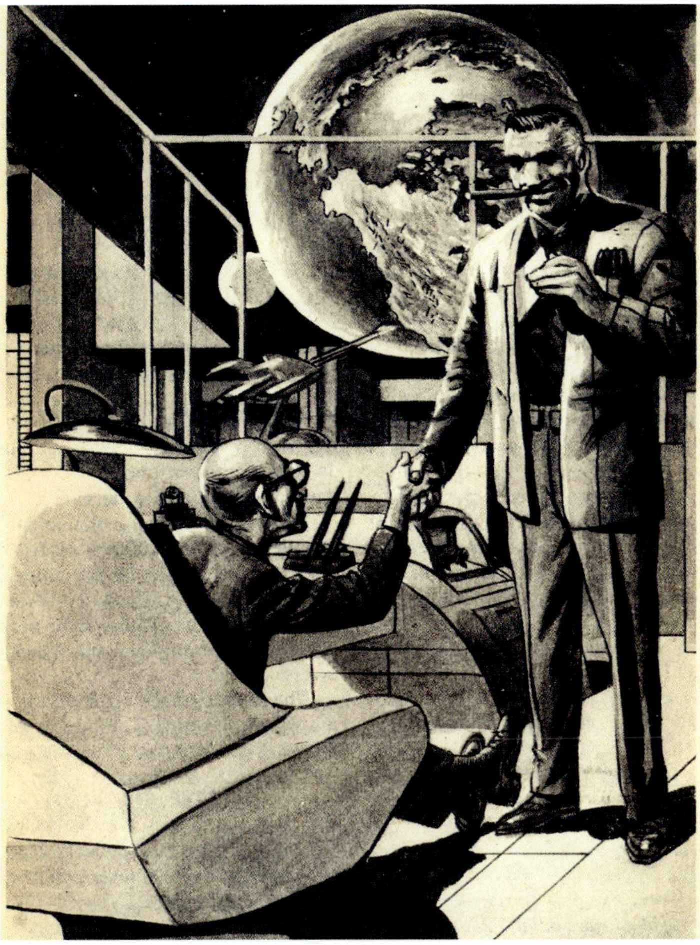

Clockwise from upper left:
Galaxy Science Fiction December 1959
Illustration for "Blacksword" by A. J. Offutt

Galaxy Science Fiction February 1960
Illustration for "Crying Jag" by Clifford D. Simak

Galaxy Science Fiction February 1960
Illustration for "The Nuse Man" by Margaret St. Clair

Opposite, clockwise from upper left:
Galaxy Science Fiction August 1959
Illustration for "Lex" by W. T. Haggert

Galaxy Science Fiction August 1959
Illustration for "The Waging of the Peace" by Frederik Pohl

Galaxy Science Fiction December 1959
Illustration for "The Snowmen" by Frederik Pohl

Galaxy Science Fiction December 1959
Illustration for "War Game" by Philip K. Dick

Next spread:
Page 110, top:
Galaxy Science Fiction April 1960
Illustration for "Don't Look Now" by Leonard Rubin

Page 110, bottom:
Galaxy Science Fiction February 1960
Illustration for "Crying Jag" by Clifford D. Simak

Page 111:
Galaxy Science Fiction October 1959 (Cover)
Illustration for "Someone to Watch Over Me," by Christopher Grimm

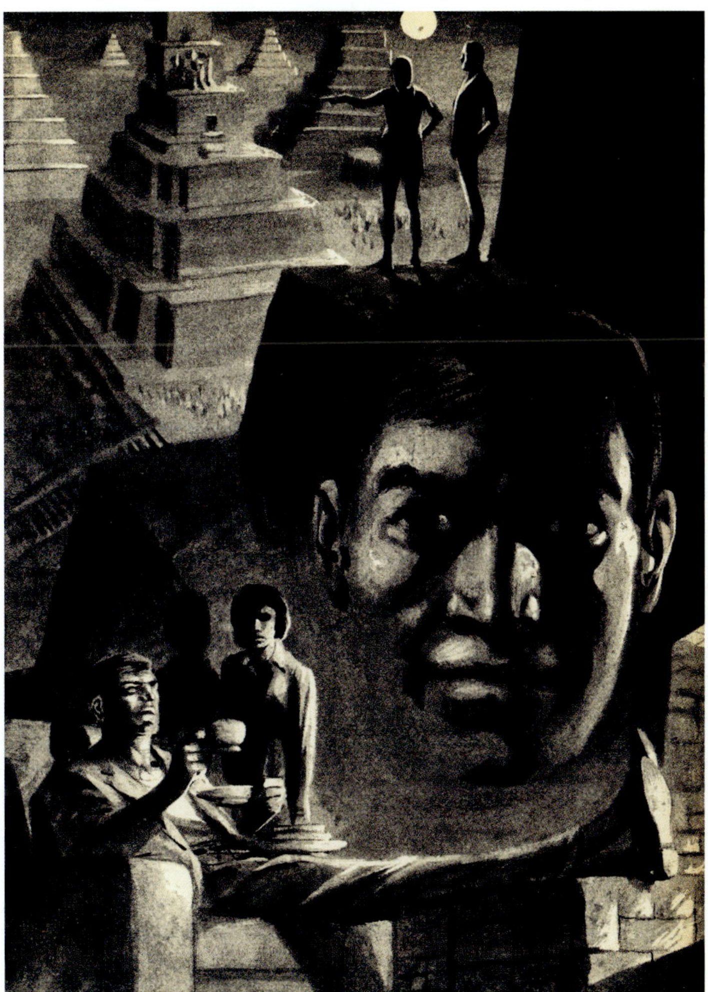

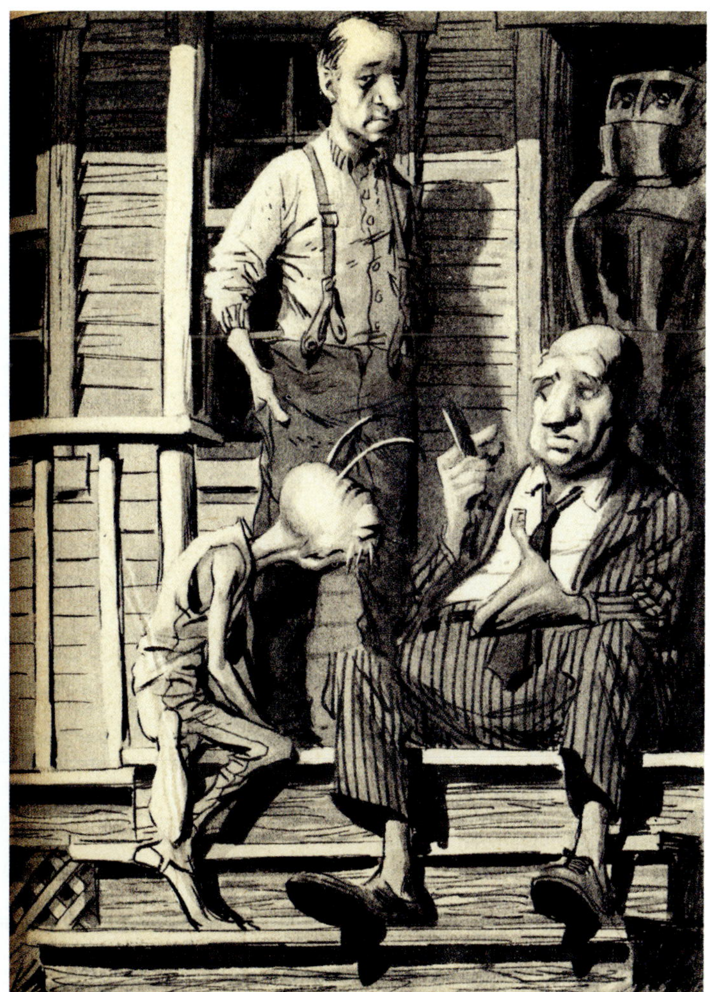

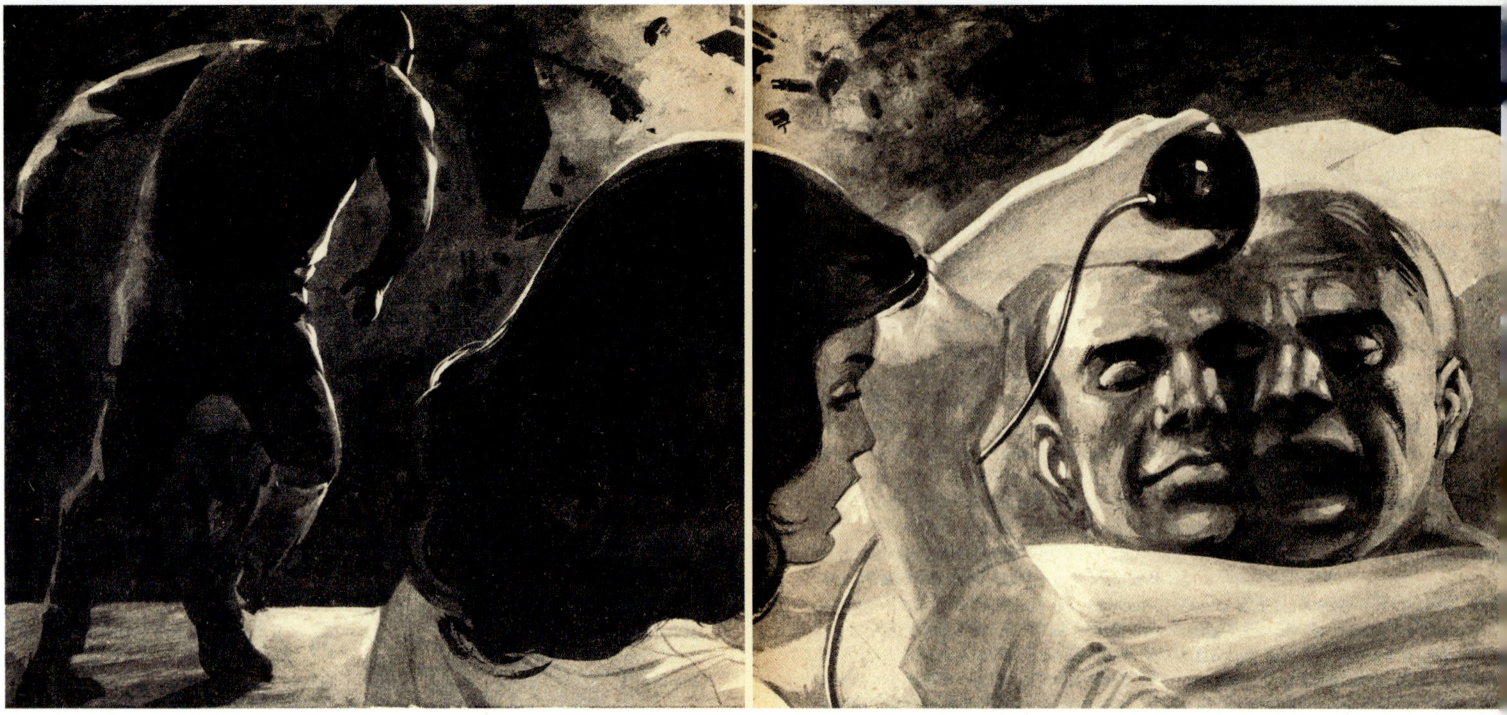

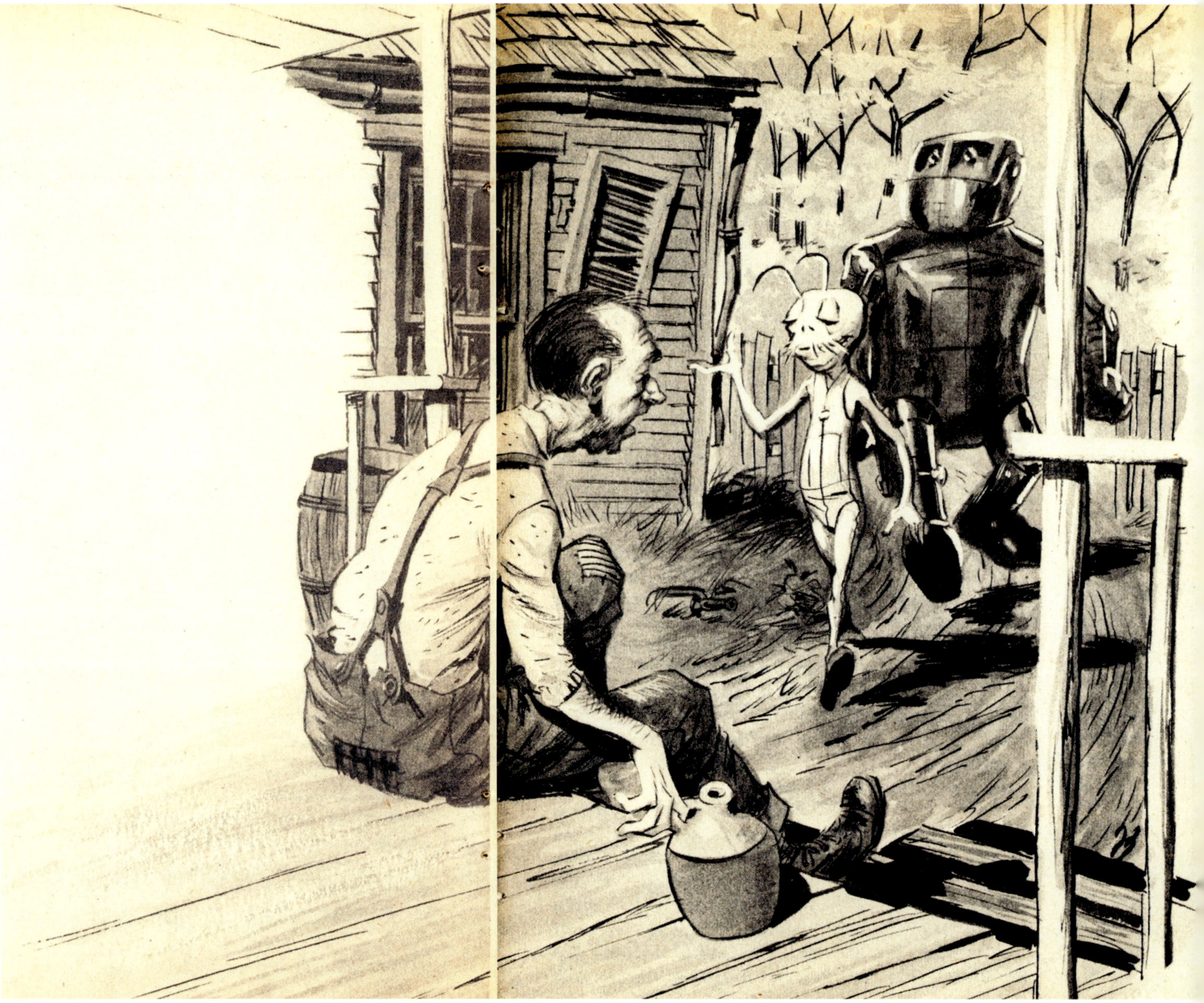

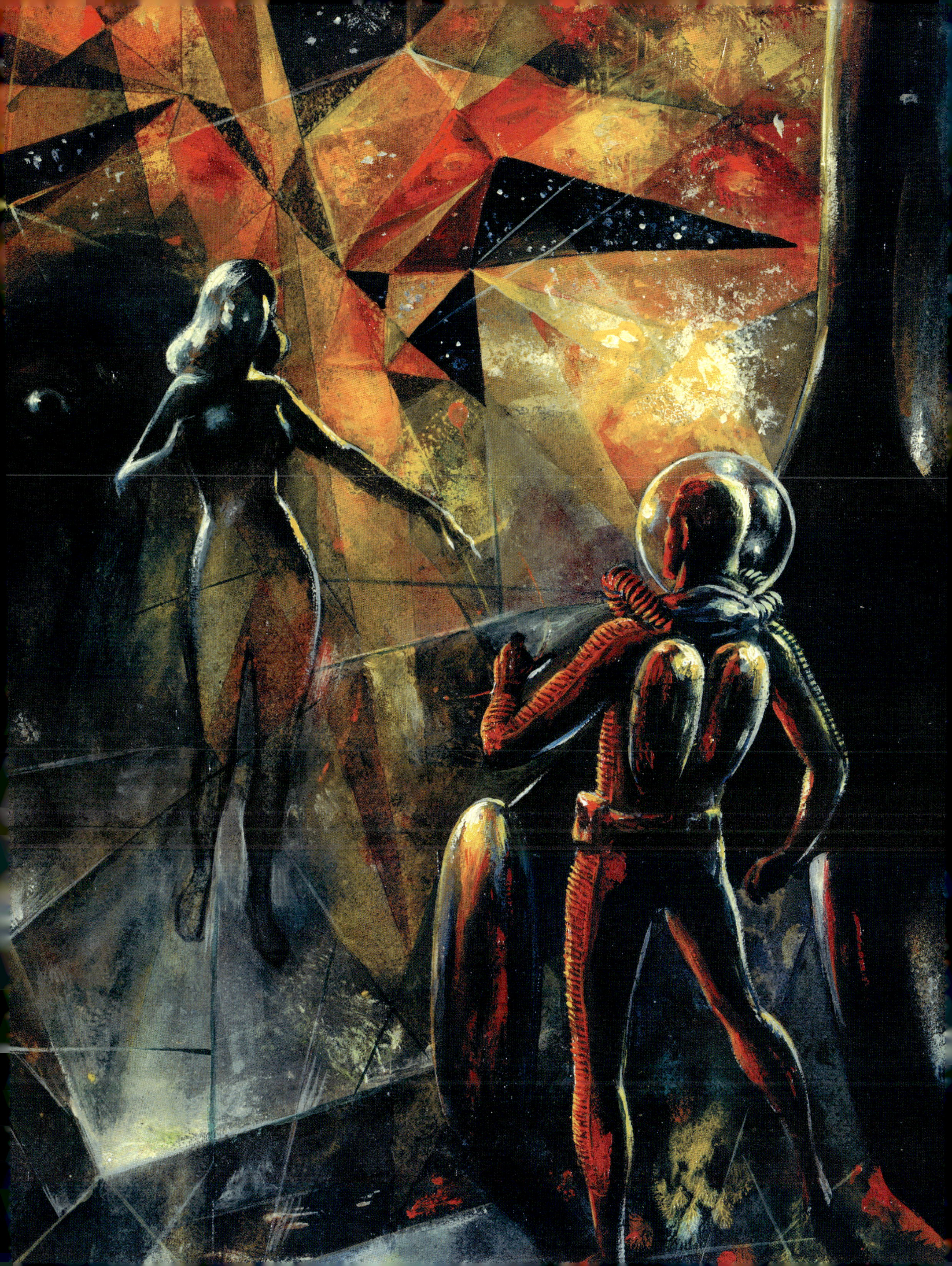

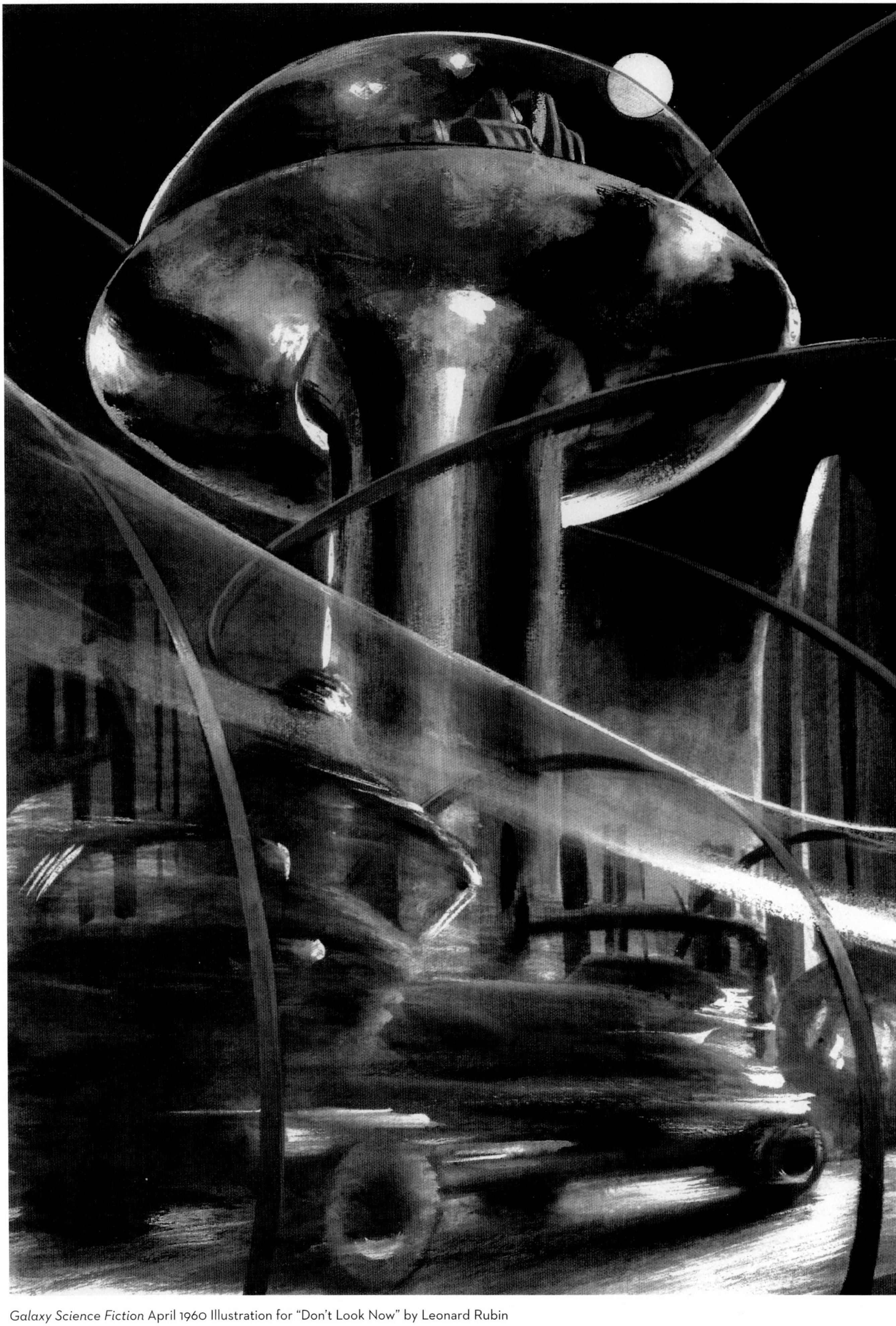

Galaxy Science Fiction April 1960 Illustration for “Don’t Look Now” by Leonard Rubin

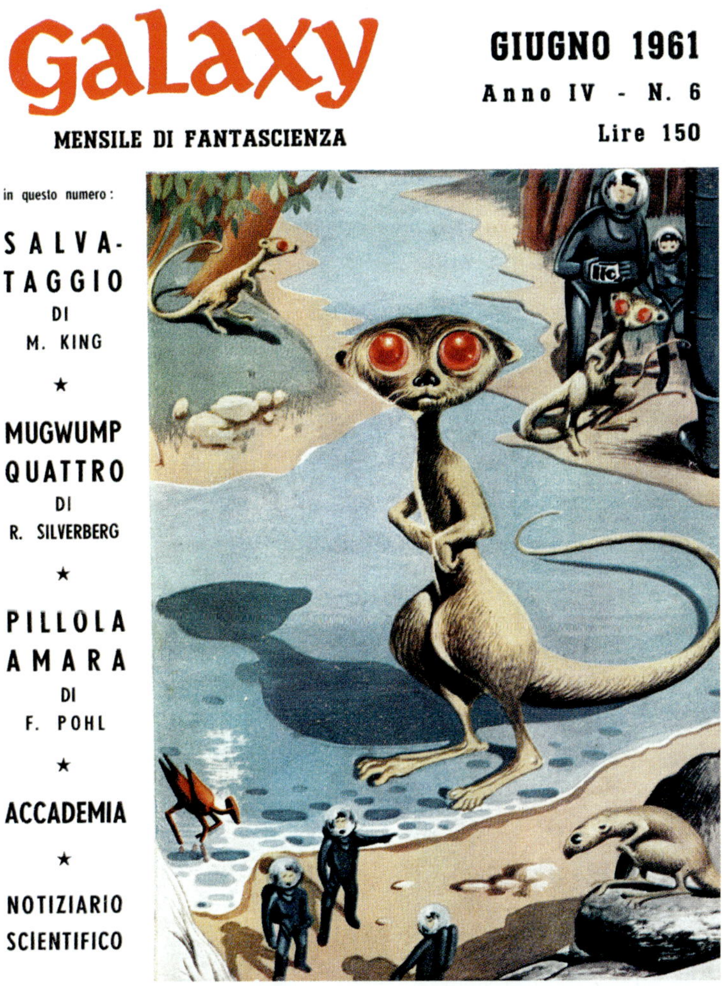

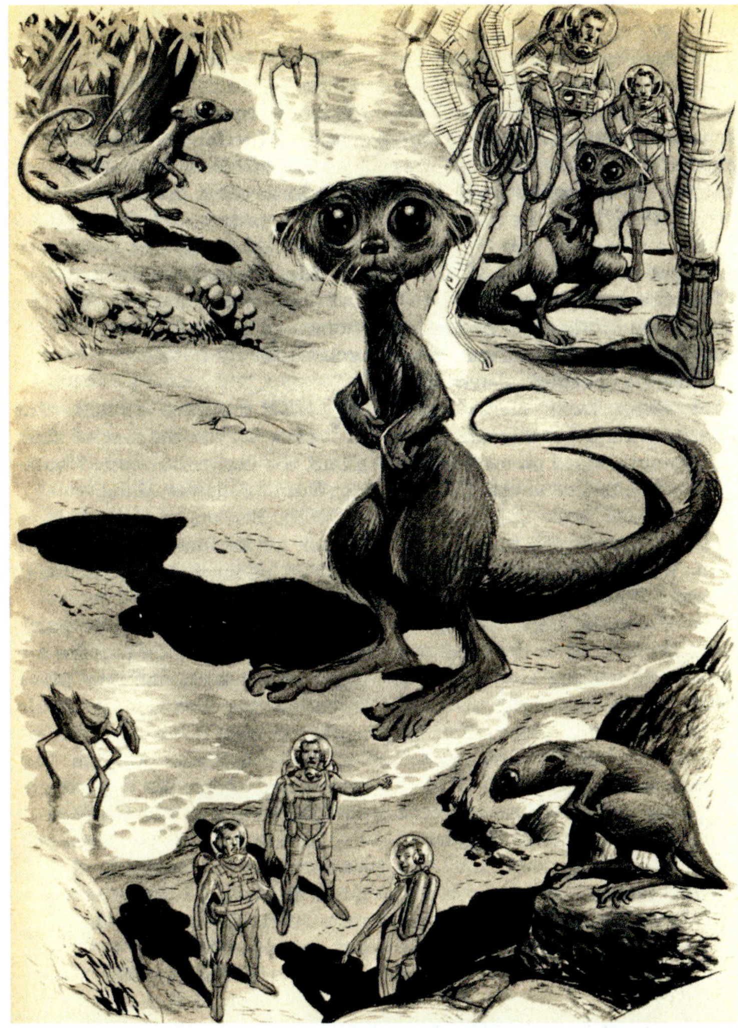

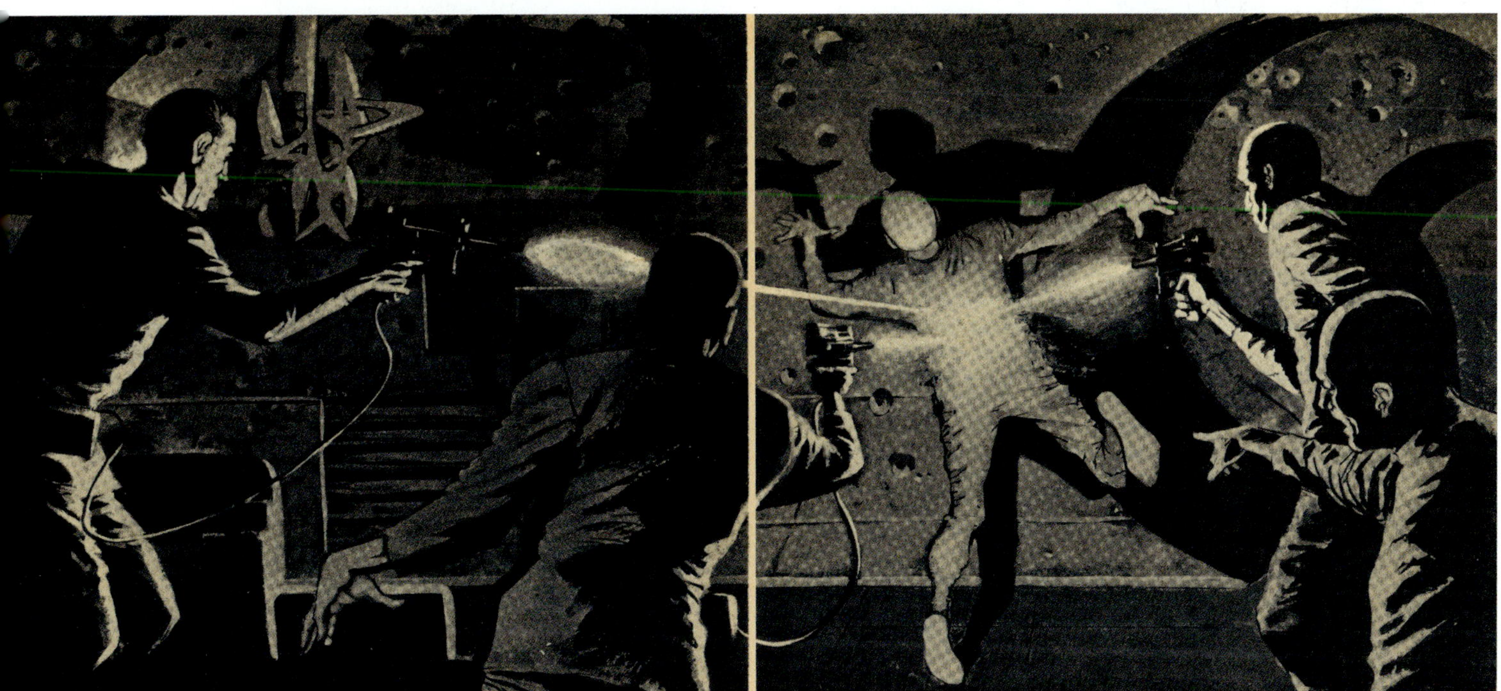

Clockwise from upper left:
Galaxy Science Fiction October 1960 (Cover - Italy)

Galaxy Science Fiction October 1960 Illustration for "Beach Scene" by Marshall King

Galaxy Science Fiction October 1960 Illustration for "The Stentorii Luggage" by Neal Barrett Jr

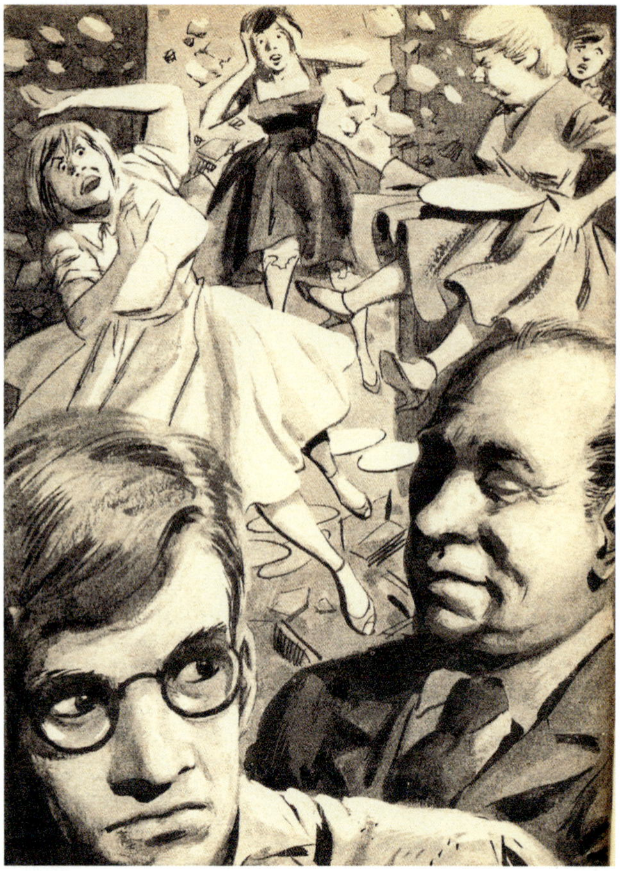

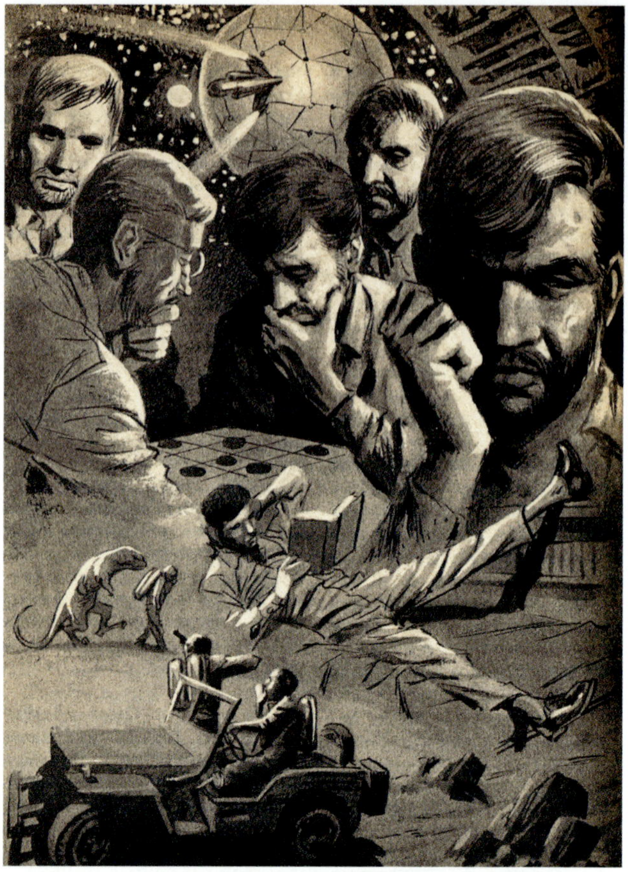

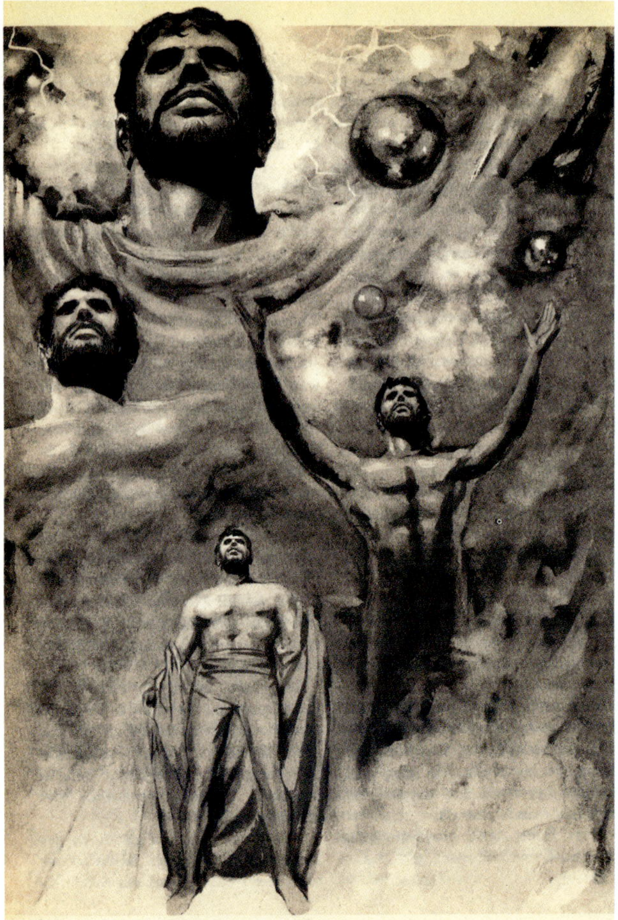

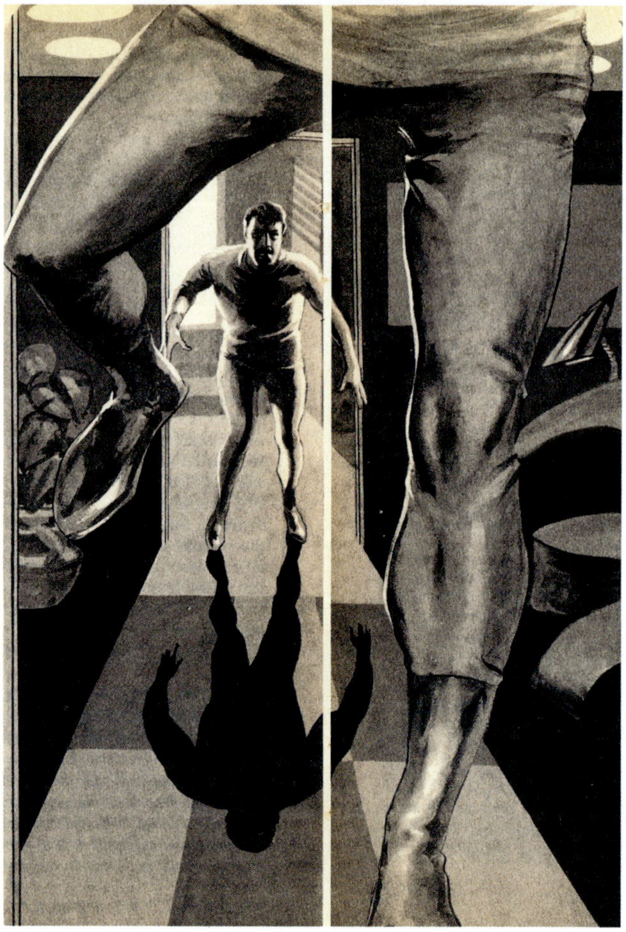

Clockwise from upper left:
Galaxy Science Fiction April 1960 Illustration for "The Airy Servitor" by Margaret St. Clair

Galaxy Science Fiction June 1960 Illustration for "The Dope on Mars" by Jack Sharkey

Both bottom illustrations: *Galaxy Science Fiction* August 1960 Illustration for "Sordman the Protector" by Tom Purdom

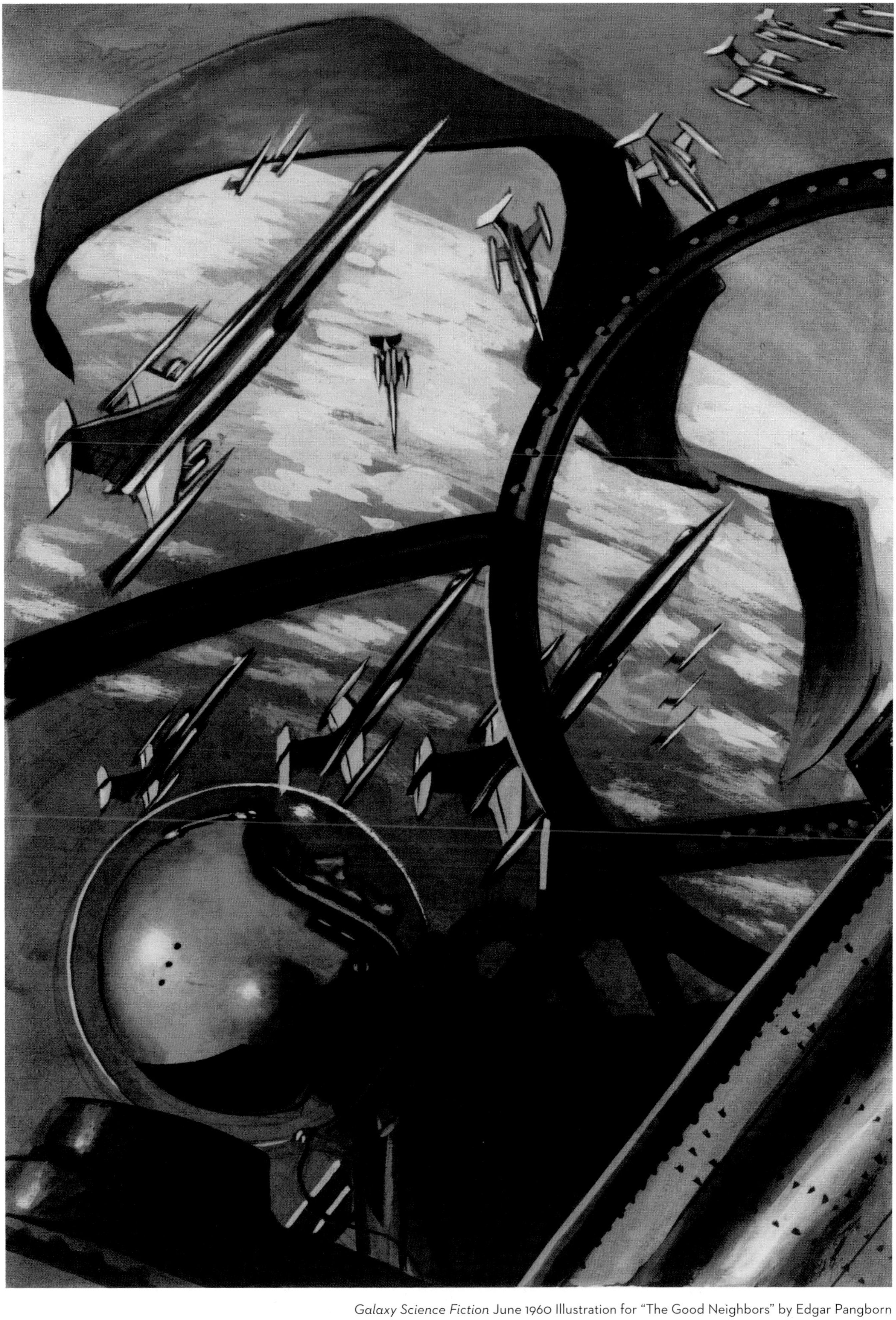

Galaxy Science Fiction June 1960 Illustration for "The Good Neighbors" by Edgar Pangborn

Top and lower left:
Galaxy Science Fiction October 1960
Illustration for "The Stentorii Luggage"
by Neal Barrett Jr.

Lower right:
Galaxy Science Fiction December 1960
Illustration for "Round-And-Round Trip" by H. B. Fyfe

Opposite:
Top to bottom:
Galaxy Science Fiction
December 1960
Illustration for "Round-And-Round Trip" by H. B. Fyfe

Galaxy Science Fiction August 1963
Illustration for "The Great Nebraska Sea"
by Allan Danzig

Bottom two illustrations:
Galaxy Science Fiction June 1963
Illustration for "Here Gather the Stars" part 1
by Clifford D. Simak

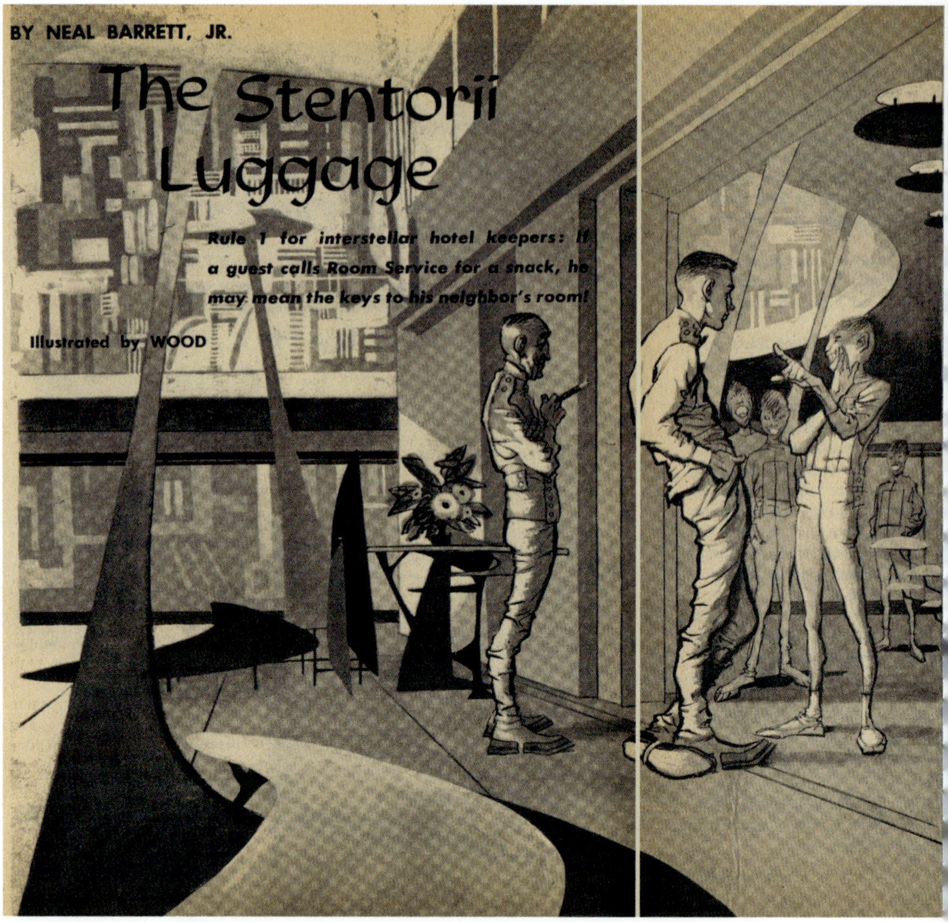

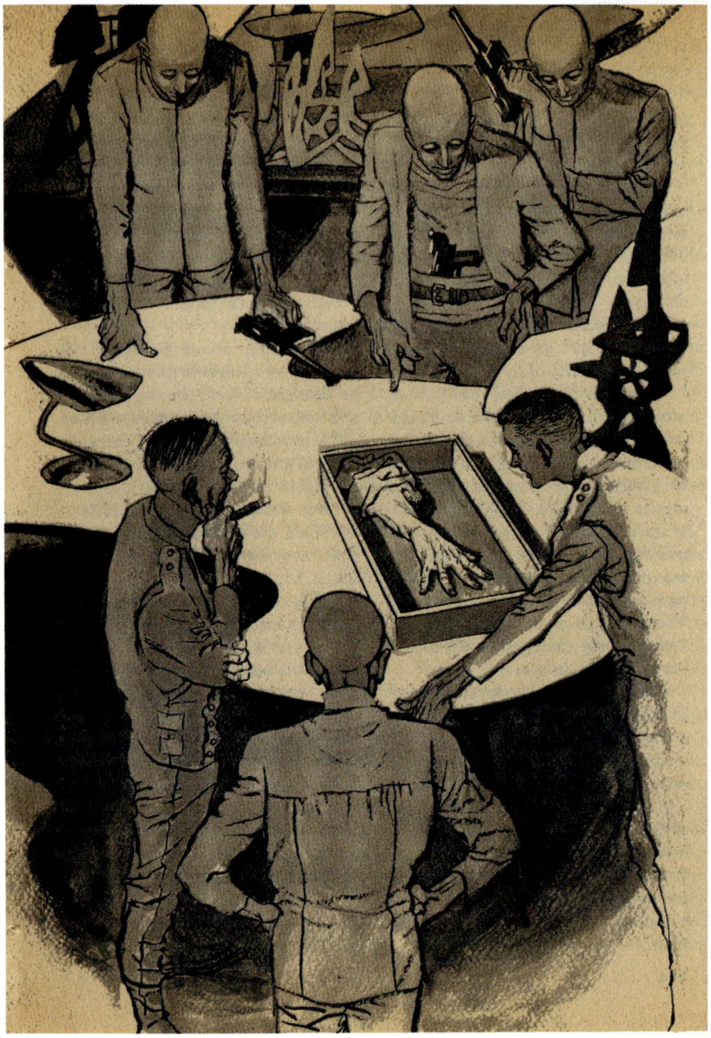

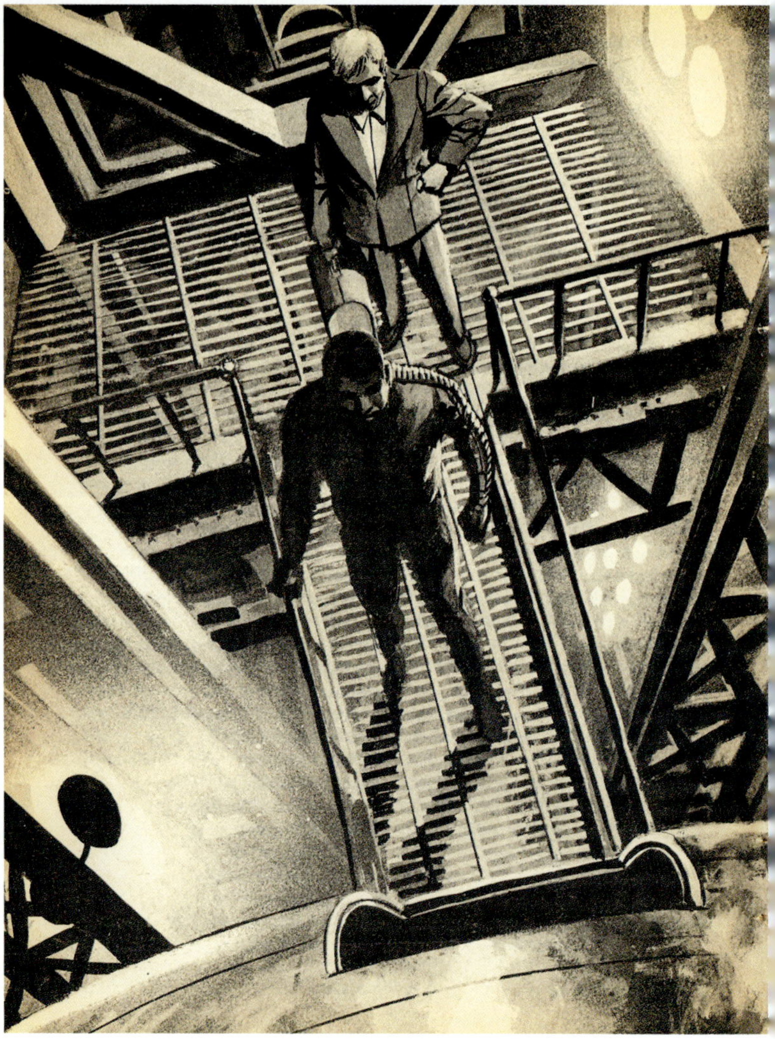

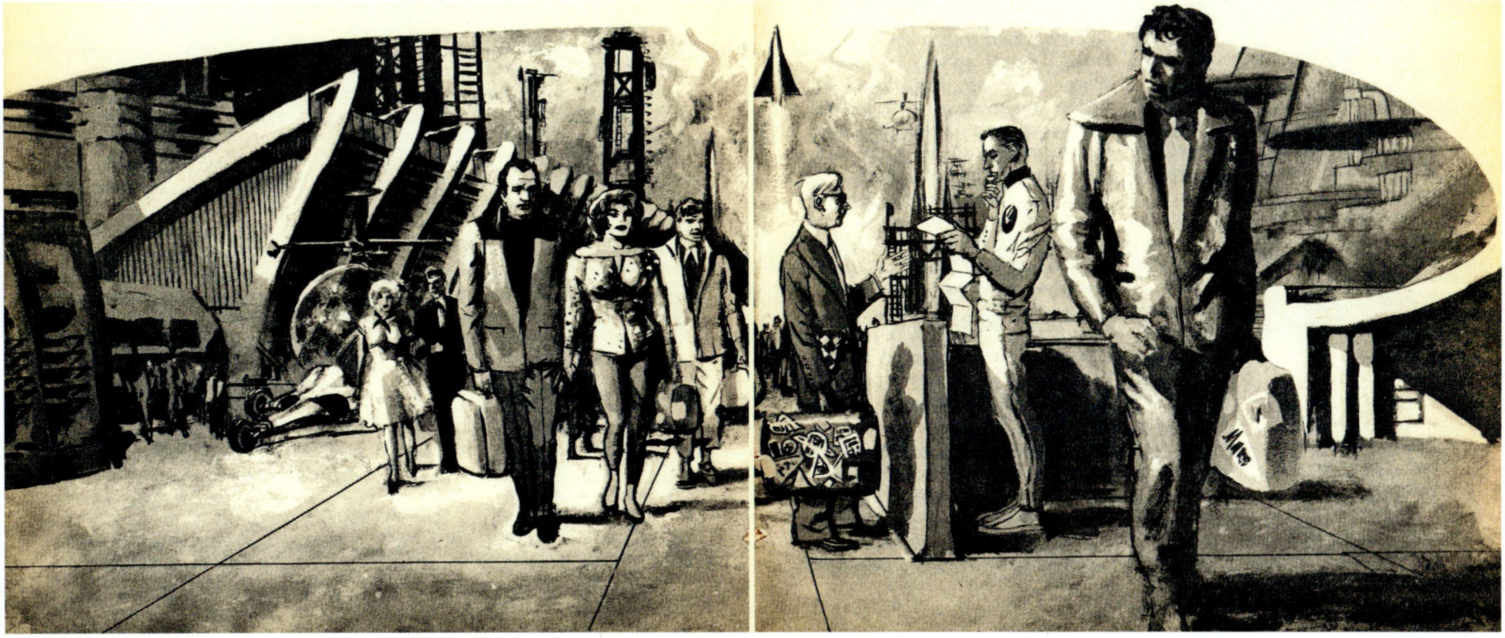

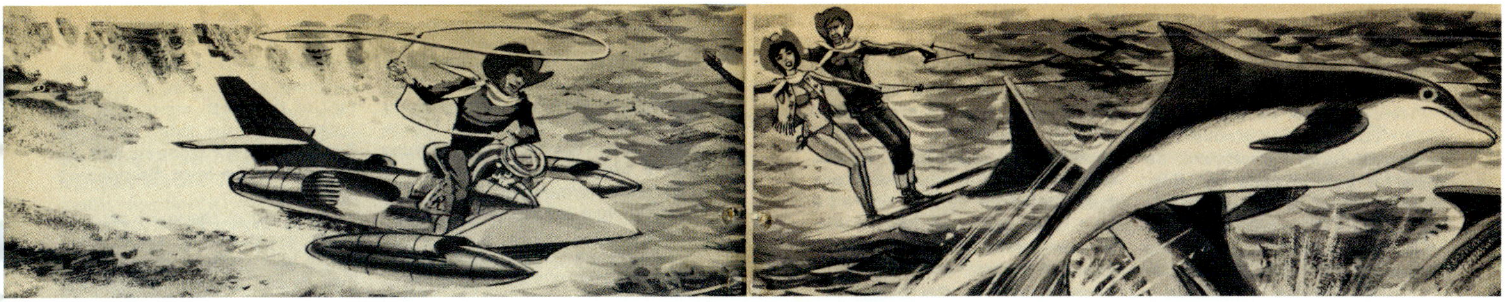

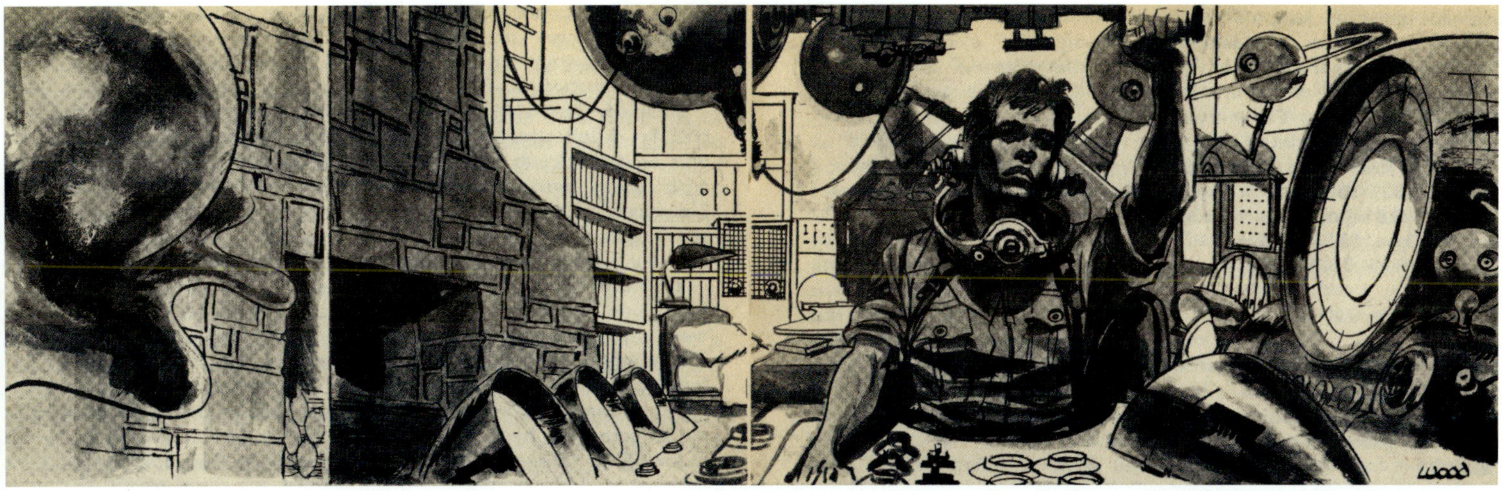
WOOD

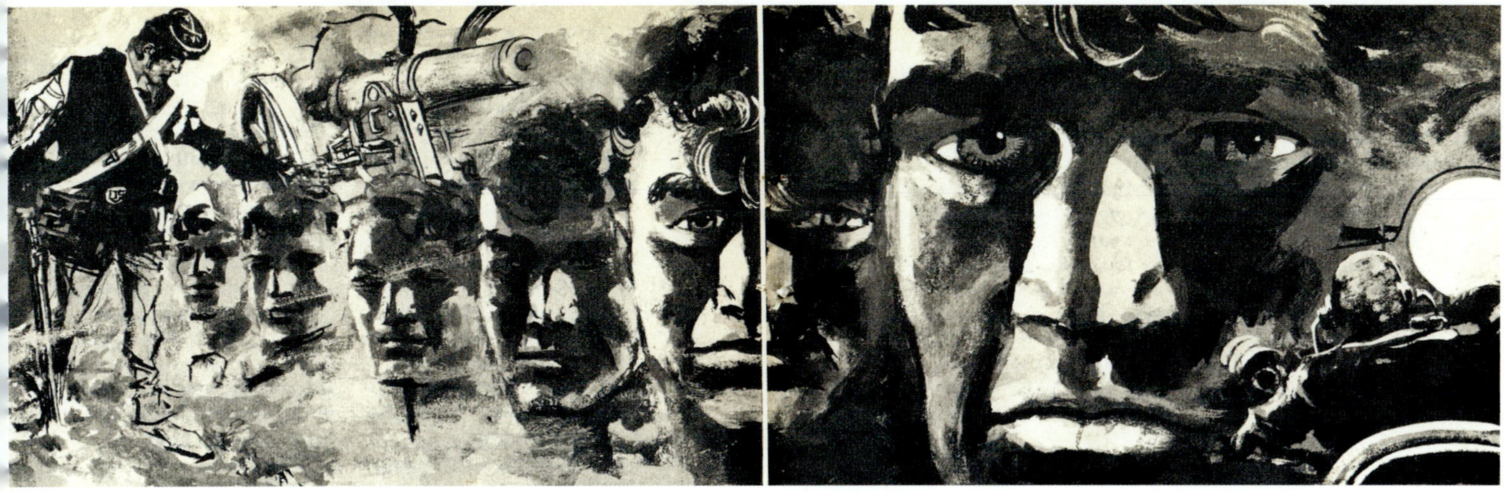

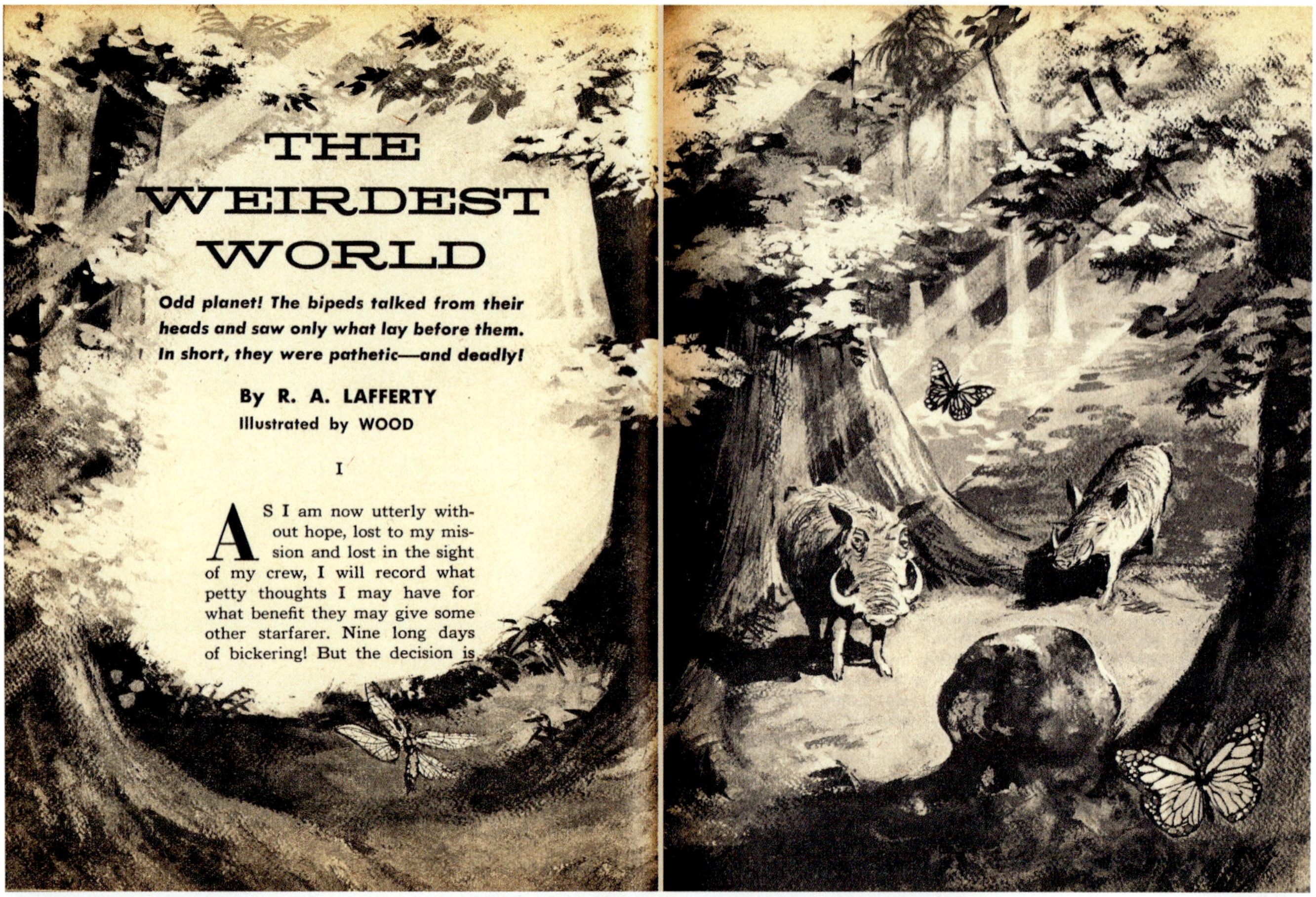
THE WEIRDEST WORLD

Odd planet! The bipeds talked from their heads and saw only what lay before them. In short, they were pathetic—and deadly!

By R. A. LAFFERTY
Illustrated by WOOD

I

AS I am now utterly without hope, lost to my mission and lost in the sight of my crew, I will record what petty thoughts I may have for what benefit they may give some other starfarer. Nine long days of bickering! But the decision is

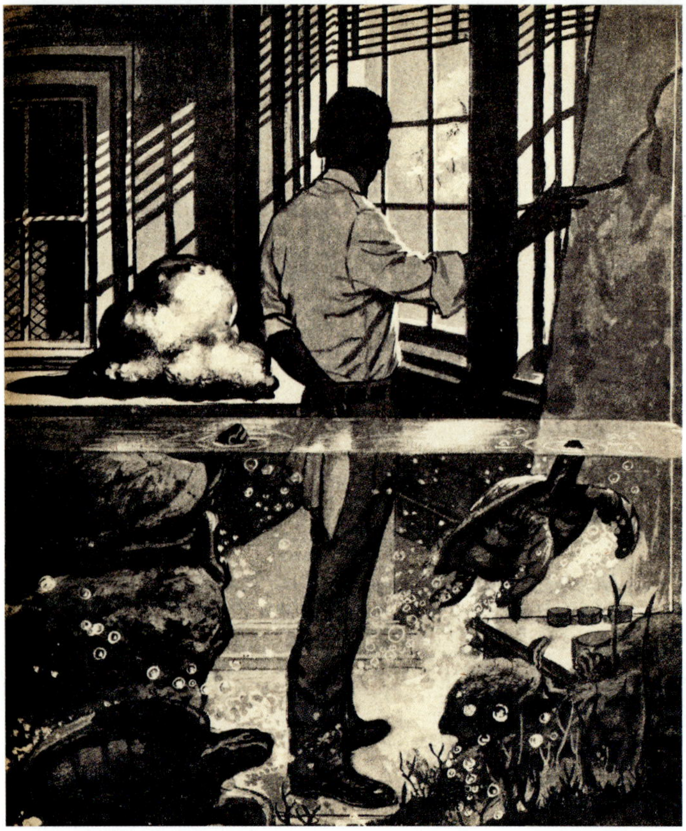

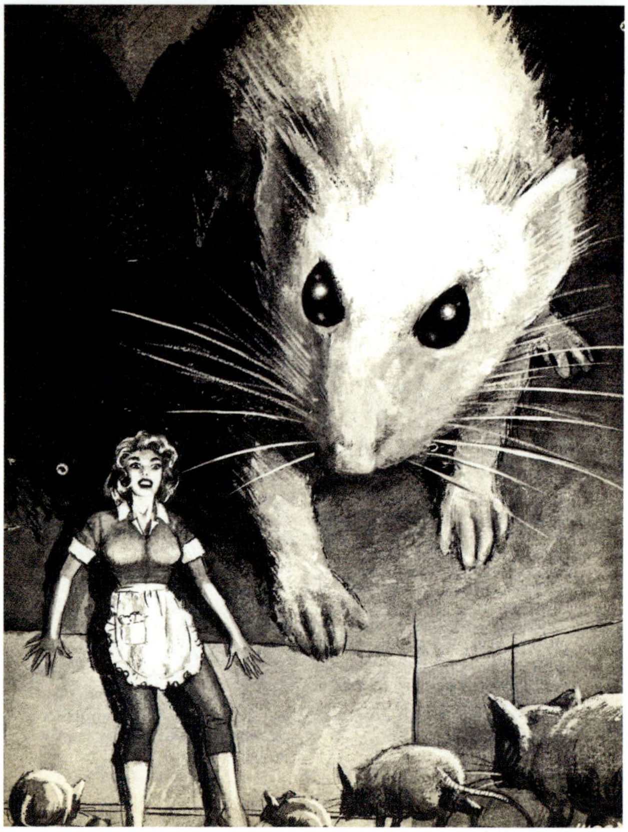

Top and lower left: *Galaxy Science Fiction* June 1961
Illustration for "The Weirdest World" by R. F. Lafferty

Lower right: *Galaxy Science Fiction* December 1961
Illustration for "Oh Rats!" by Miriam Allen Deford

Opposite: *Galaxy Science Fiction* June 1961 Illustration for "The Weirdest World" by R. F. Lafferty

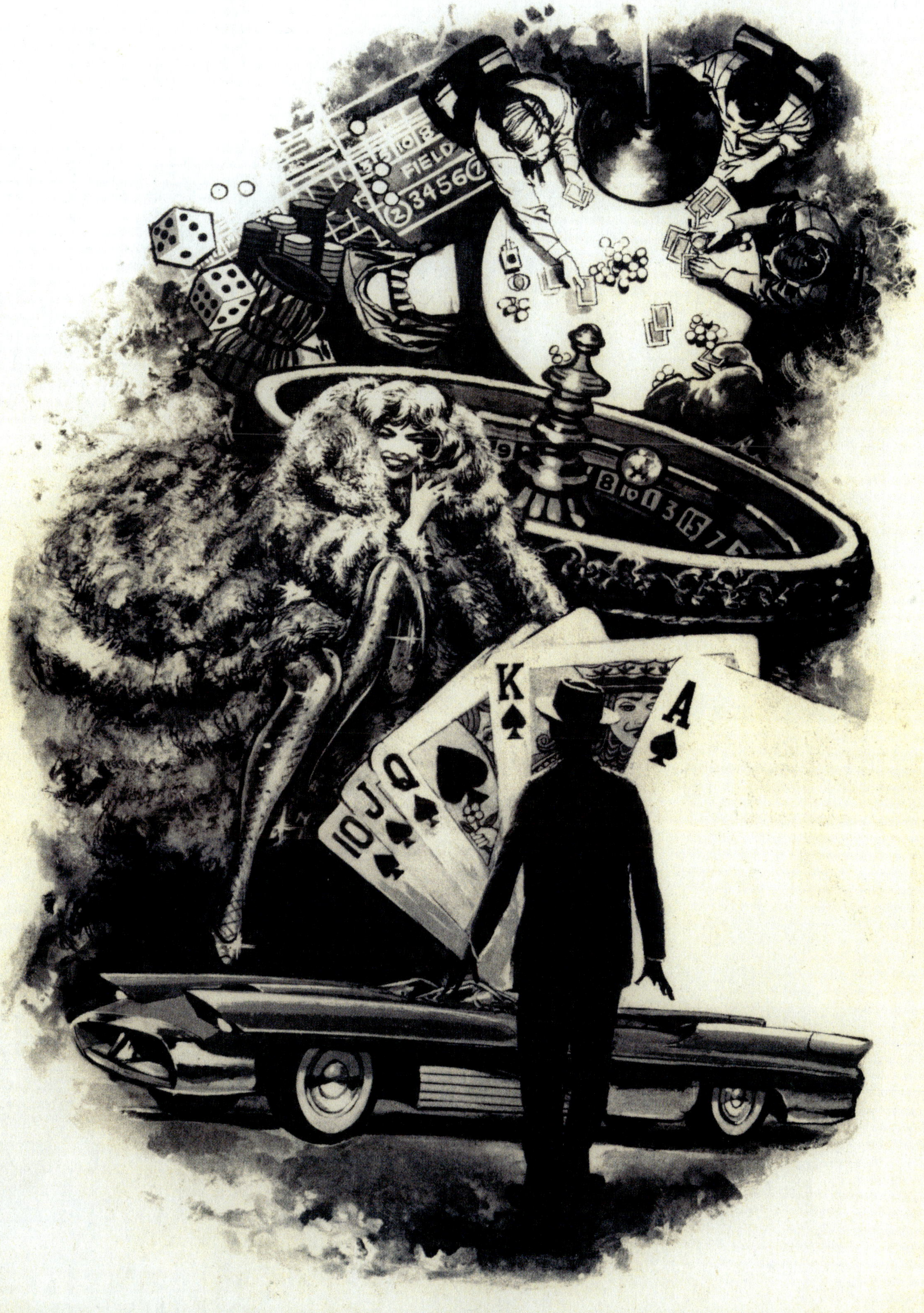
FIELD
3456
K
A
Q
J
10

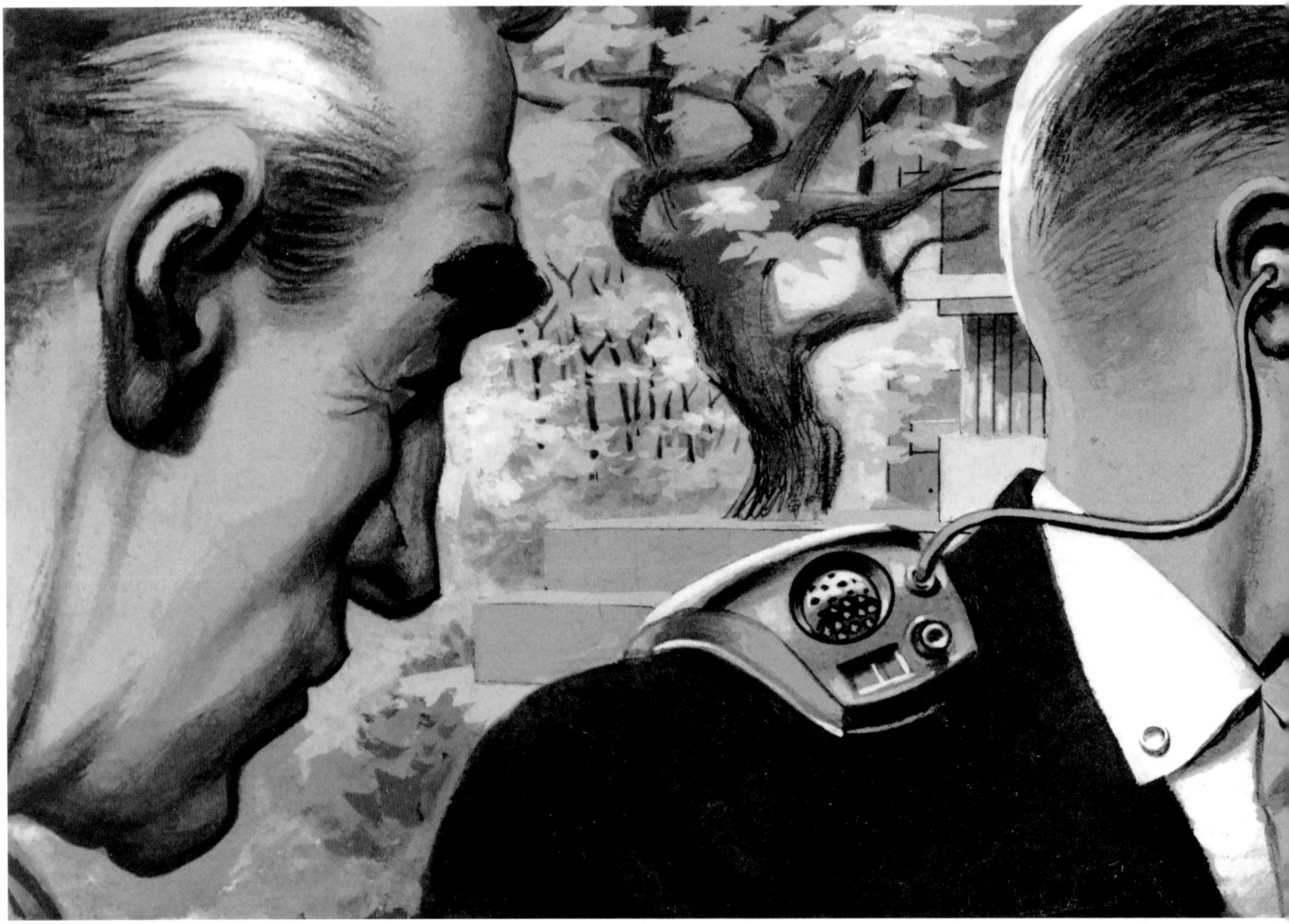

Galaxy Science Fiction December 1962
Illustration for "The Creature from the Cleveland Depths" by Fritz Leiber

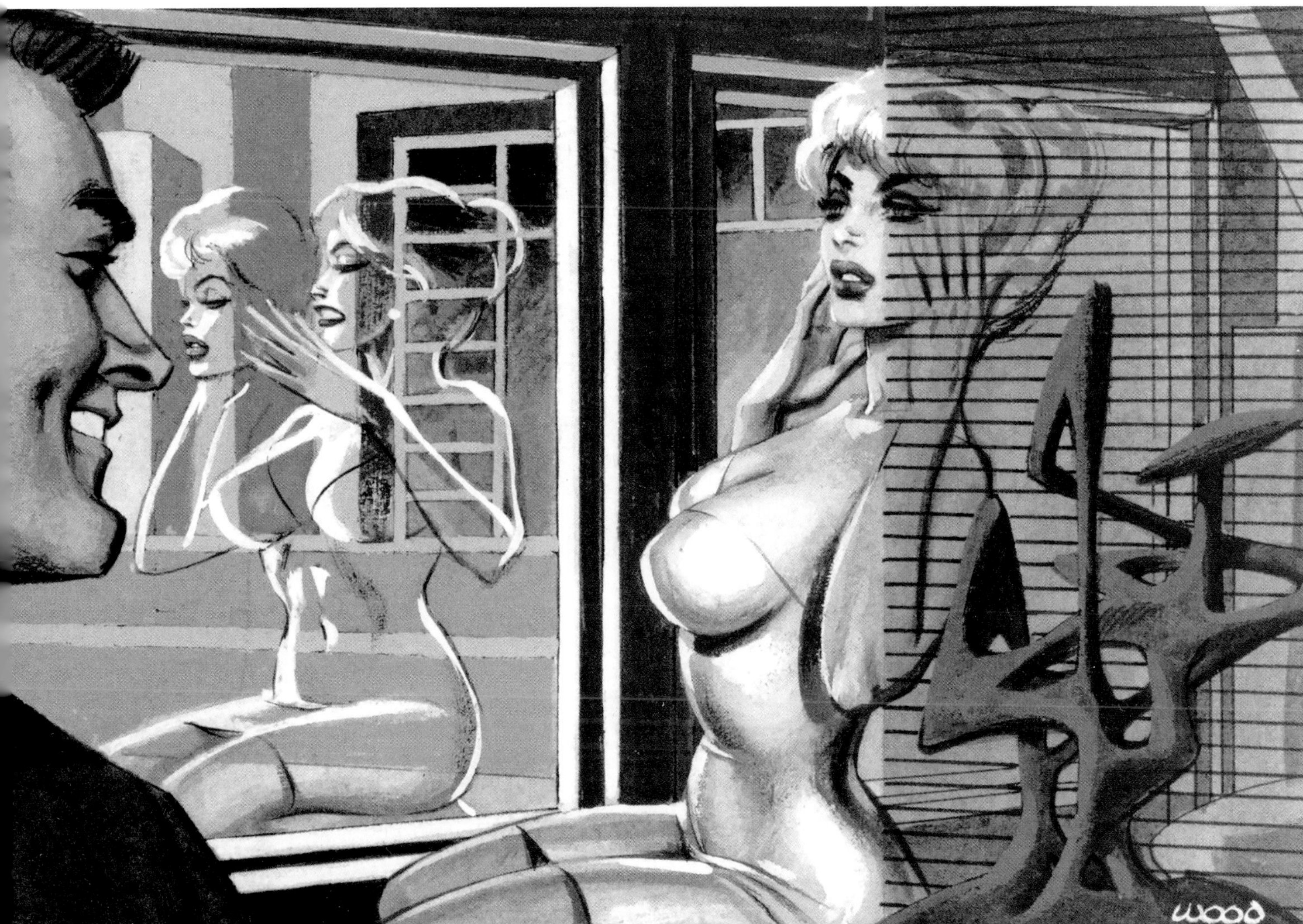
WOOD

Clockwise from top: *Galaxy Science Fiction* August 1963 Illustration for "The Great Nebraska Sea" by Allan Danzig

The Survivors by Tom Godwin, Gnome Press, 1958 (Cover)

The Shrouded Planet by Robert Randall, Gnome Press, 1957 (Cover)

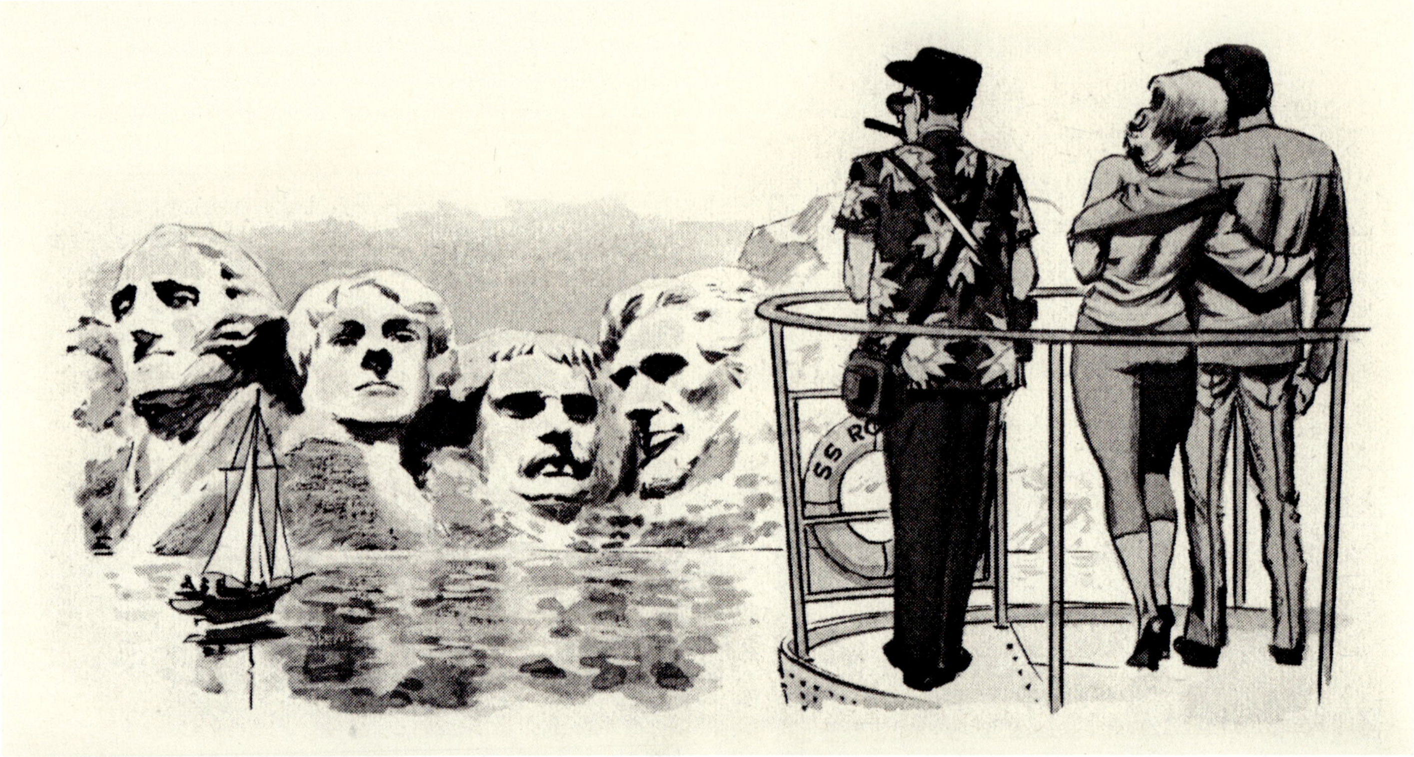

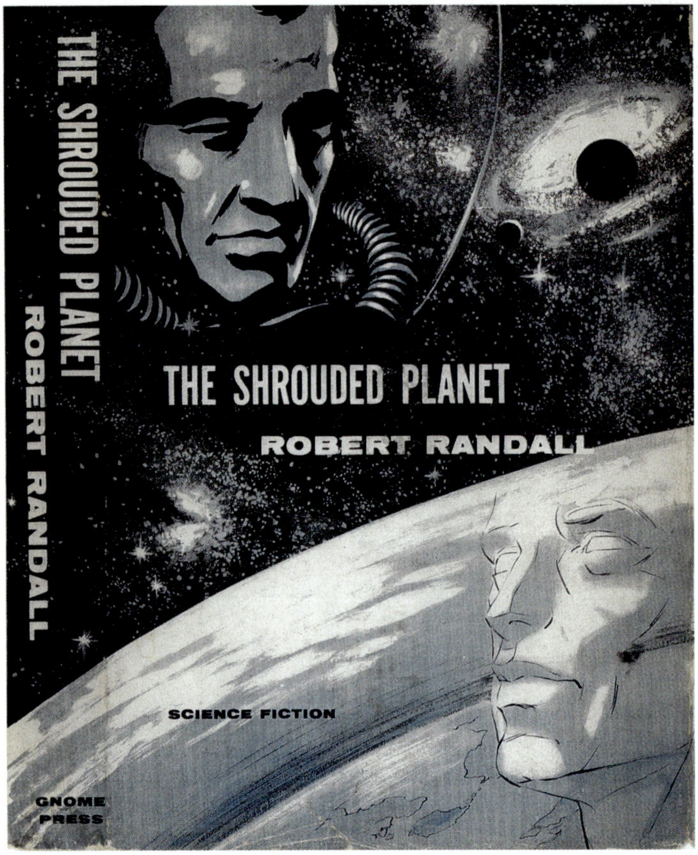

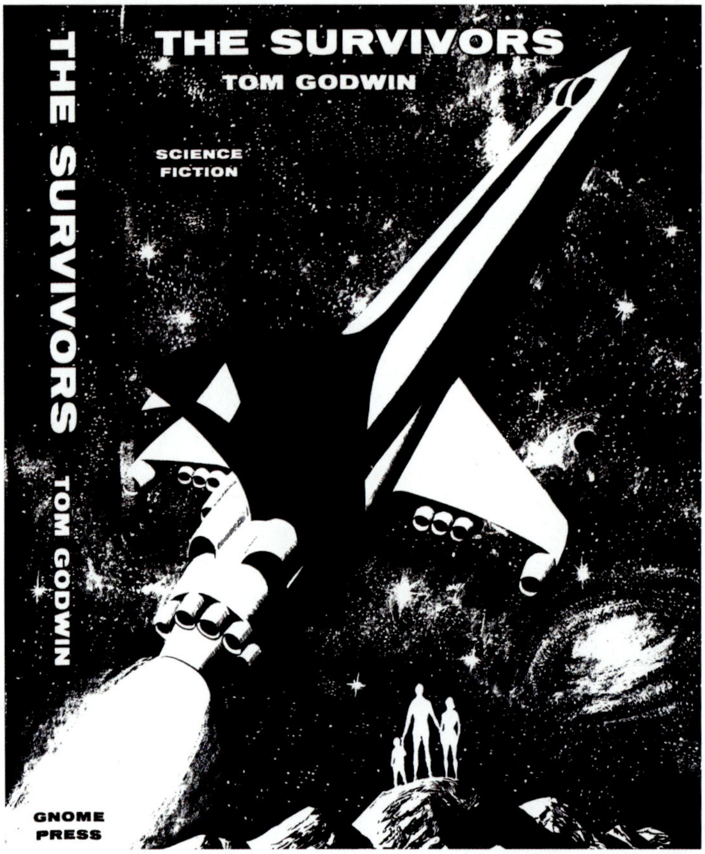

Opposite, clockwise from upper left:
Galaxy Science Fiction December 1962 Illustration for "The Creature from the Cleveland Depths" by Fritz Leiber

Galaxy Science Fiction October 1963 Illustration for "On the Gem Planet" by Cordwainer Smith

Galaxy Science Fiction June 1963 Illustration for "Here Gather the Stars" part 1 by Clifford D. Simak

Galaxy Science Fiction August 1963 Illustration for "Here Gather the Stars" part 2 by Clifford D. Simak

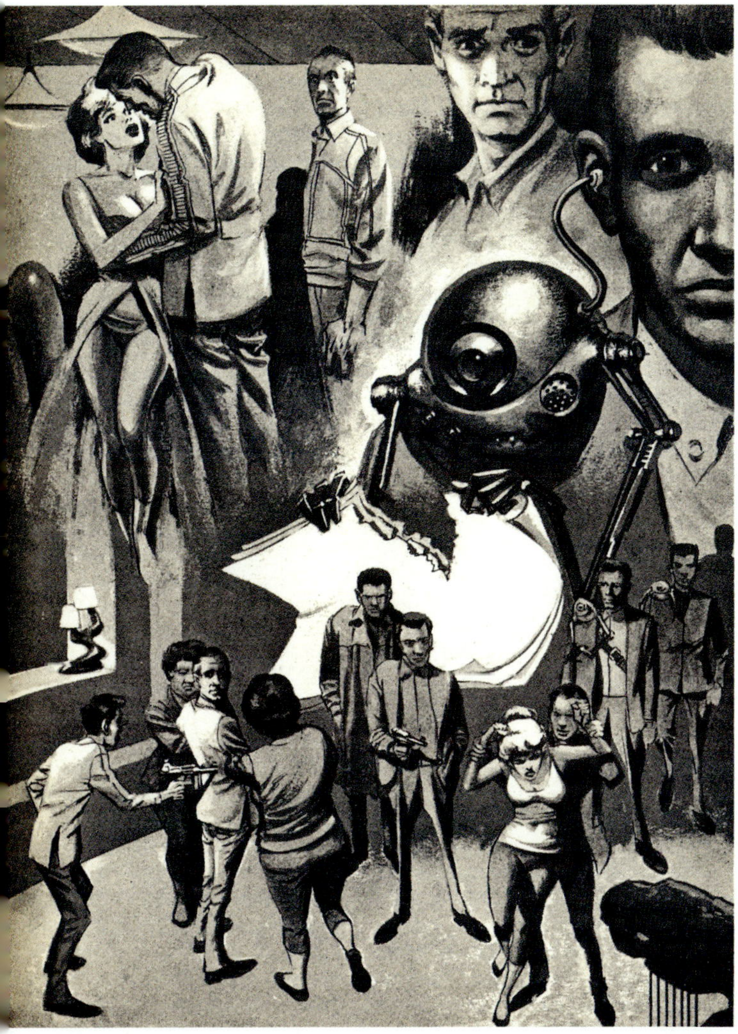

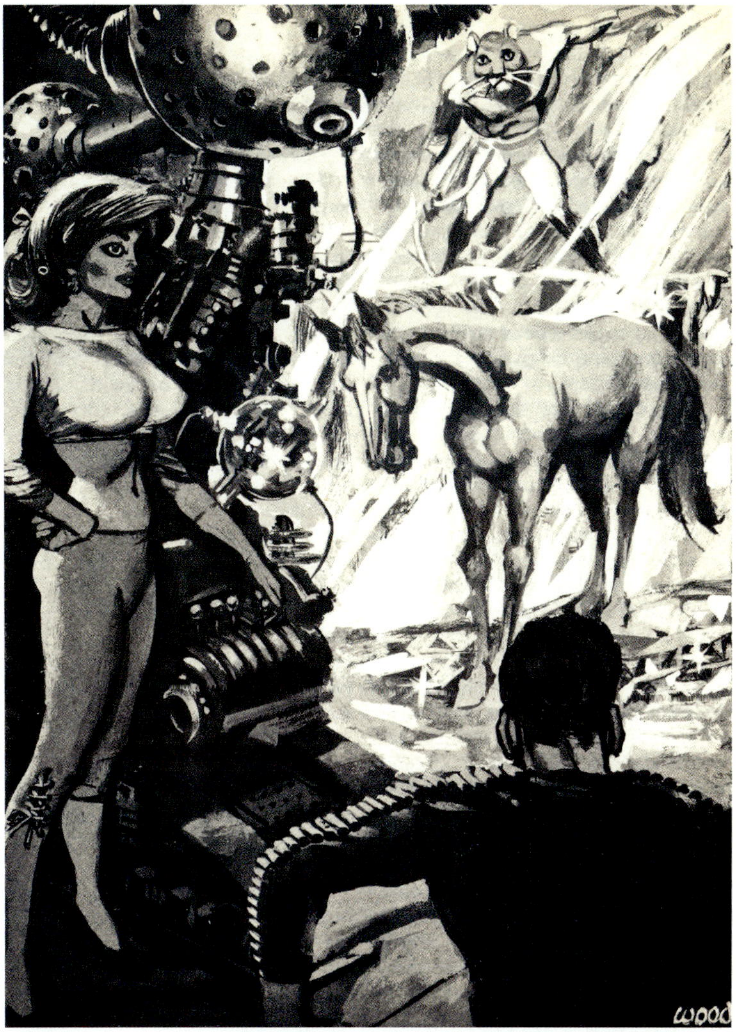
wood

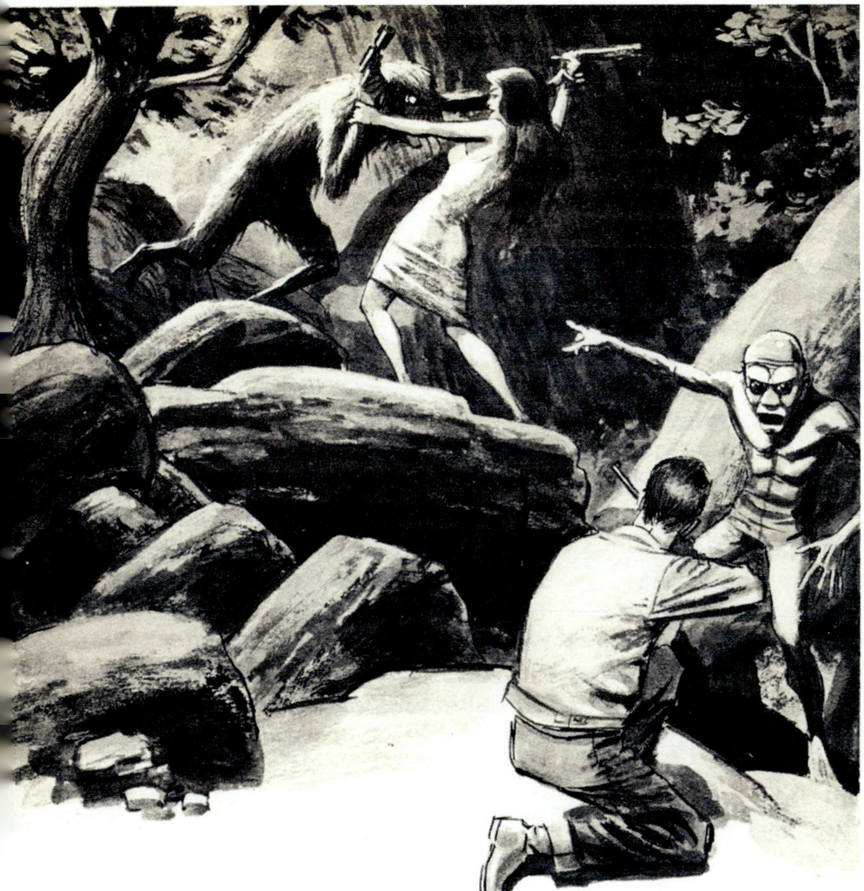

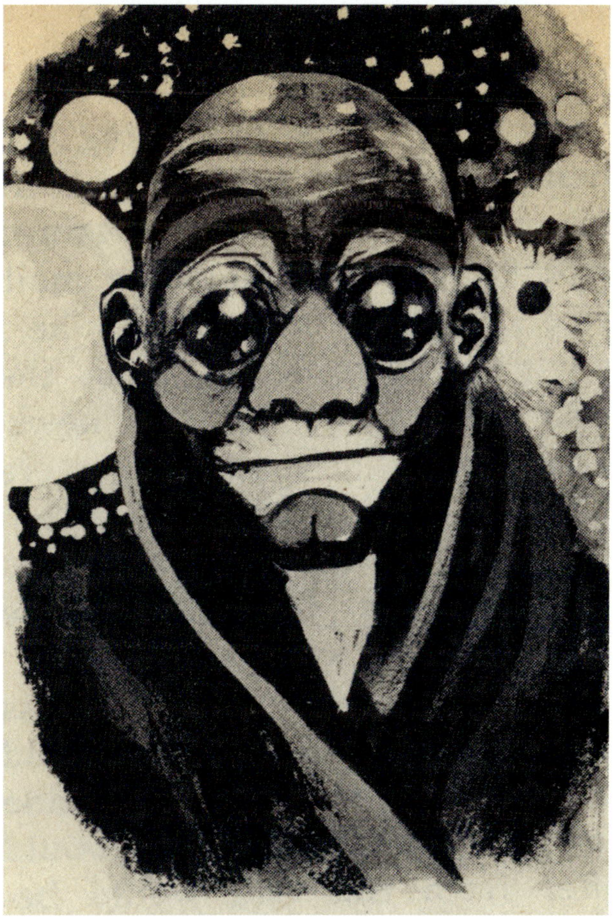

Upper left:
Galaxy Science Fiction
June 1963
Illustration for "Here Gather the Stars"
part 1 by Clifford D. Simak

Lower left:
Galaxy Science Fiction
August 1963
Illustration for "Here Gather the Stars"
part 2 by Clifford D. Simak

Lower right:
Galaxy Science Fiction
December 1965
Illustration for "The Age of the Pussyfoot"
part 2 by Frederik Pohl

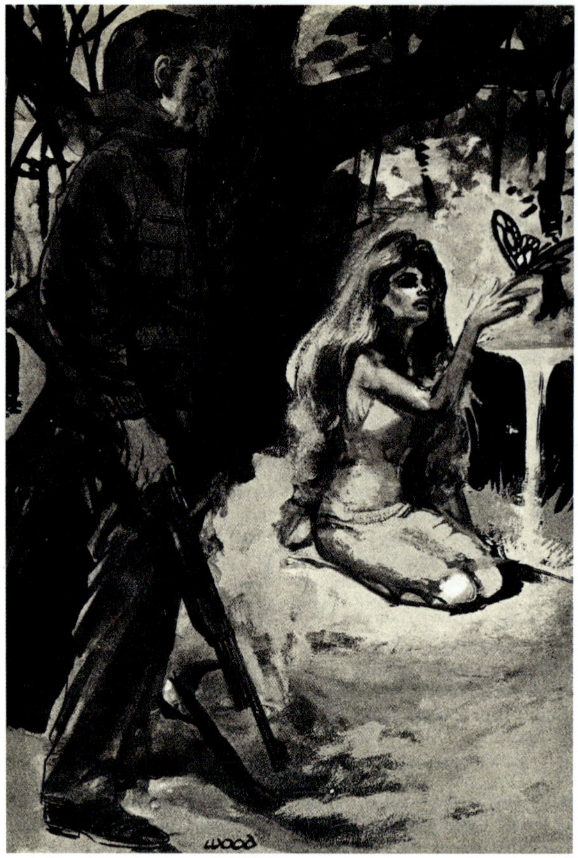

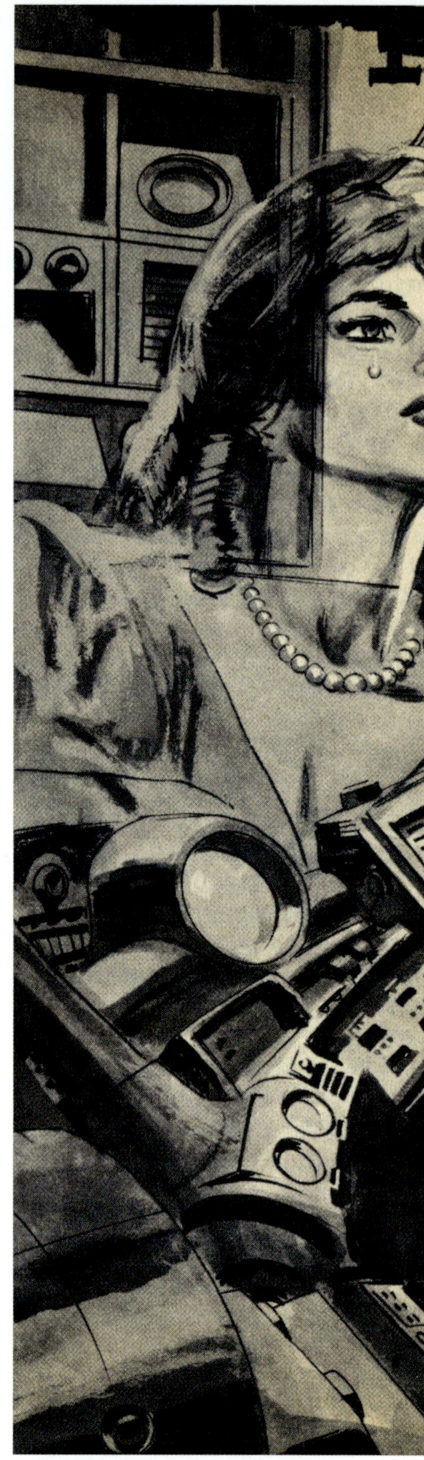

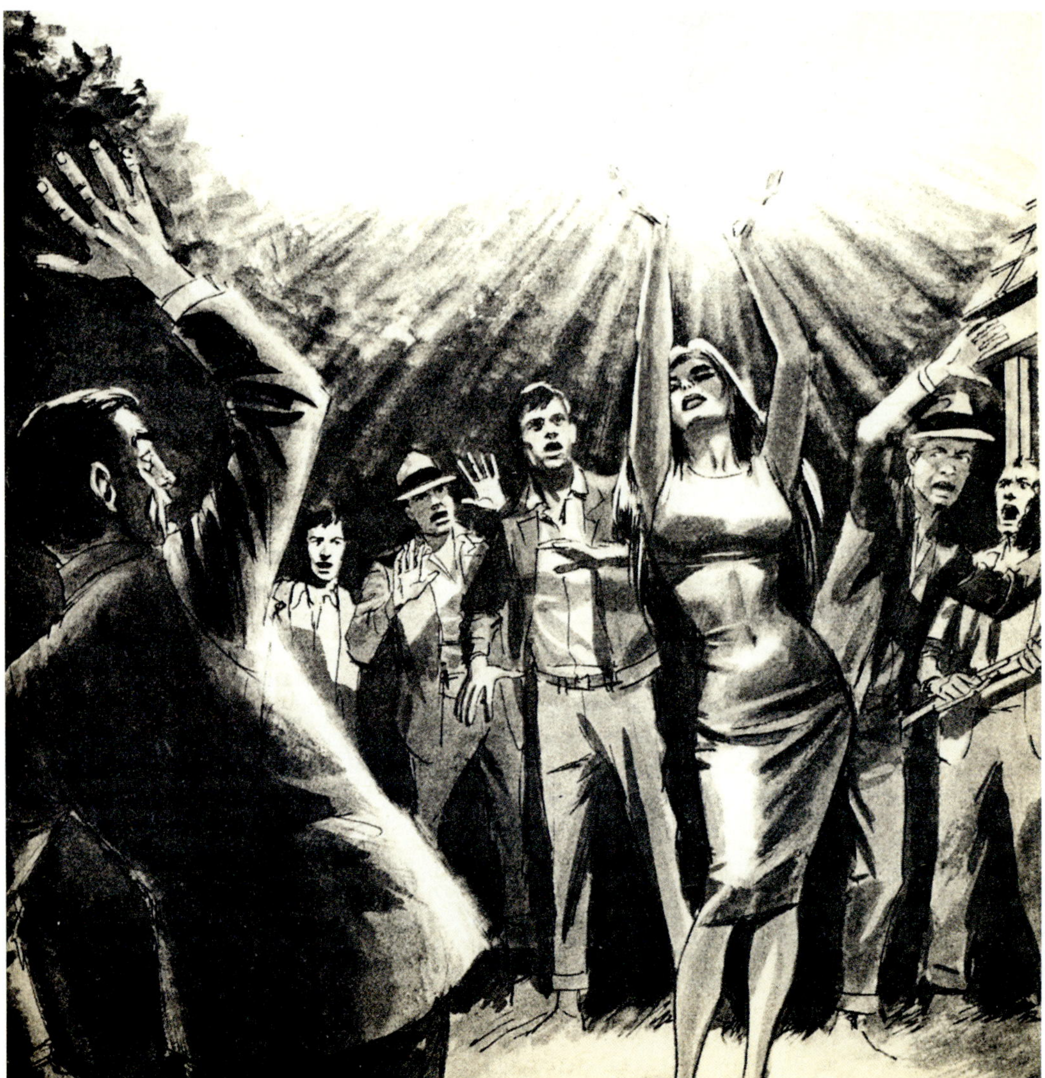

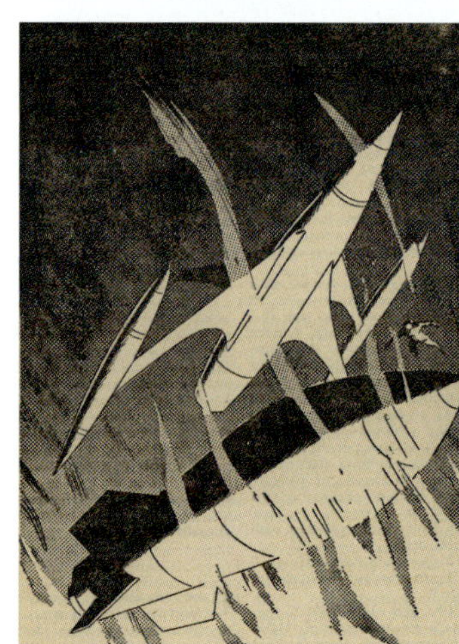

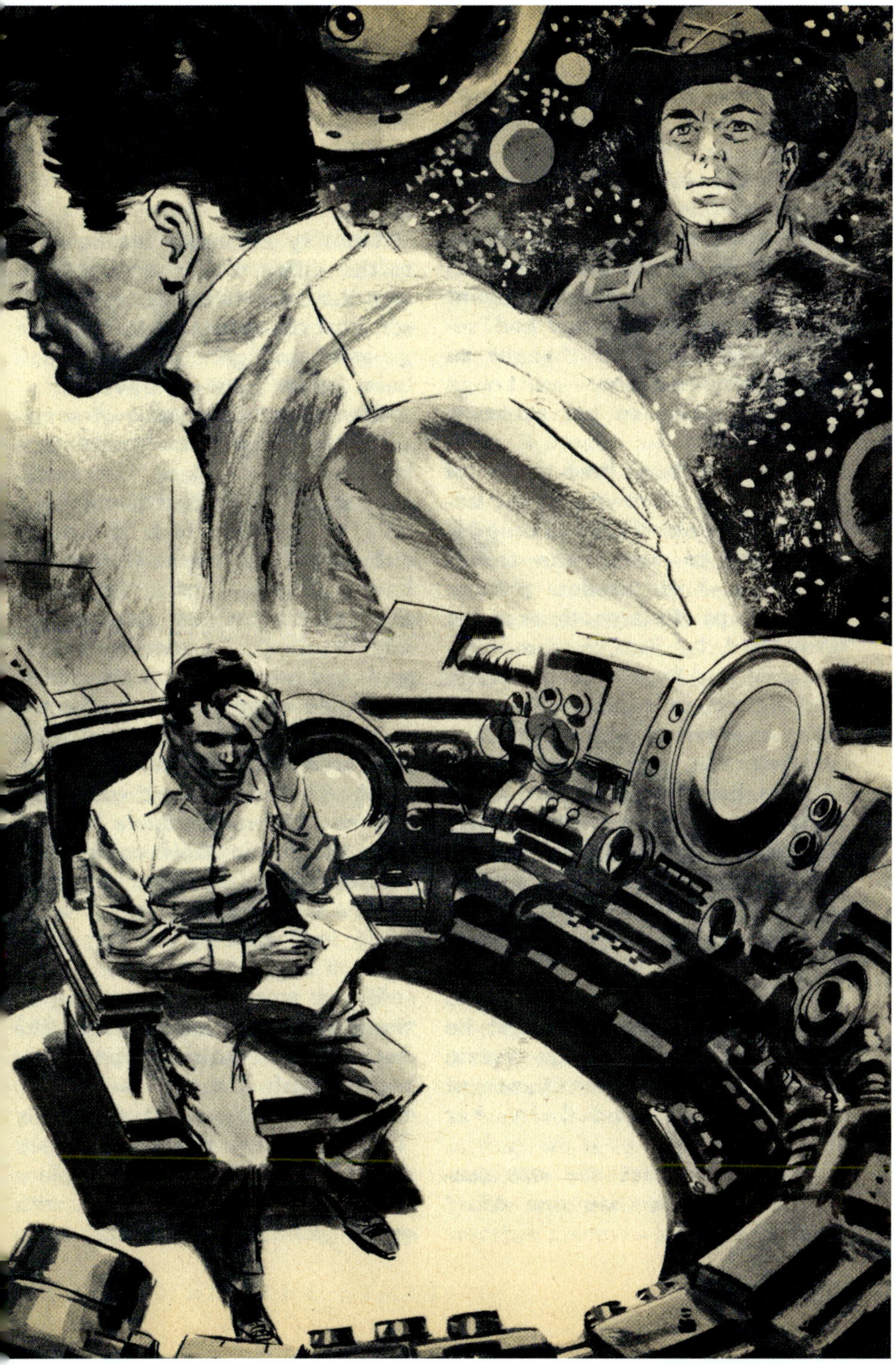

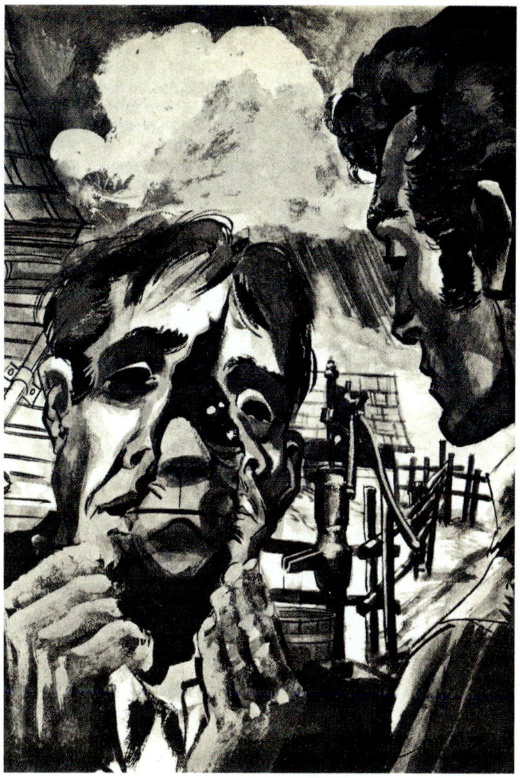

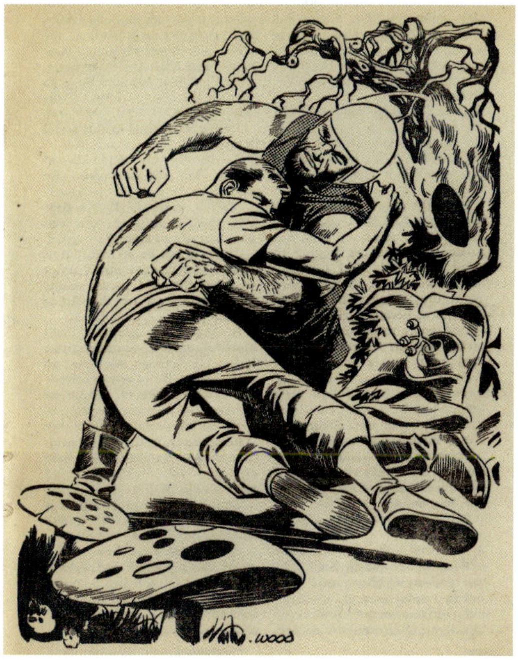

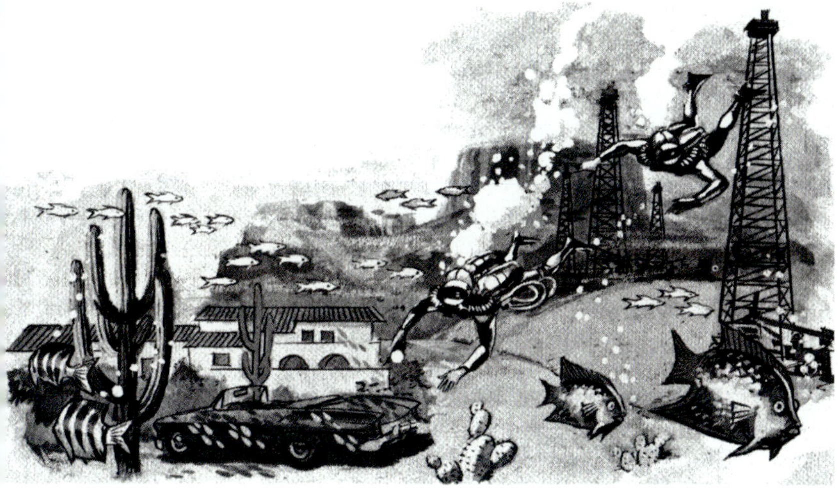

Clockwise from upper left:
Galaxy Science Fiction August 1963
Illustration for "Here Gather the Stars"
part 2 by Clifford D. Simak

Galaxy Science Fiction June 1963
Illustration for "Here Gather the Stars"
part 1 by Clifford D. Simak

Galaxy Science Fiction February 1966
Illustration for "The Age of the Pussyfoot"
part 3 by Frederik Pohl

Galaxy Science Fiction August 1963
Illustration for "The Great Nebraska Sea"
by Allan Danzig

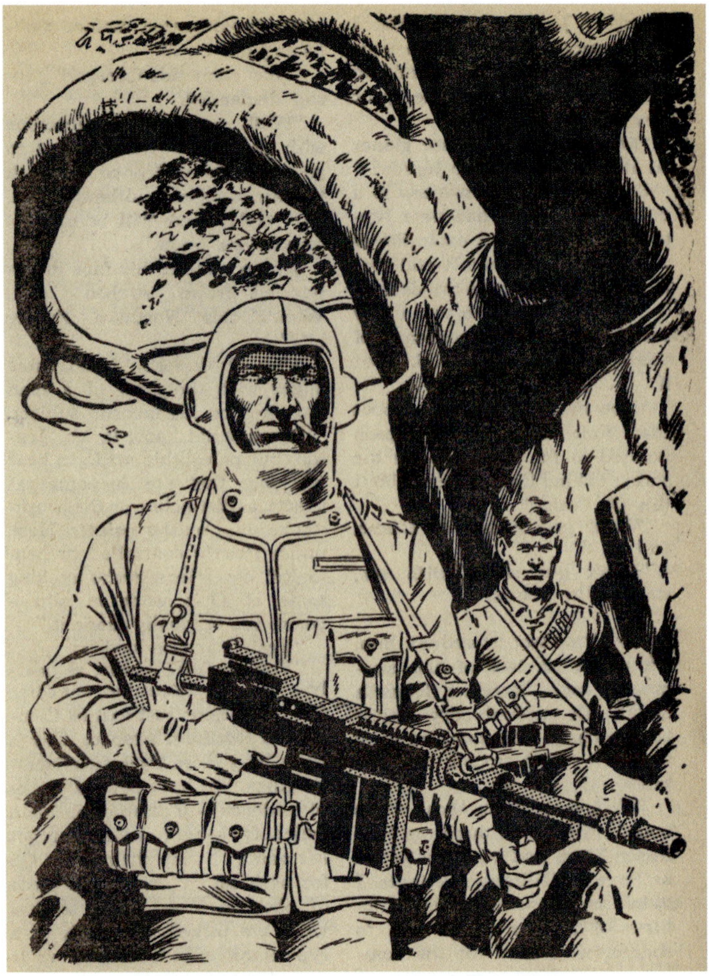

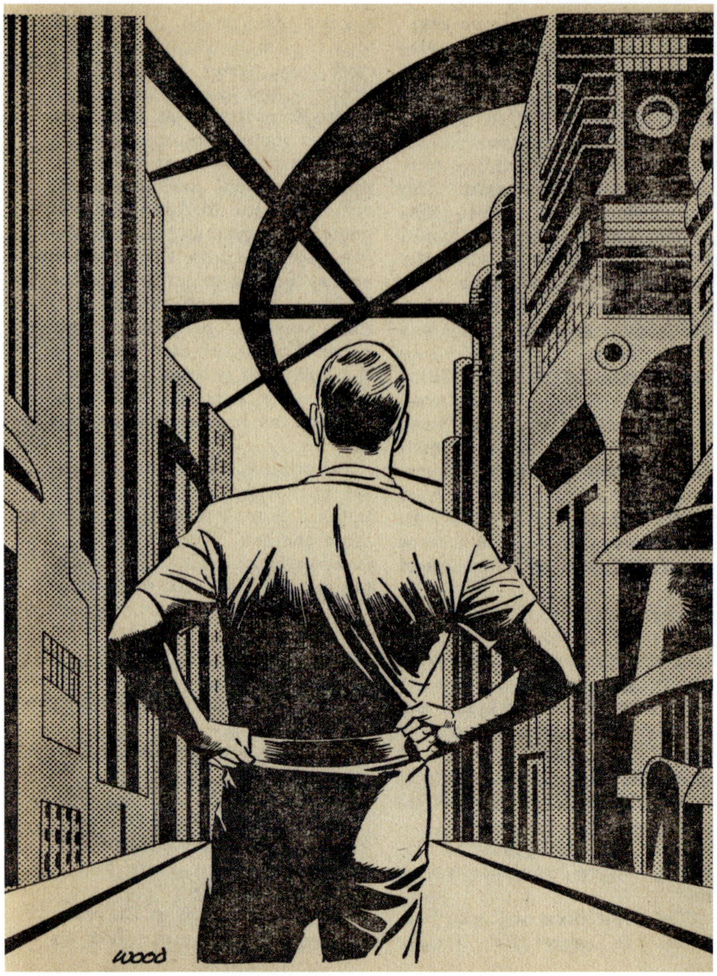
wood

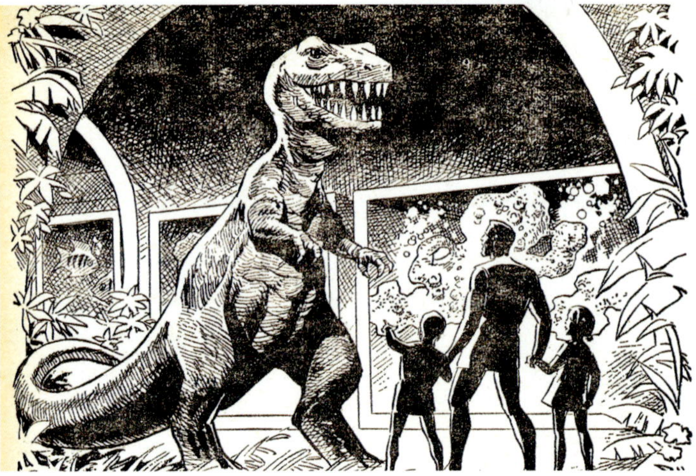

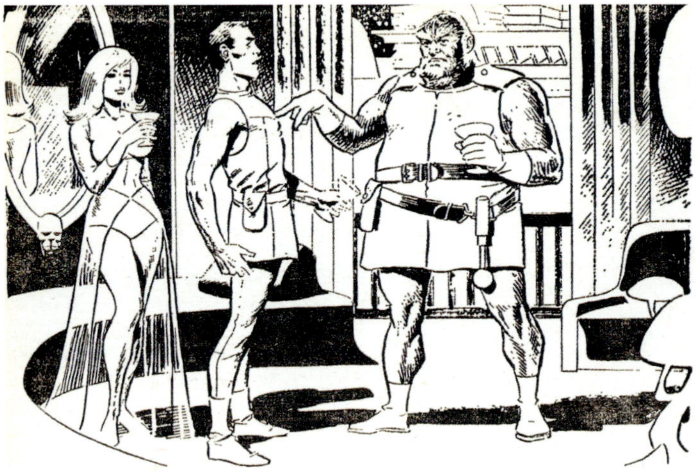

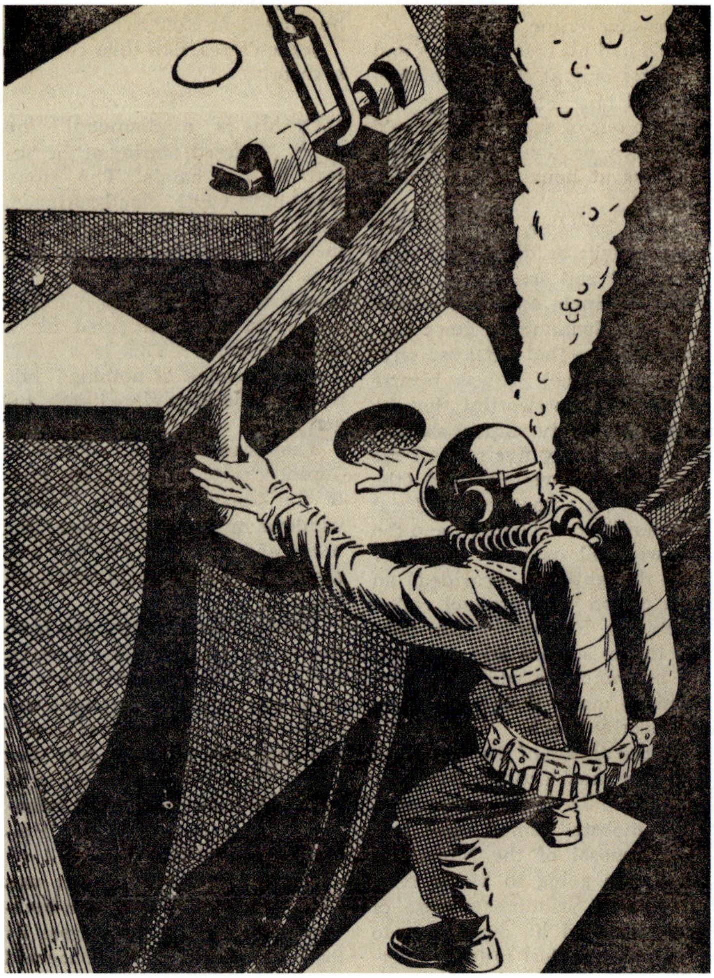

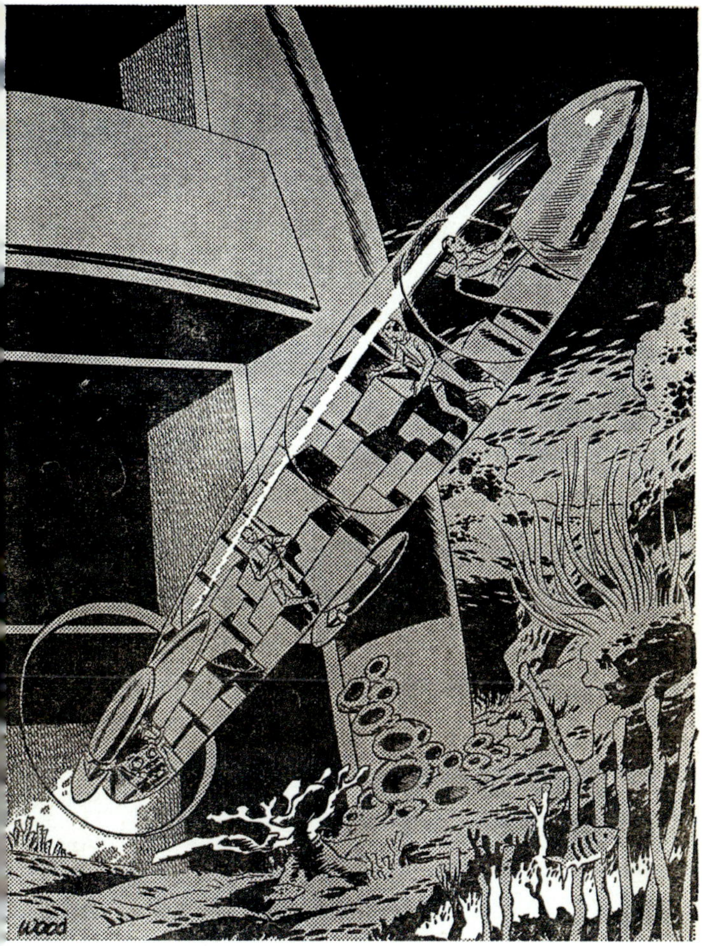

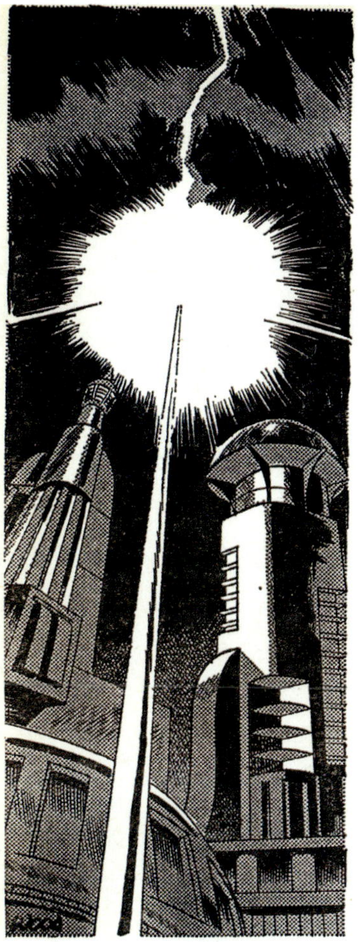

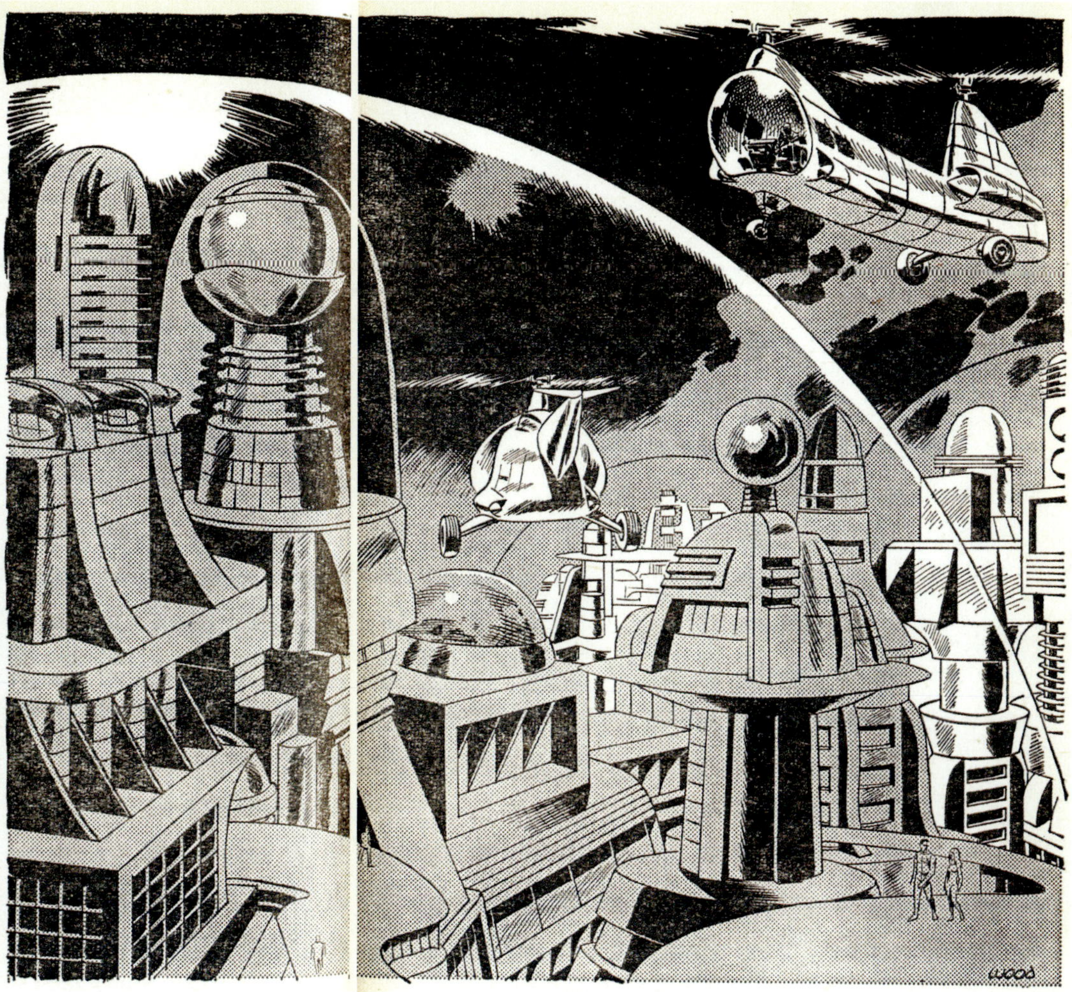

This page, all:
Galaxy Science Fiction
August 1967
Illustration for “Travelers Guide to Megahouston” by H. H. Hollis

Opposite:
Upper left and lower right:
Galaxy Science Fiction
April 1966
Illustration for “The Primitives” by Frank Herbert

Upper right:
Galaxy Science Fiction
February 1966
Illustration for “The Age of the Pussyfoot” part 3 by Frederik Pohl

Both lower left illustrations:
Galaxy Science Fiction
October 1965
Illustration for “The Age of the Pussyfoot” part 1 by Frederik Pohl

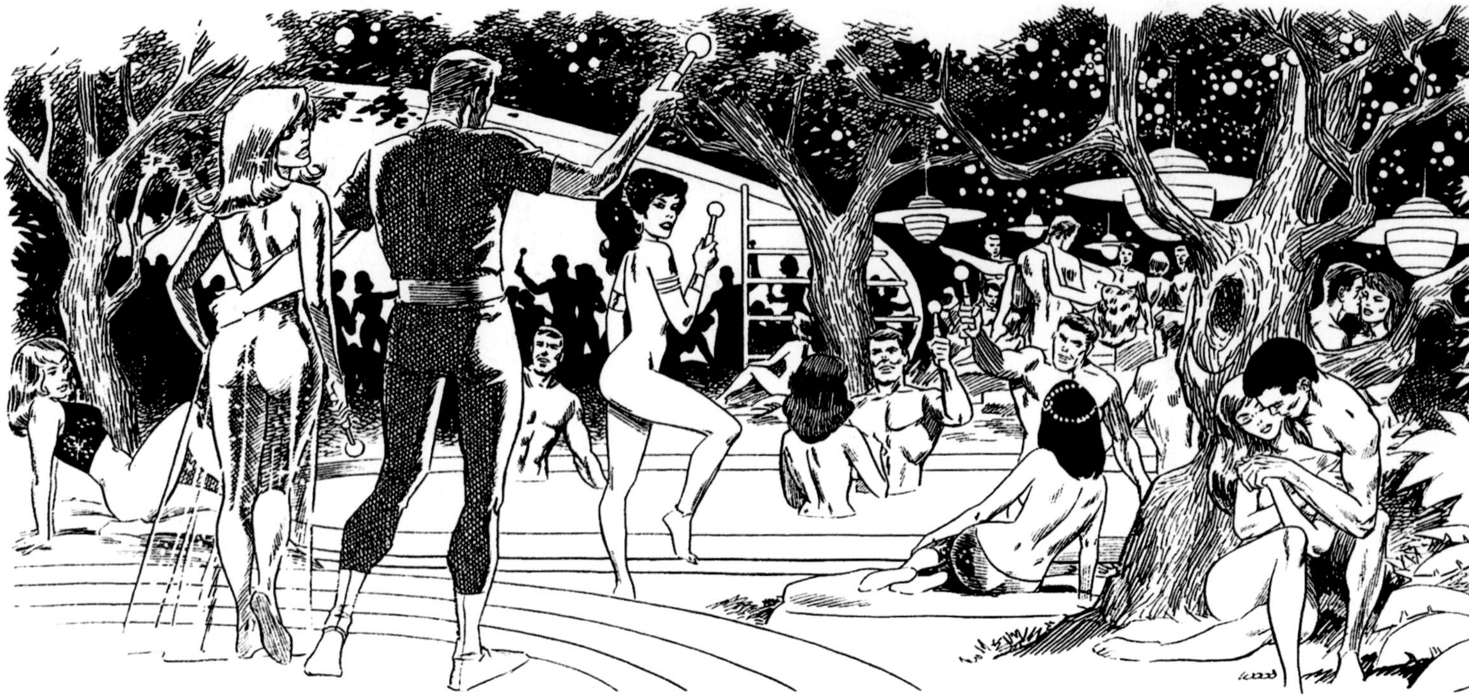

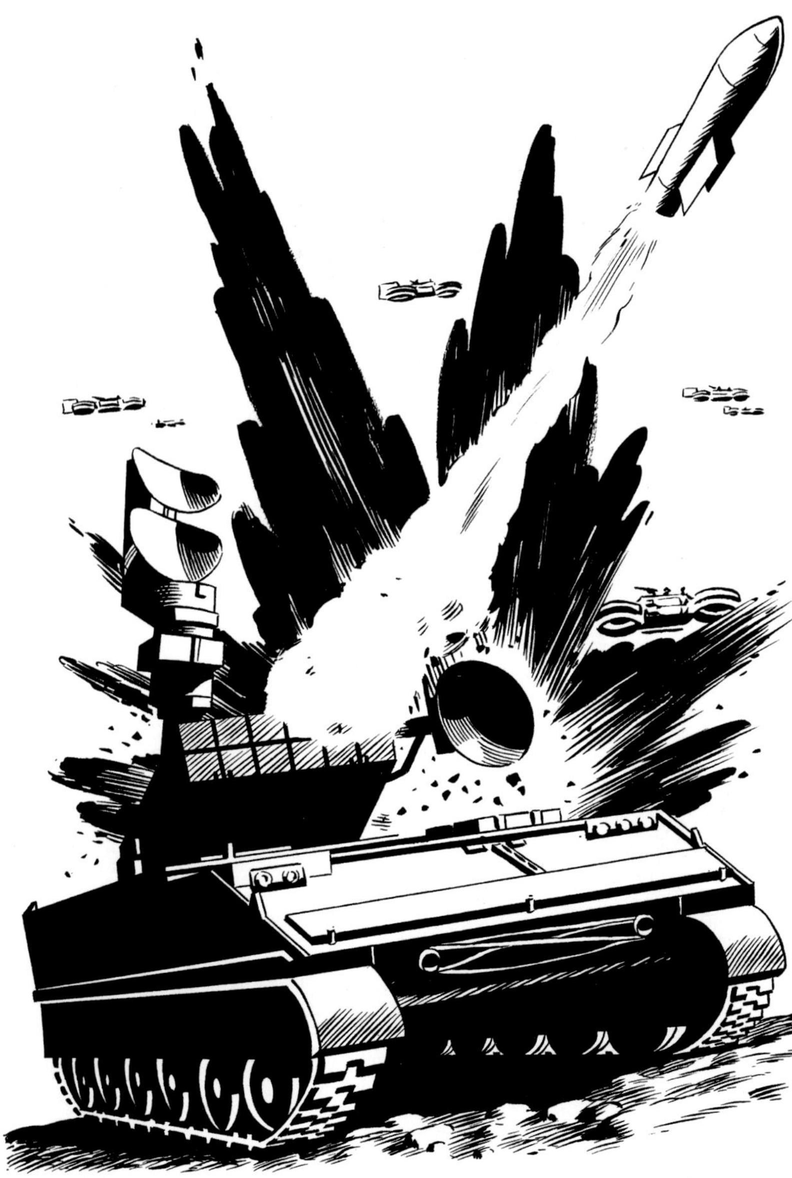

128

Top: *Galaxy Science Fiction* October 1965
Illustration for "The Age of the Pussyfoot" part 1
by Frederik Pohl

Bottom: *Galaxy Science Fiction* April 1966
Illustration for "The Primitives" by Frank Herbert

Opposite: *Galaxy Science Fiction* December 1965
Illustration for "The Age of the Pussyfoot" part 2
by Frederik Pohl

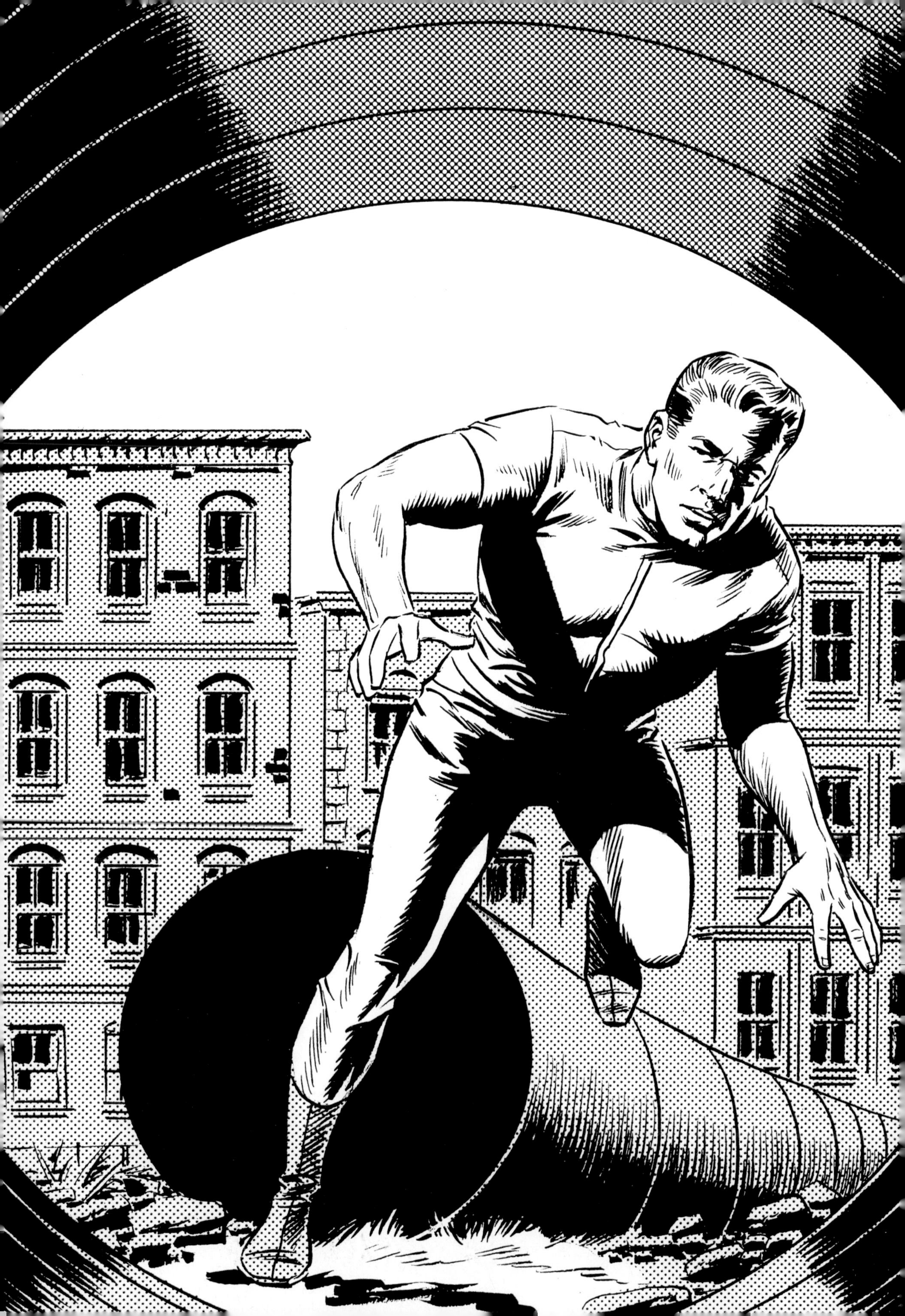

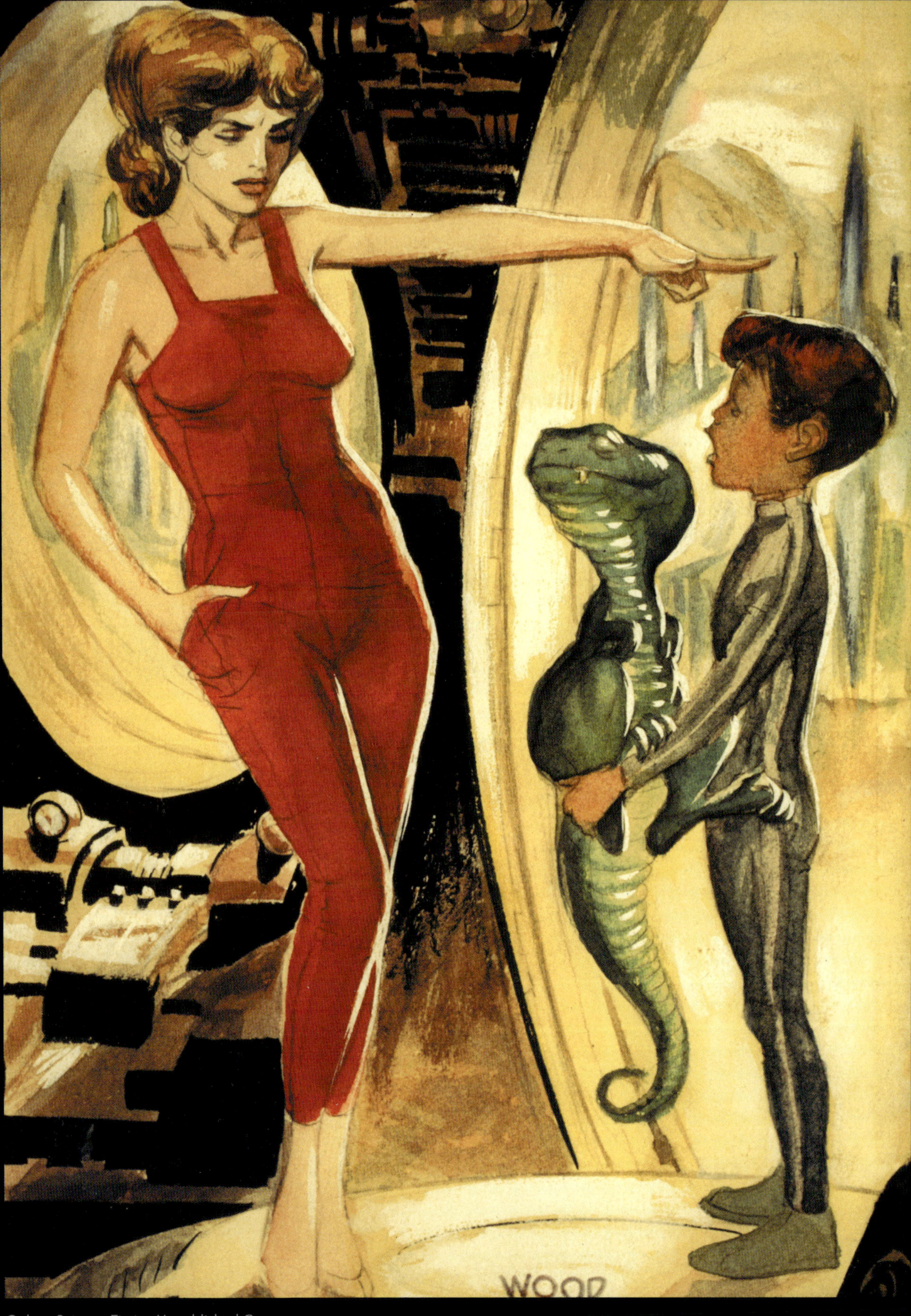

Galaxy Science Fiction Unpublished Cover

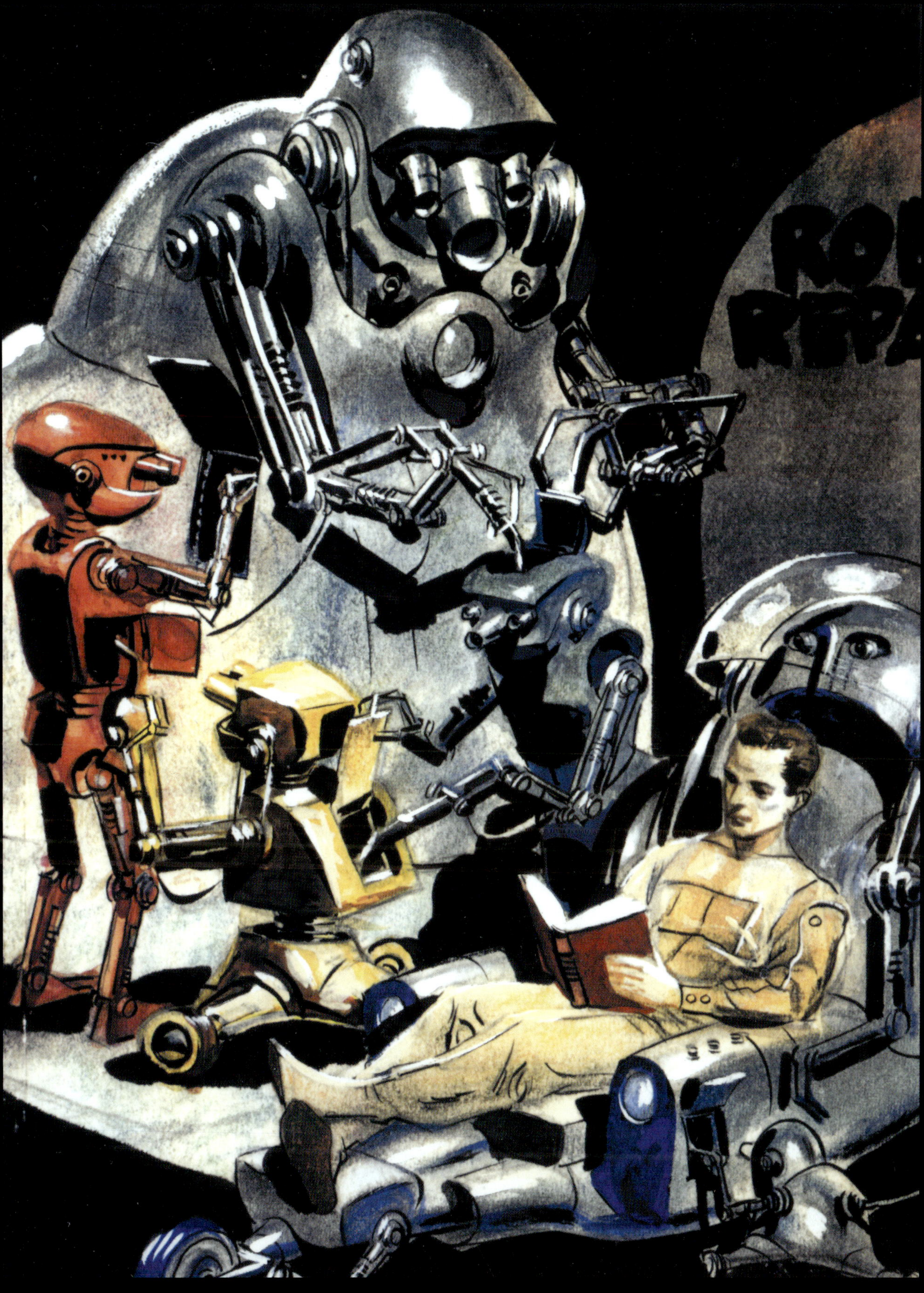

Galaxy Science Fiction Unpublished Cover

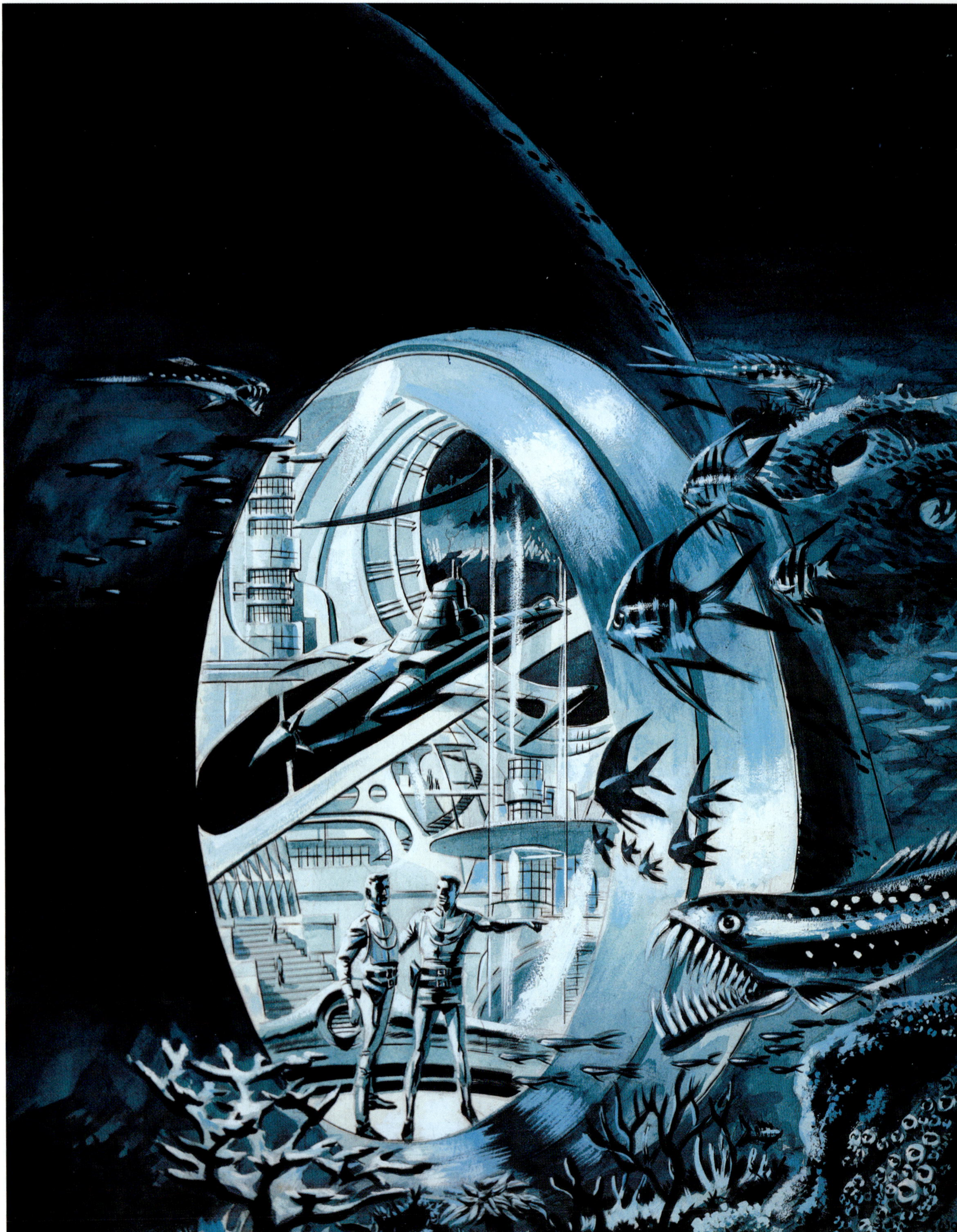

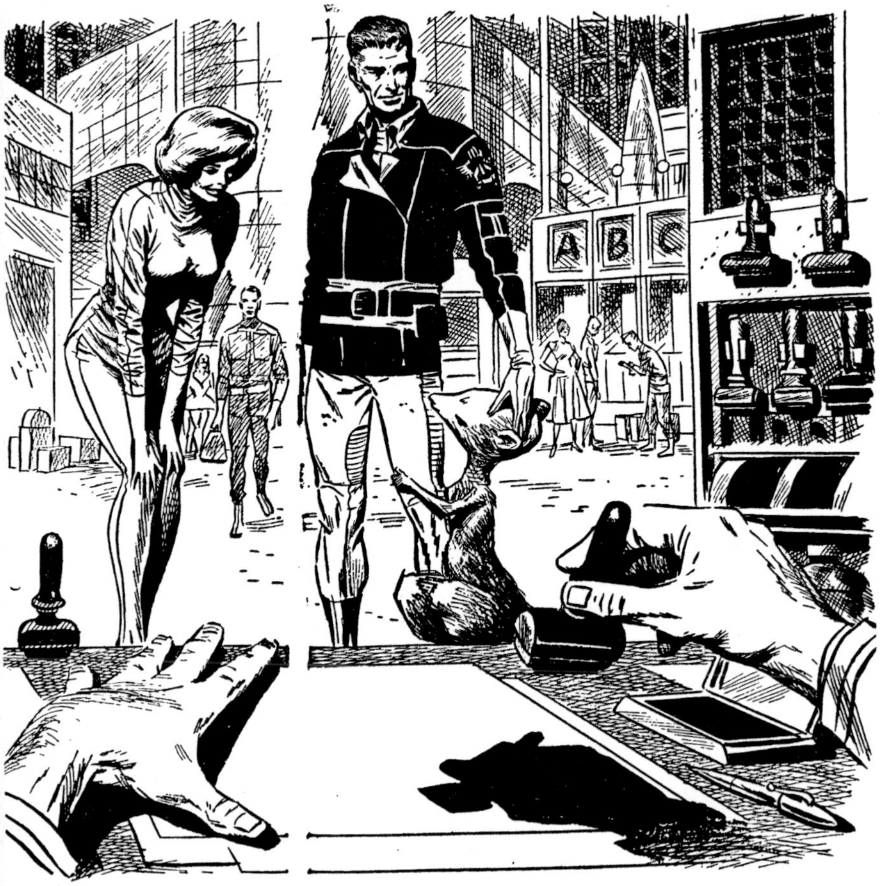

Clockwise from upper left:
If Science Fiction September 1959
Illustration for "Homecoming" by Gordon R. Dickson

If Science Fiction December 1968
Illustration for "The Starman of Pritchard's Creek" by Julian F. Grow

If Science Fiction January 1960
Illustration for "The Good Seed" by Mark Mallory

Opposite:
Undersea City by Frederik Pohl and Jack Williamson
Gnome Press, 1958 (Cover)

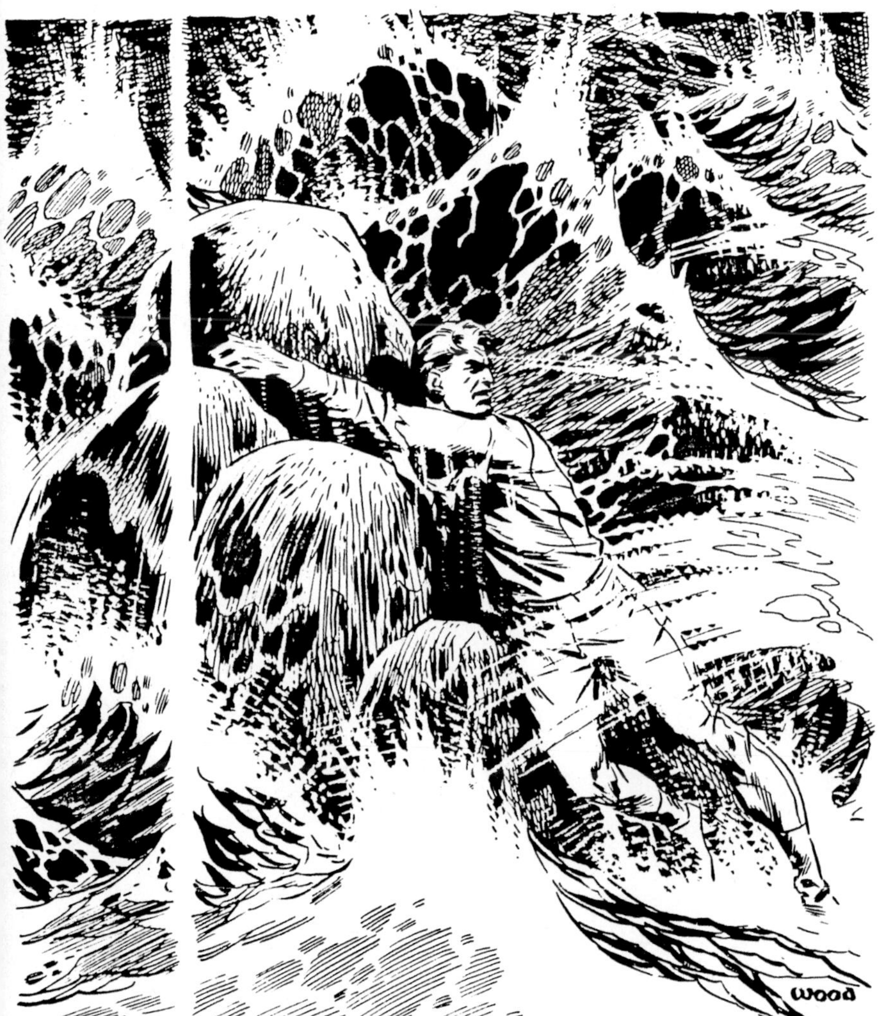

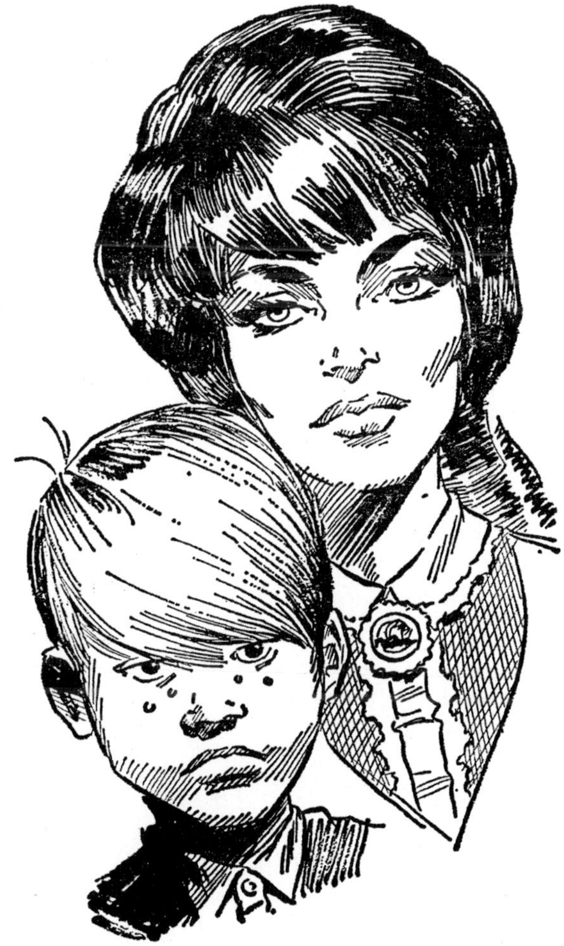

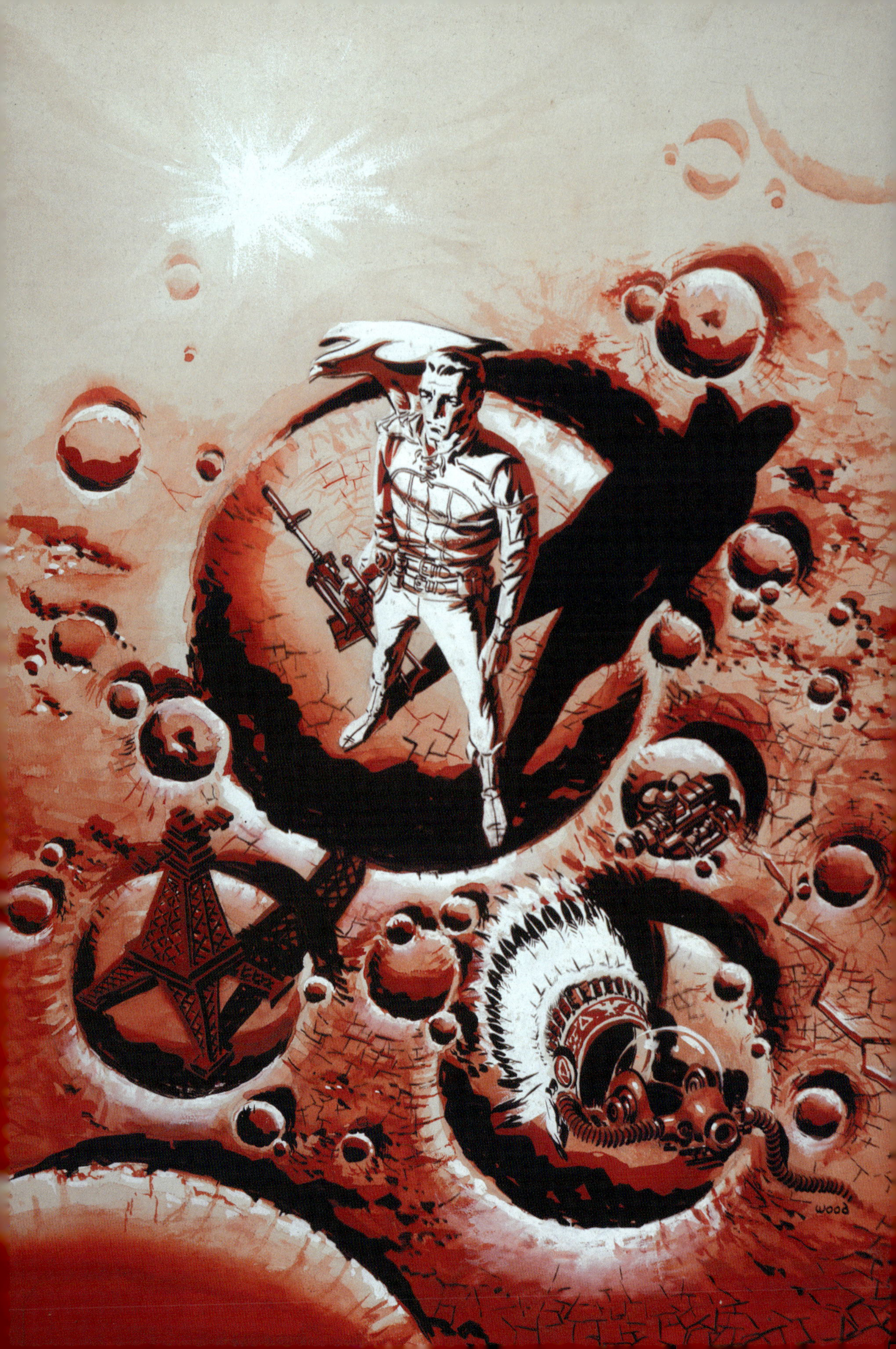
wood

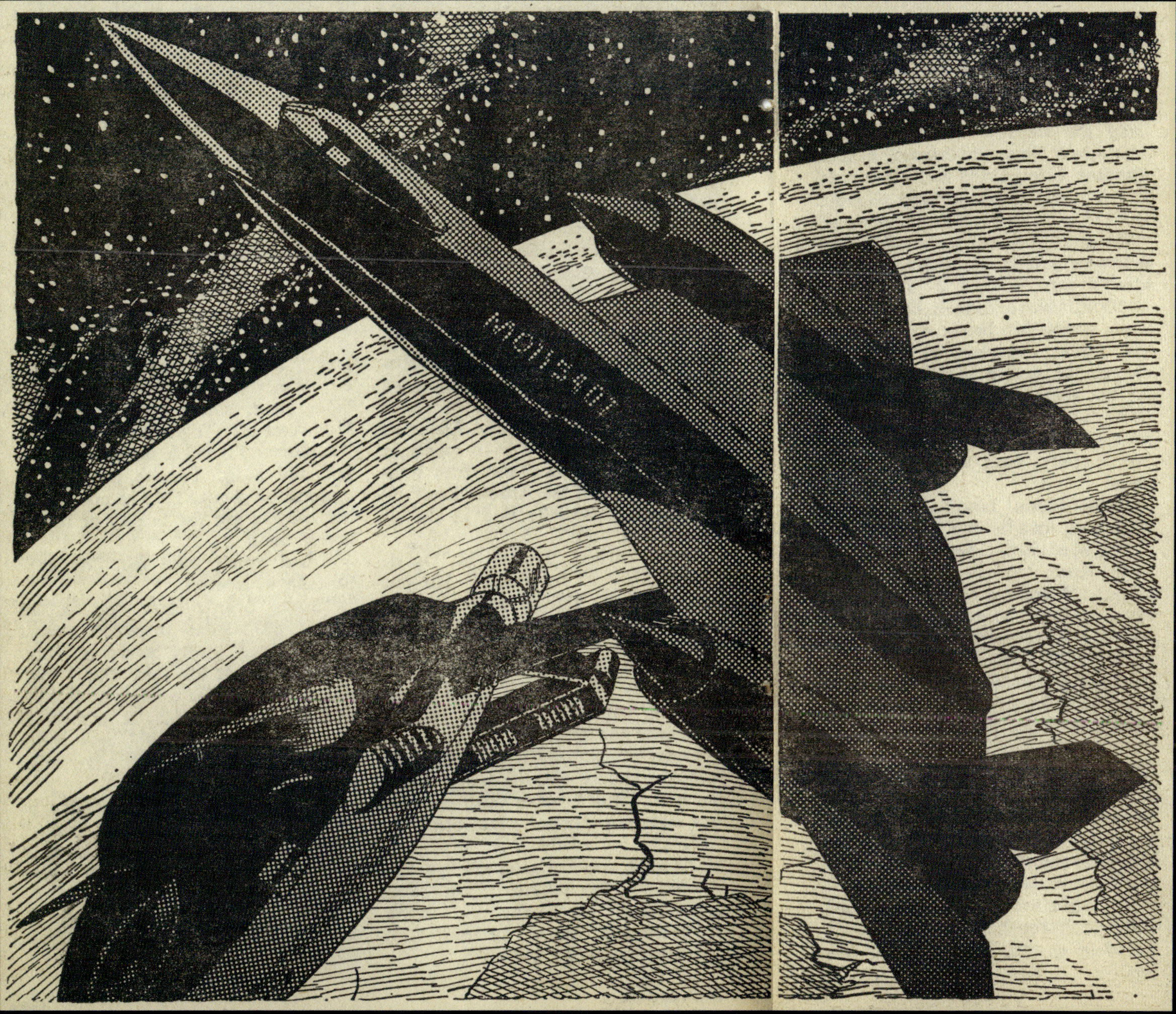

If Science Fiction July 1966
Illustration for "Earthblood" part 4 by Keith Laumer and Rosel George Brown

Opposite: *Colonial Survey* by Murray Leinster
Gnome Press, 1957 (Cover)

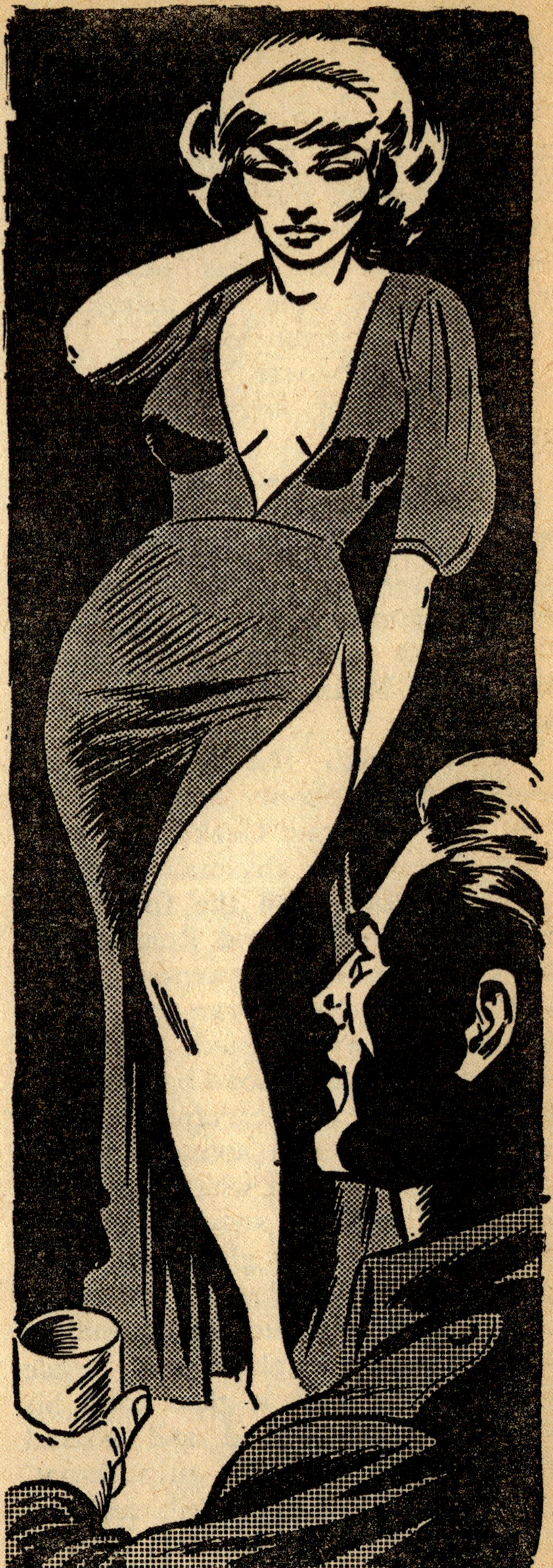

If Science Fiction June 1966
Illustration for "Earthblood" part 3
Keith Laumer and Rosel George Brown

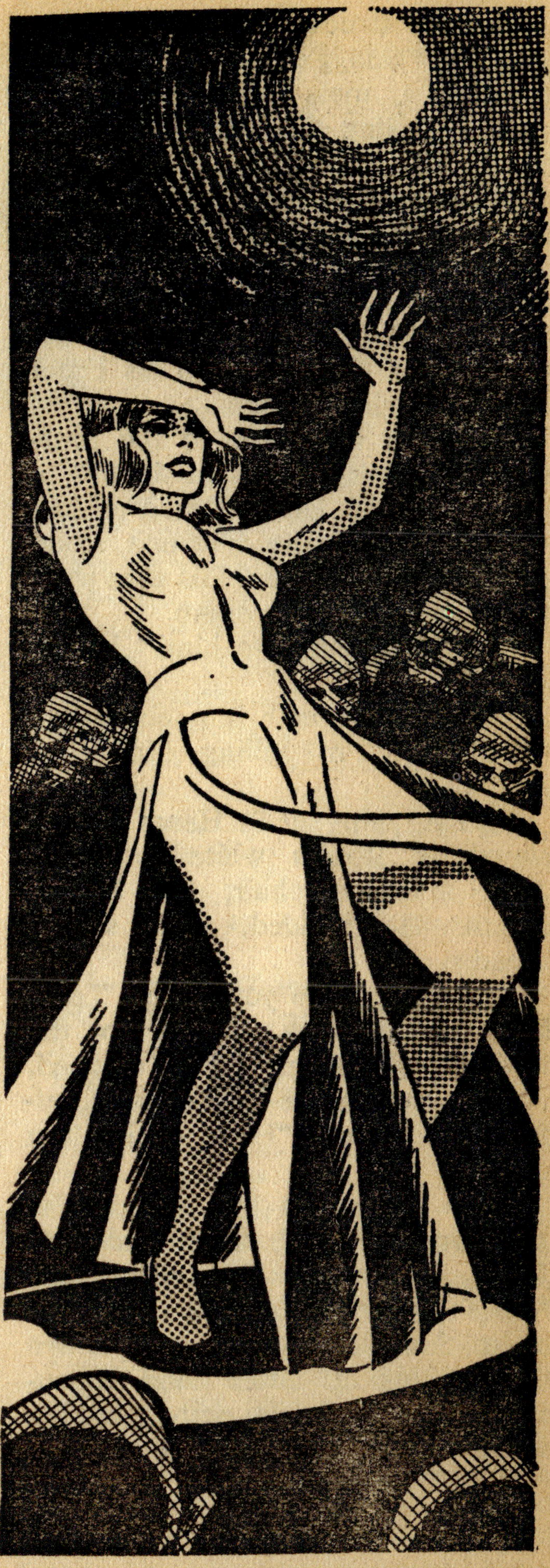

If Science Fiction
Illustration for "Ea
by Keith Laumer a

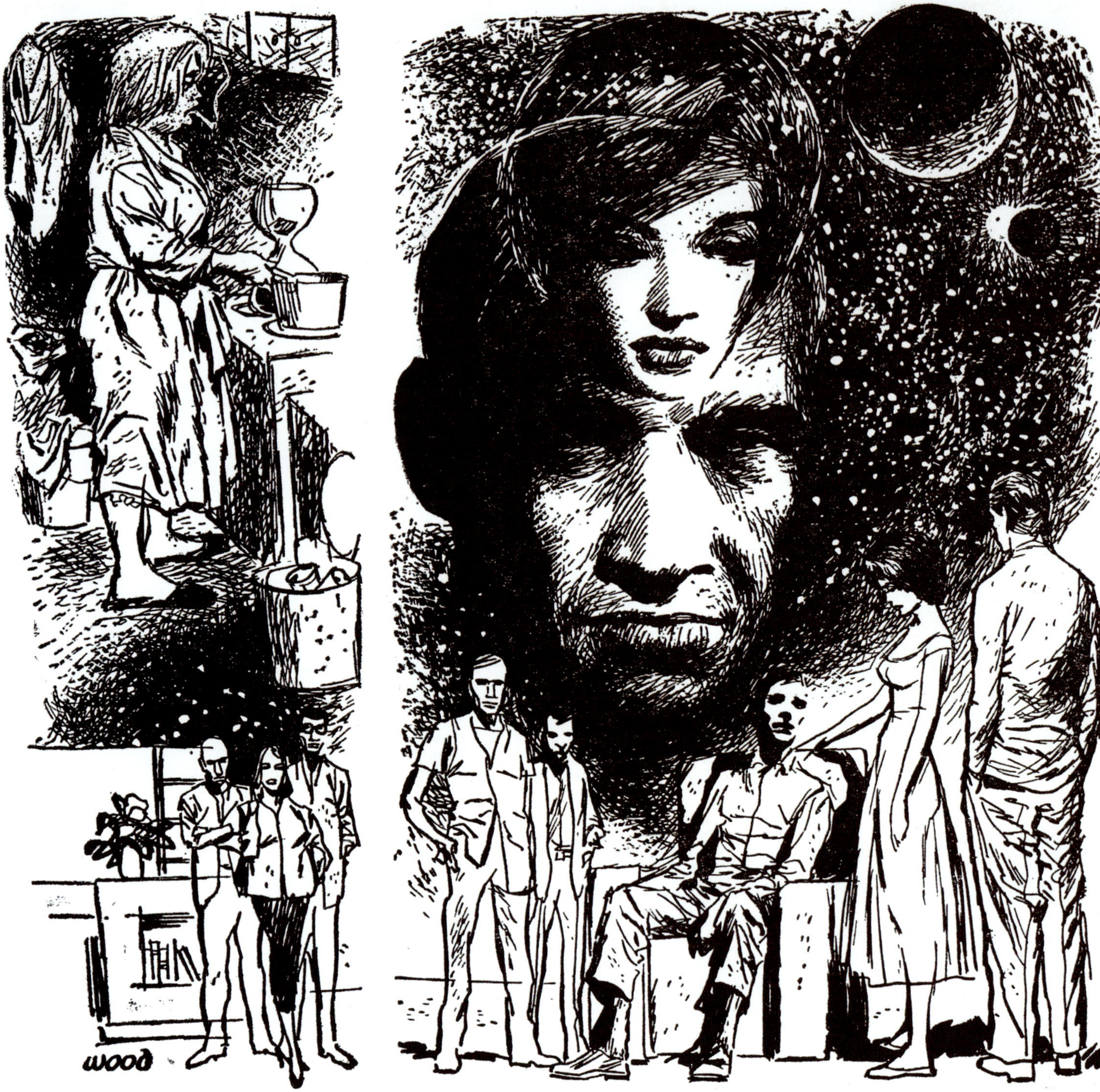

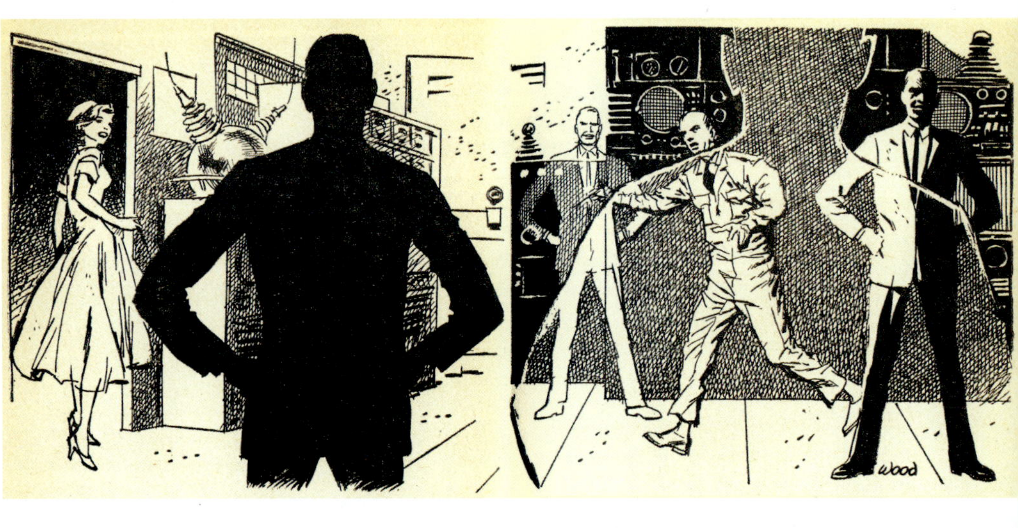

Top: *If Science Fiction* January 1960
Illustration for "The Divers"
by James Stamers

Bottom: *If Science Fiction* January 1960
Illustration for "The Last Leap"
by Daniel F. Galouye

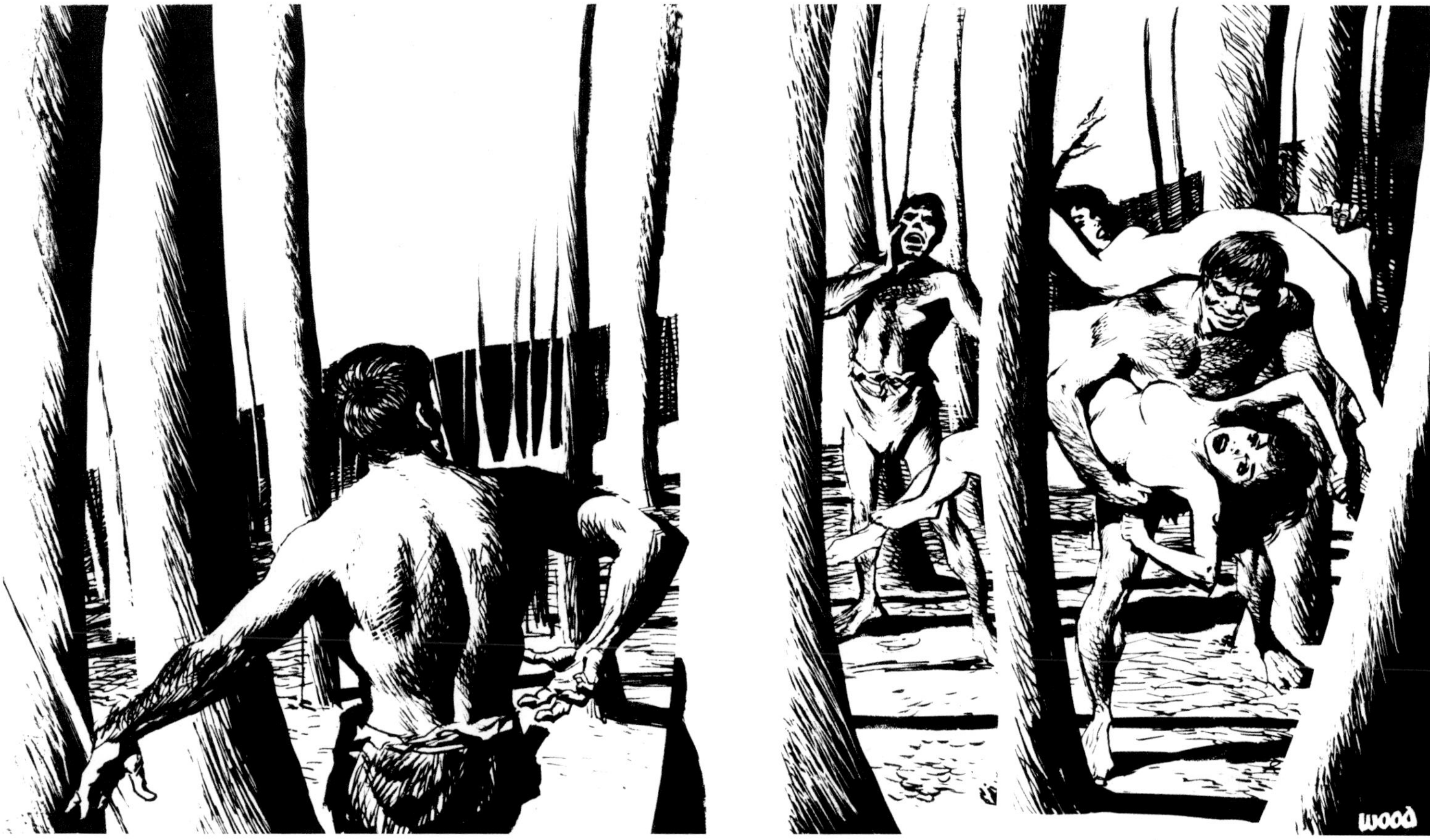

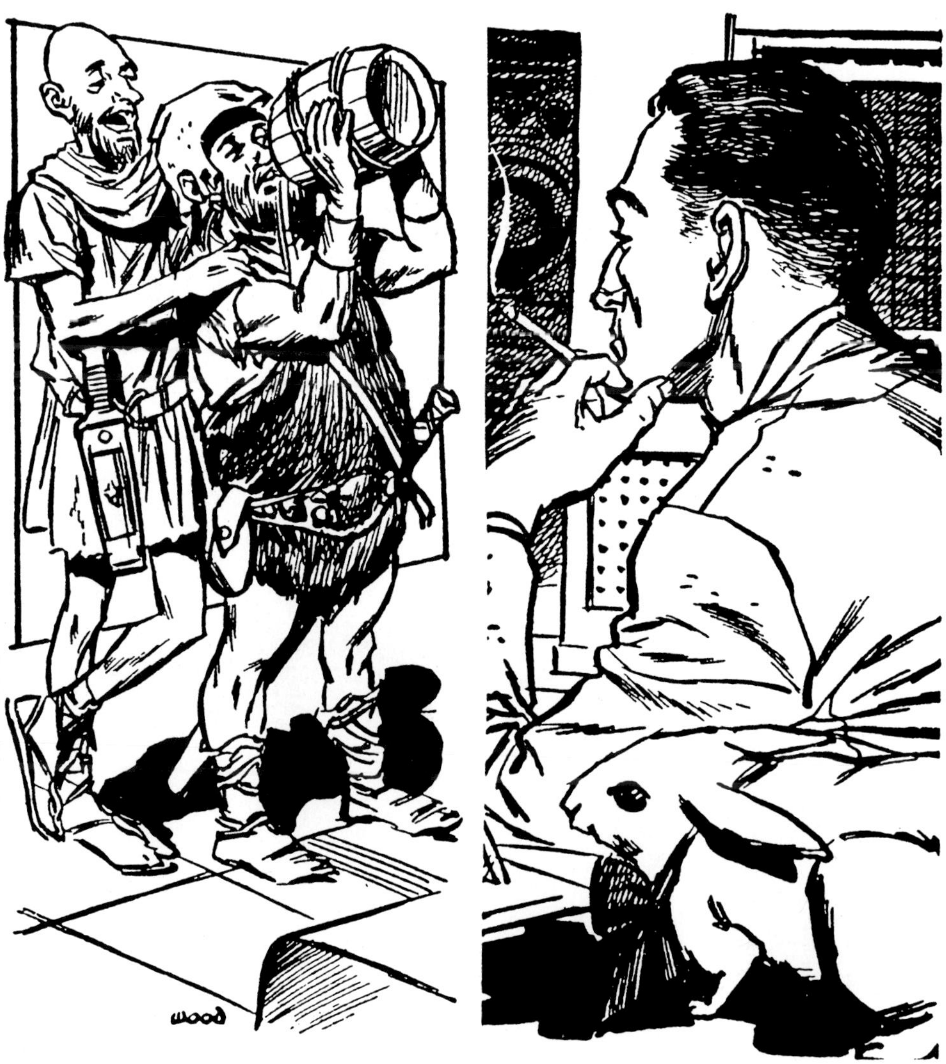

Top: *If Science Fiction* May 1960
Illustration for "A Pride of Islands"
by C. C. MacApp

Bottom: *If Science Fiction* March 1960
Illustration for "Cleaners"
by Clifford D. Simak

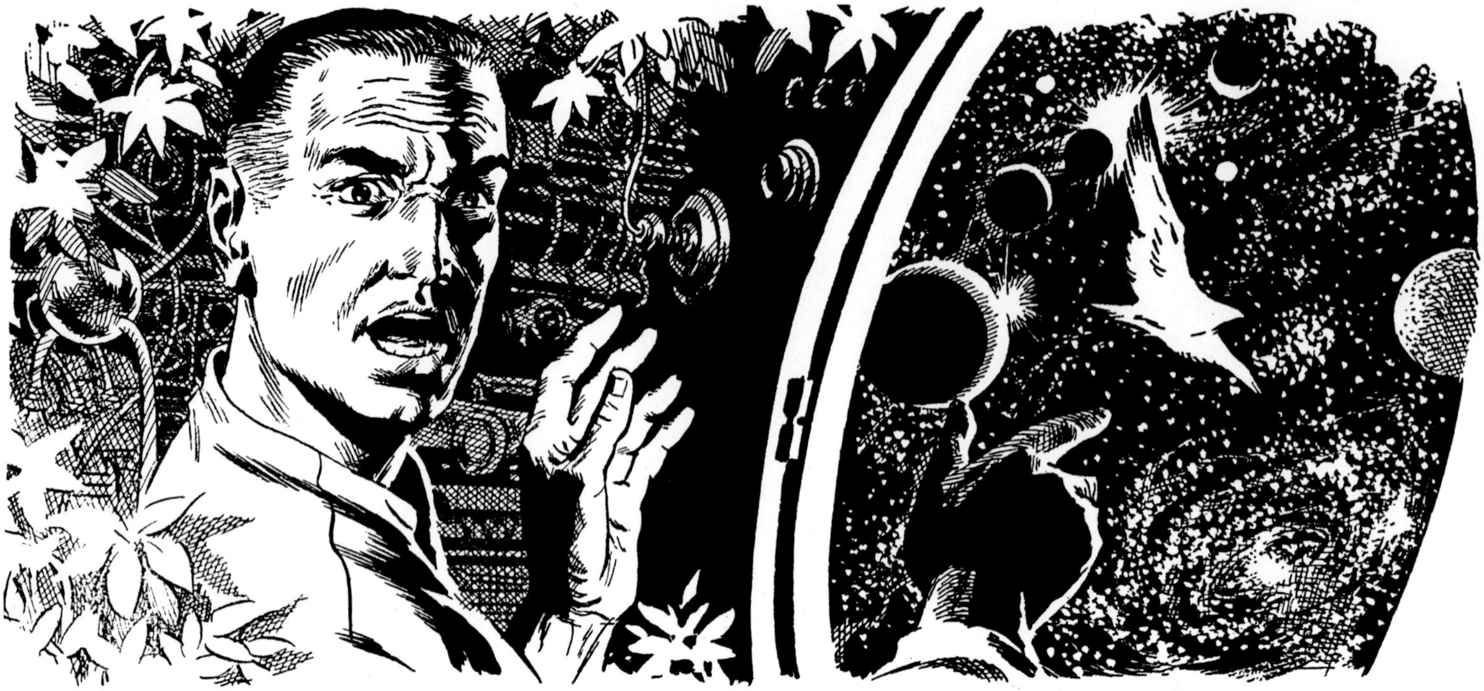

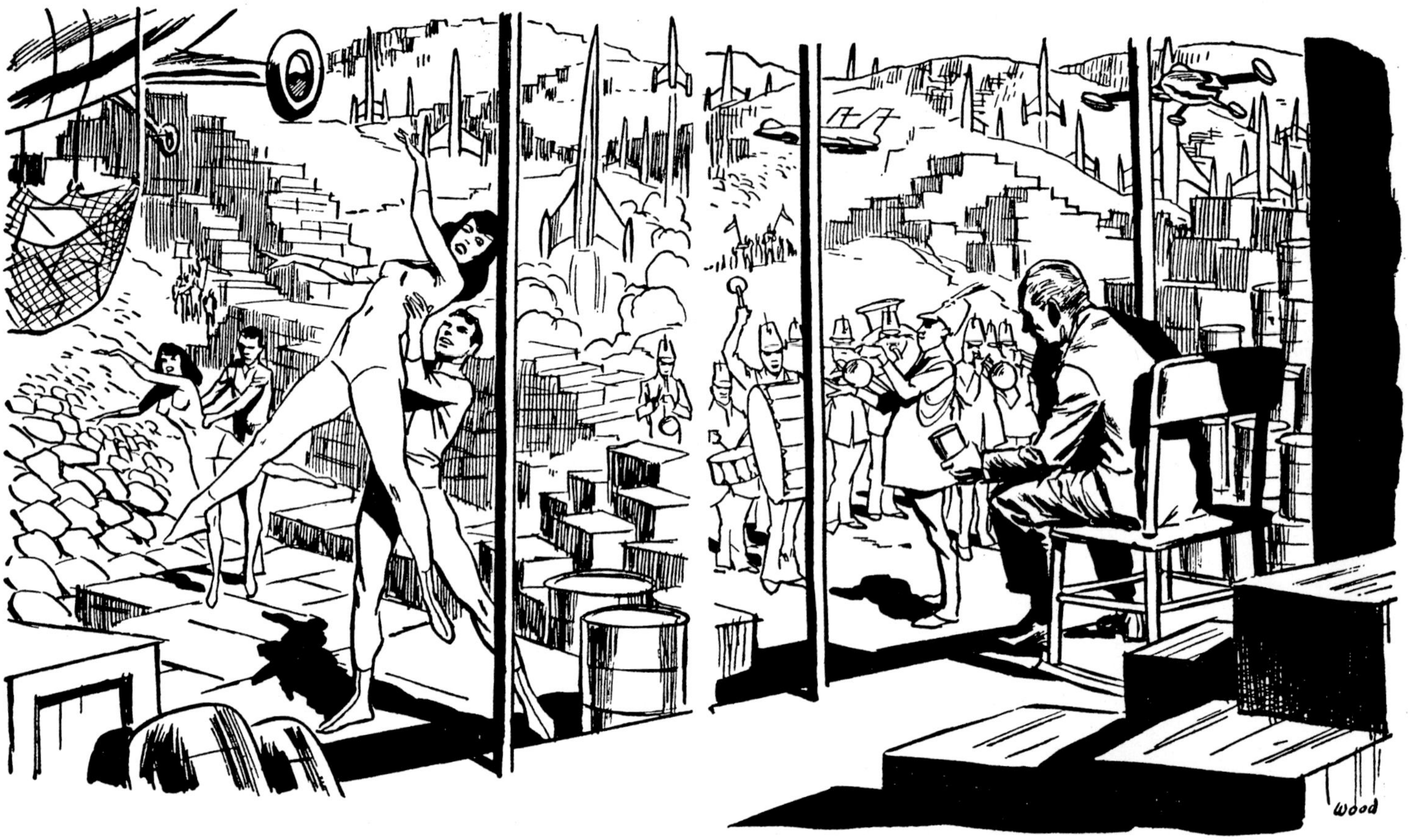

Top: *If Science Fiction* July 1959
Illustration for "Growing Season" by F. L. Wallace

Bottom: *If Science Fiction* March 1960
Illustration for "Gravy Train" by Daniel F. Galouye

Top: *If Science Fiction* September 1959
Illustration for "Summer Guests" by James H. Schmitz

Bottom: *If Science Fiction* May 1960
Illustration for "A Tourist Named Death" by Christopher Anvil

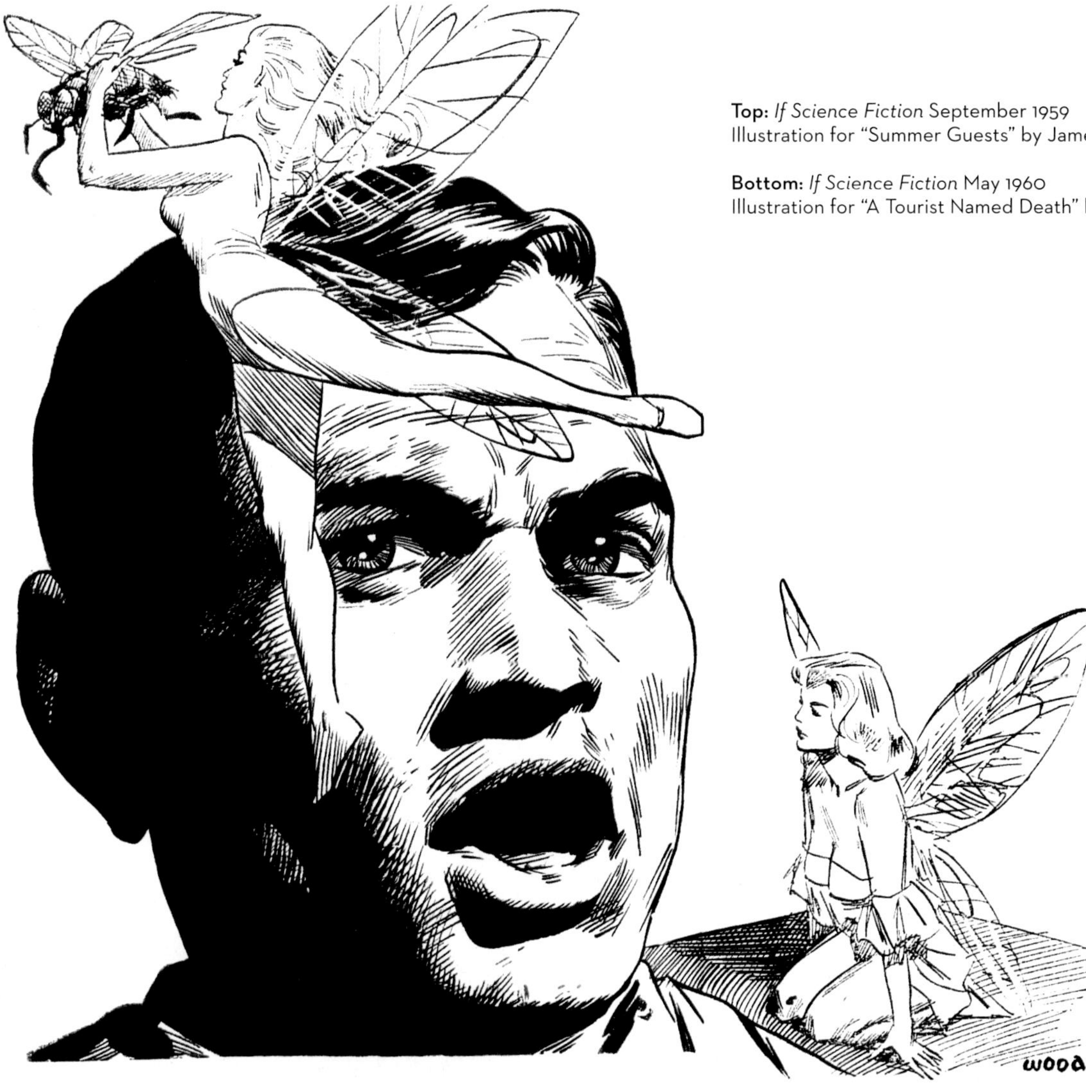

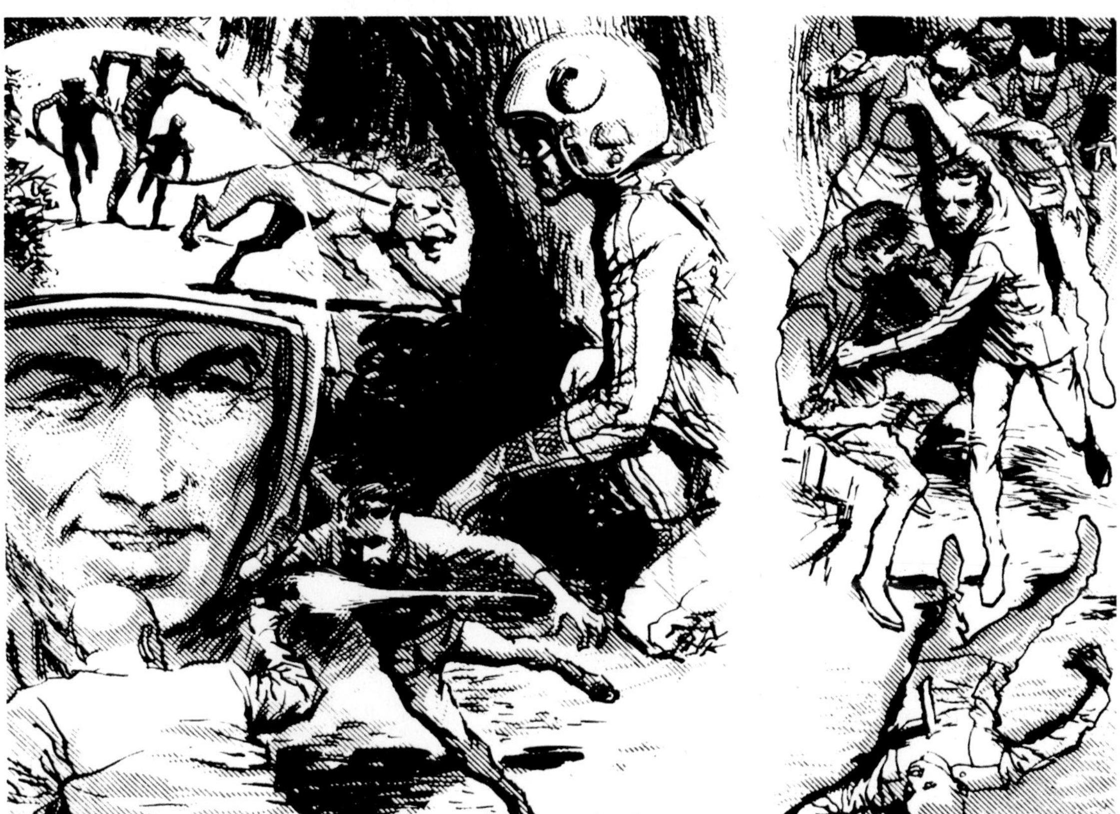

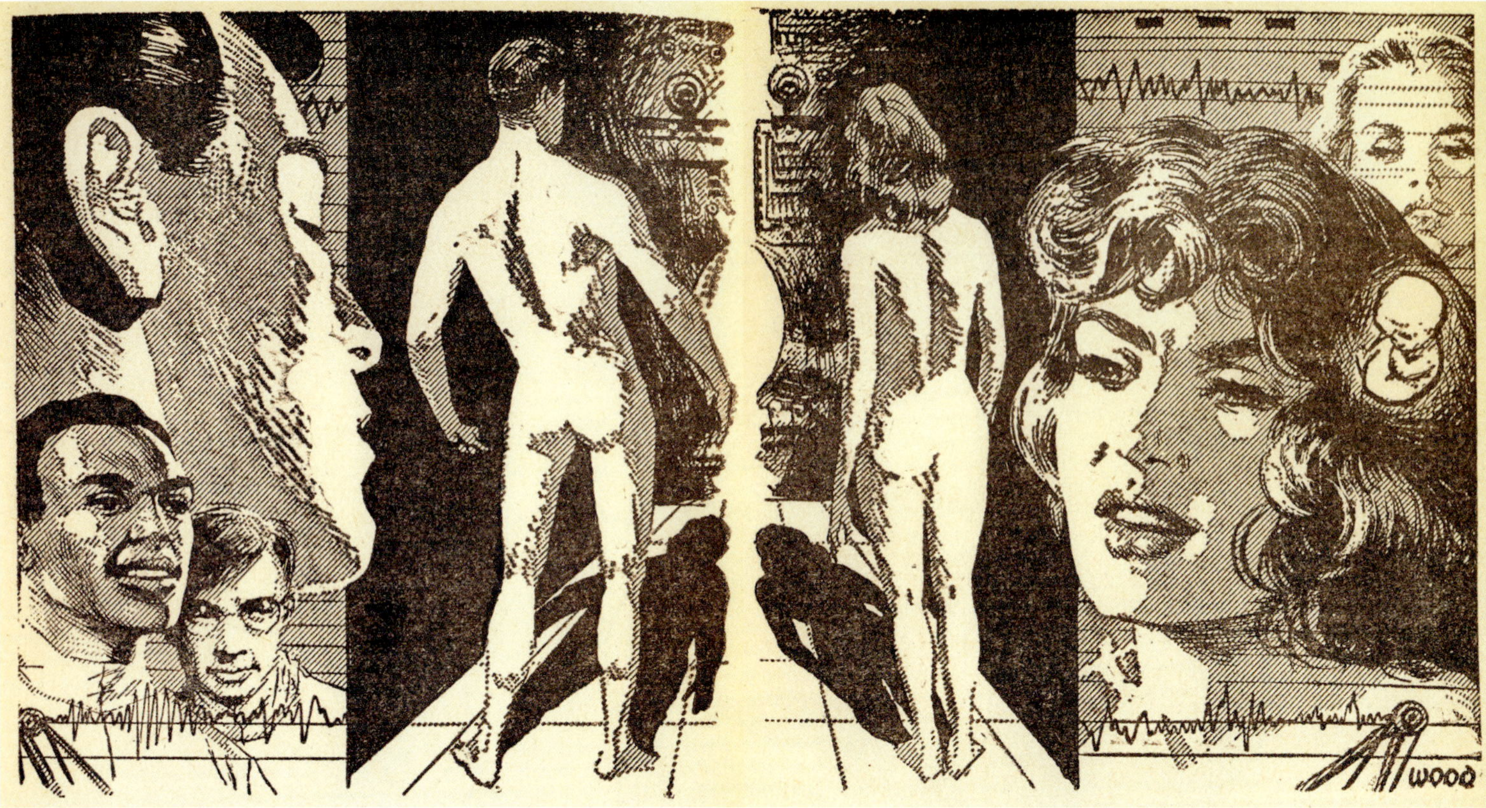

142

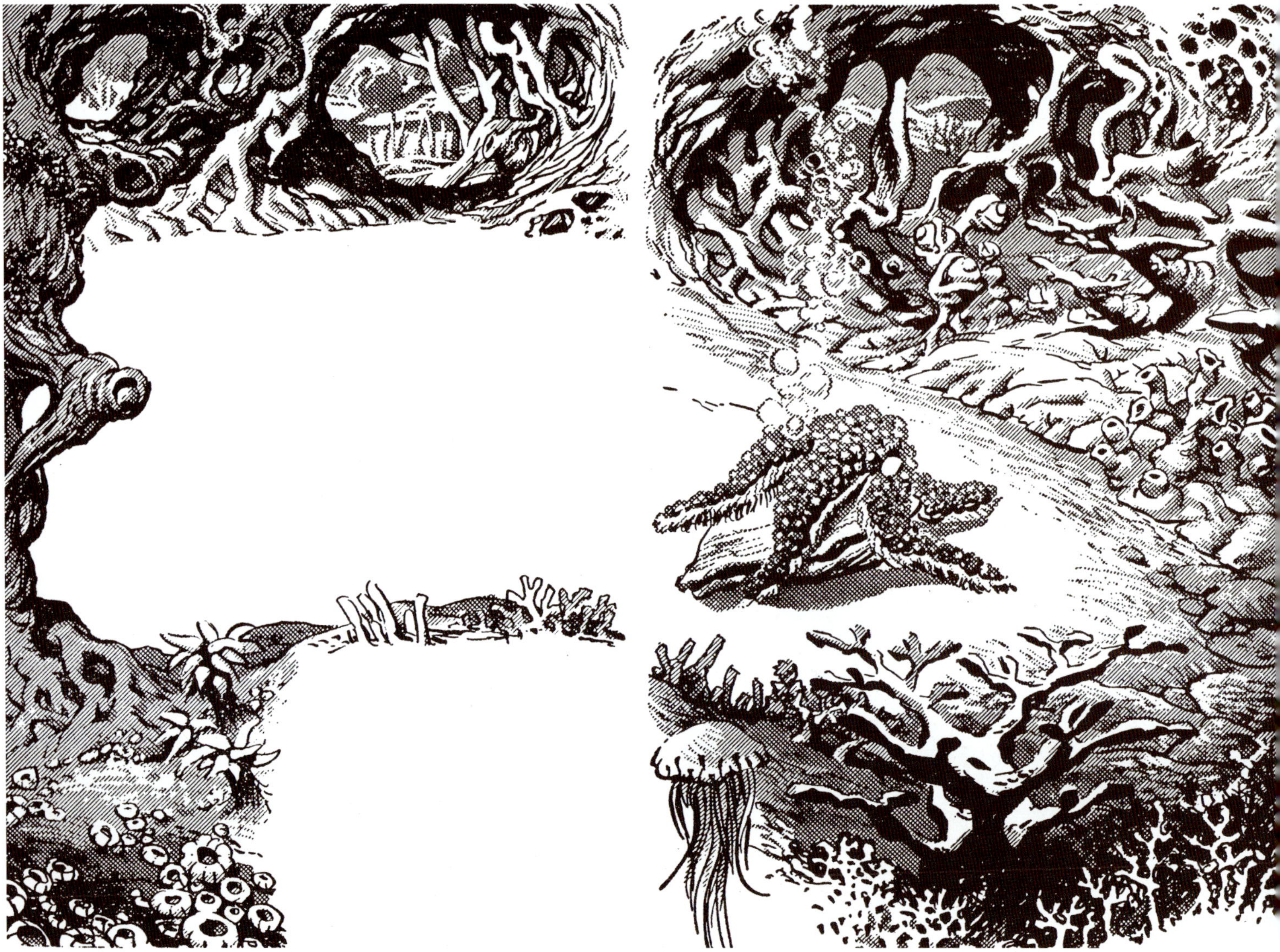

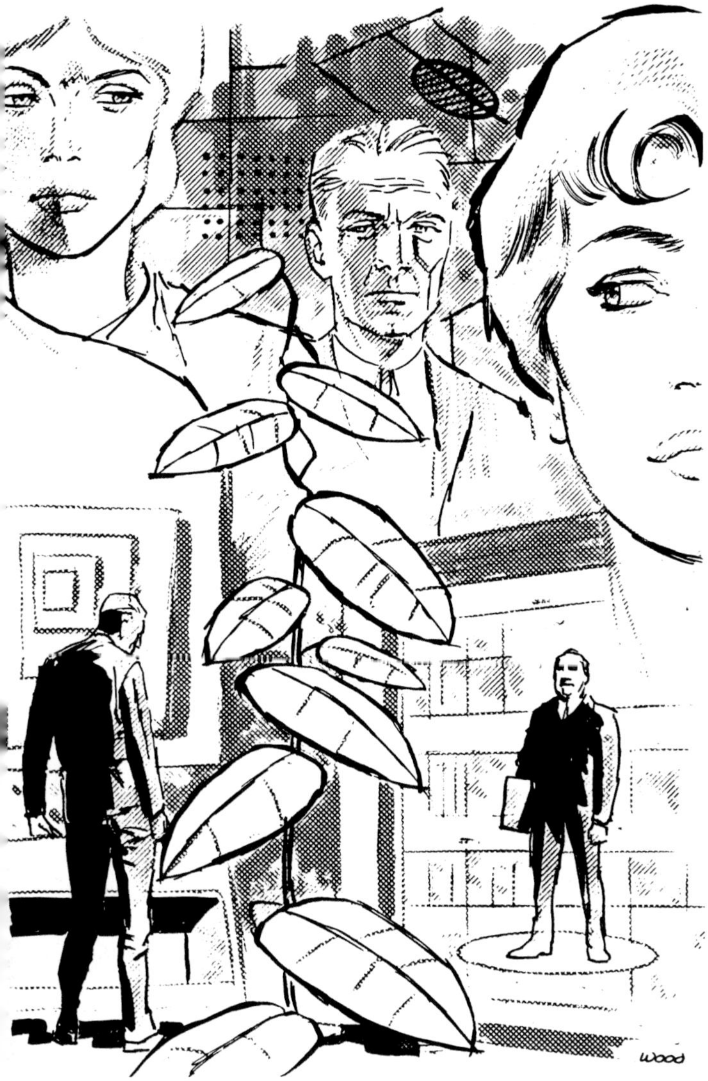

Top: *If Science Fiction* July 1961
Illustration for "The Bem Called Windigo" by Theodore Sturgeon

Bottom: *If Science Fiction* September 1960
Illustration for "Parallel Beans" by Margaret St. Clair

Opposite:
Top: *If Science Fiction* May 1960
Illustration for "Matchmaker" by Charles L. Fontenay

Bottom: *If Science Fiction* September 1960
Illustration for "To Choke an Ocean" by J. F. Bone

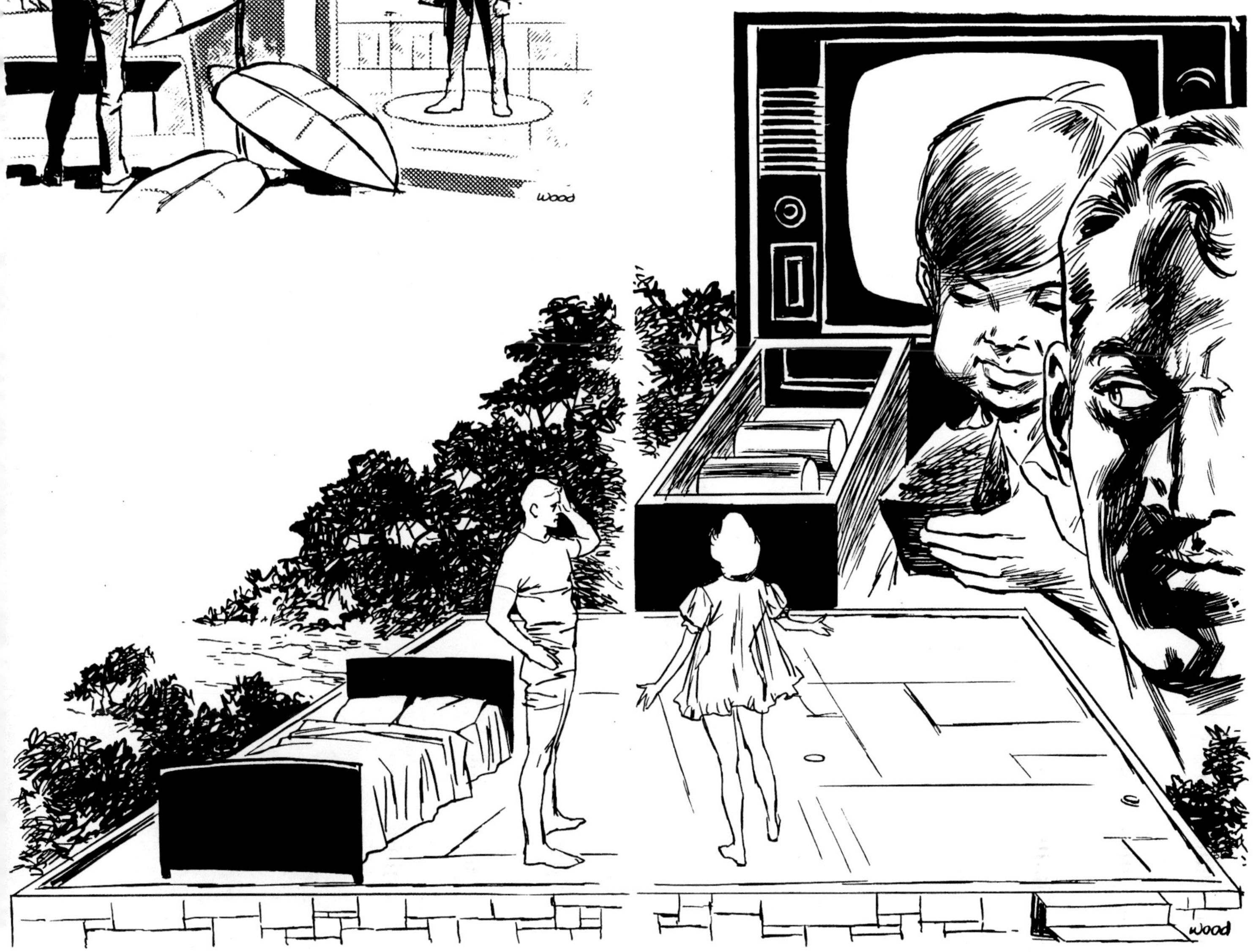

Clockwise from upper right:
If Science Fiction December 1968
Illustration for "The Starman of Pritchard's Creek"
by Julian F. Grow

If Science Fiction September 1960
Illustration for "Kangaroo Court" by Daniel F. Galouye

If Science Fiction
November 1960
Illustration for "Superjoemulloy" by Scott F. Grenville

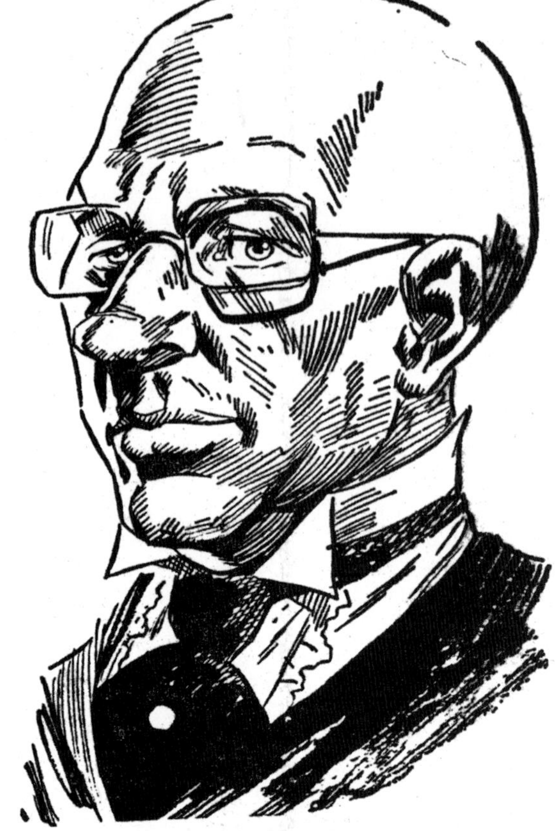

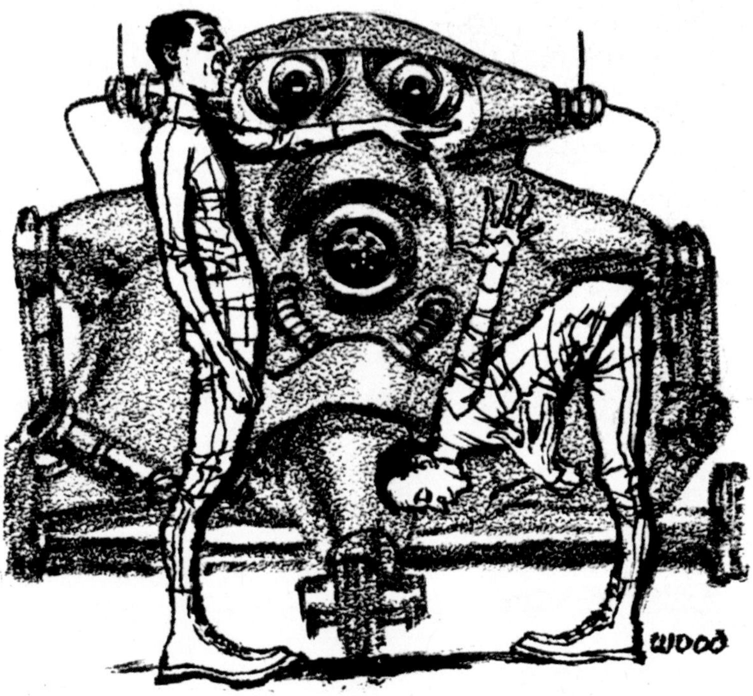

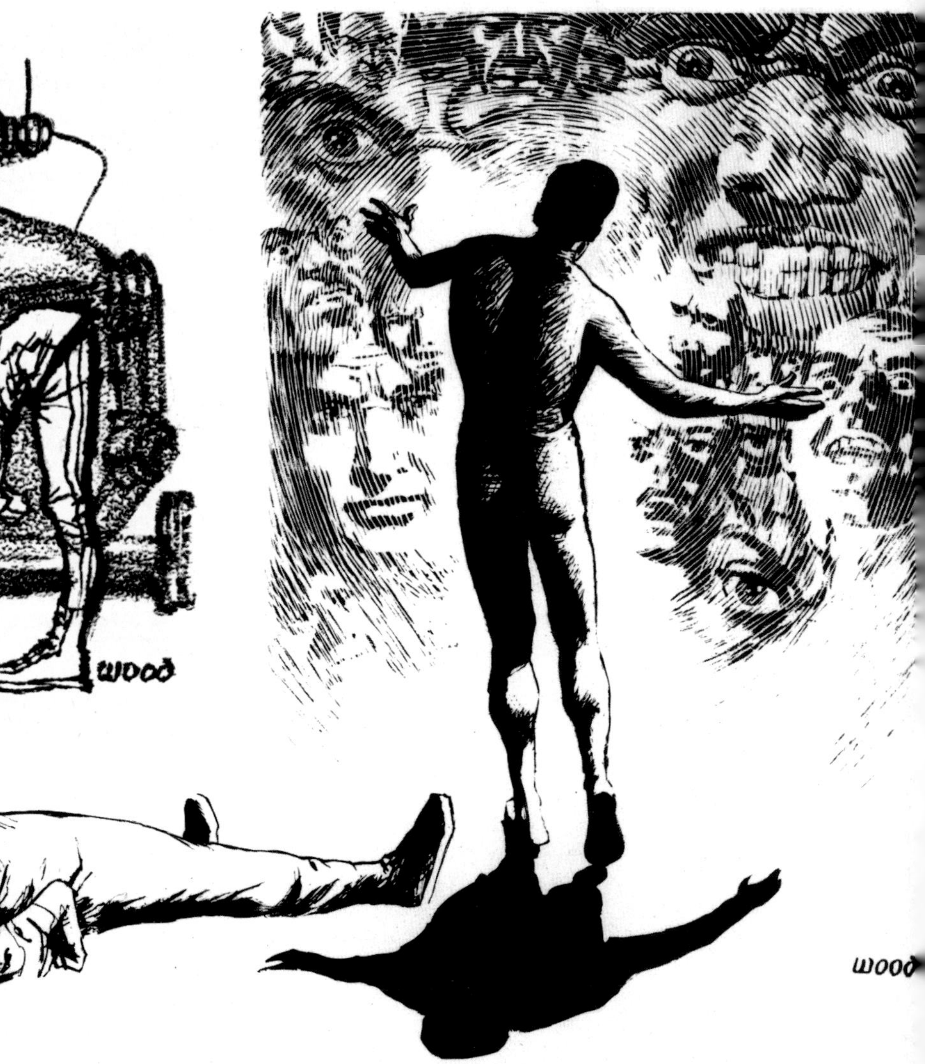

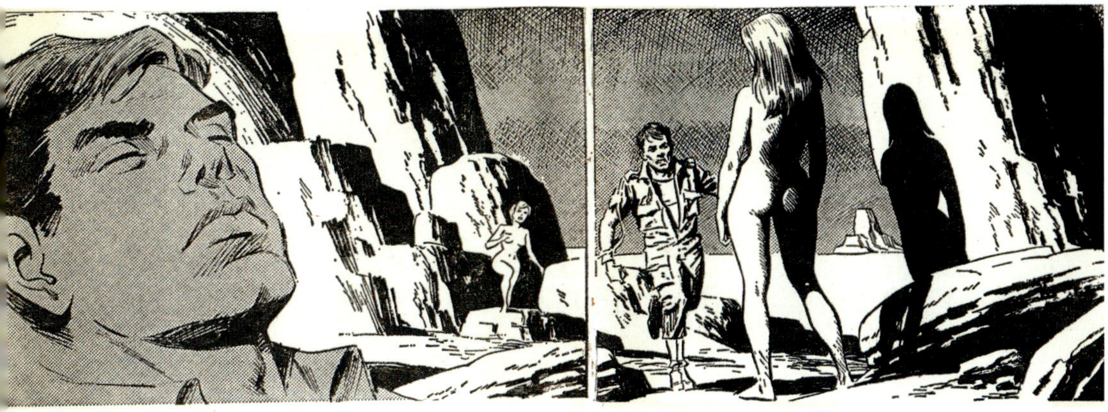

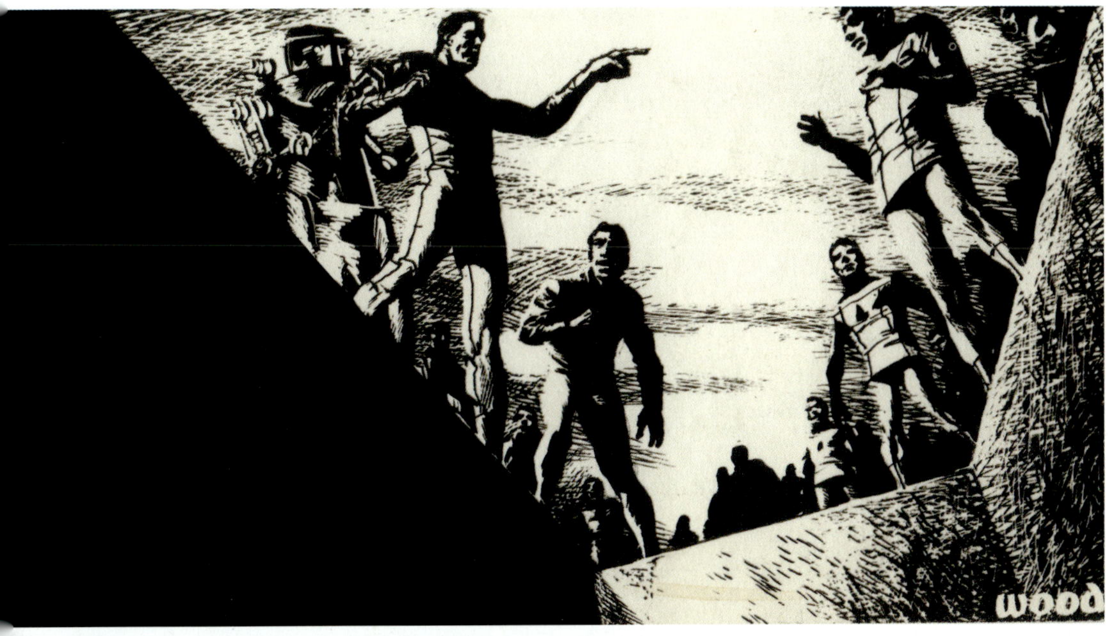

Top:
If Science Fiction November 1967
Illustration for "Dreamhouse"
by Gary Wright

Middle, previously unpublished early illustration. Bottom, final published illustration:
If Science Fiction November 1960
Illustration for "The Quality of Mercy"
by Daniel Keyes

Next page:
Upper left:
If Science Fiction April 1966 (Cover)

Both bottom illustrations and upper right:
Painting by Wood and Dan Adkins for
If Science Fiction, April 1966
(accidentally credited to Gray Morrow)
Story titled "Earthblood"
by Keith Laumer and Rosel G. Brown
(accidentally credited to Gray Morrow)

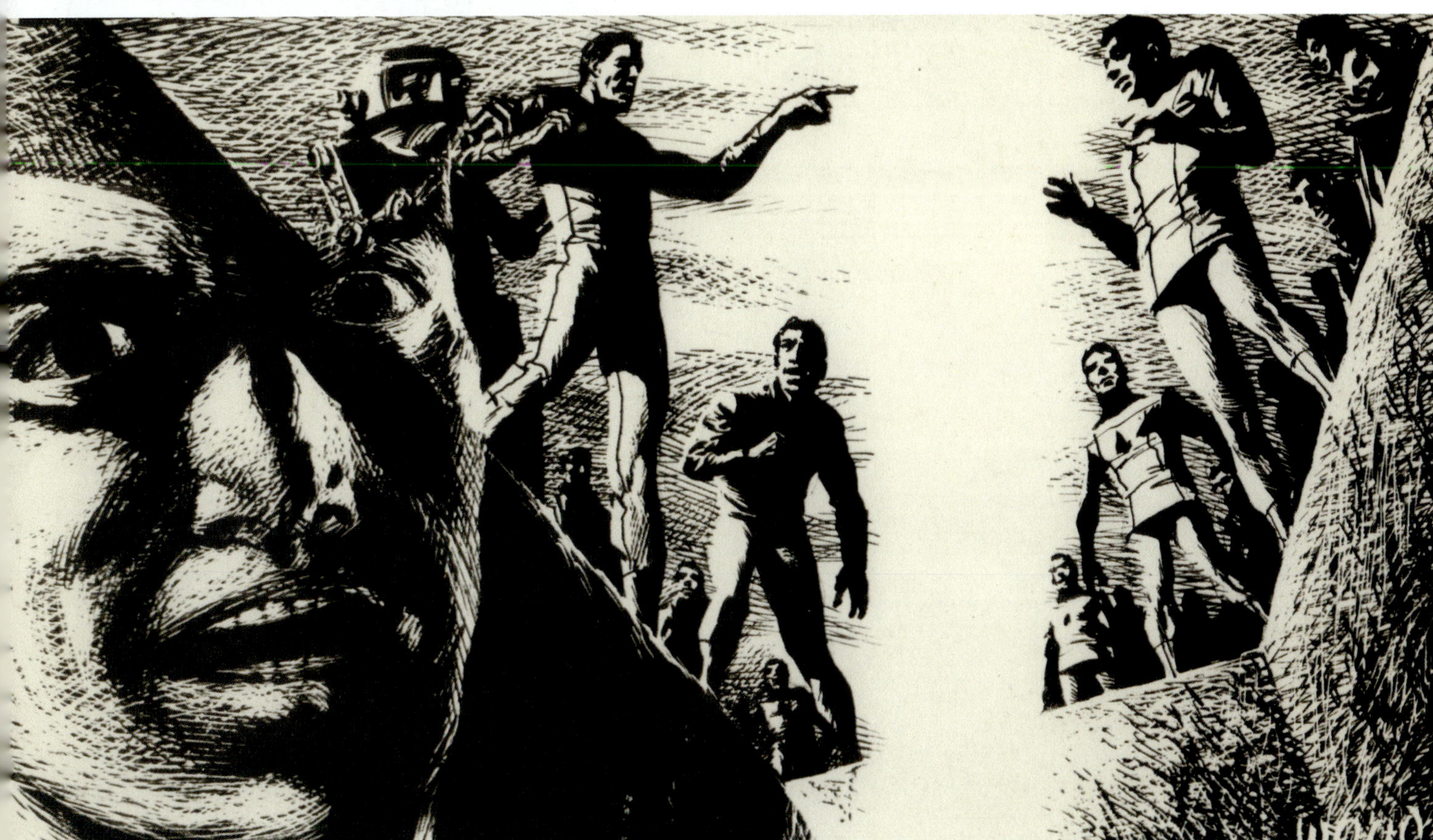

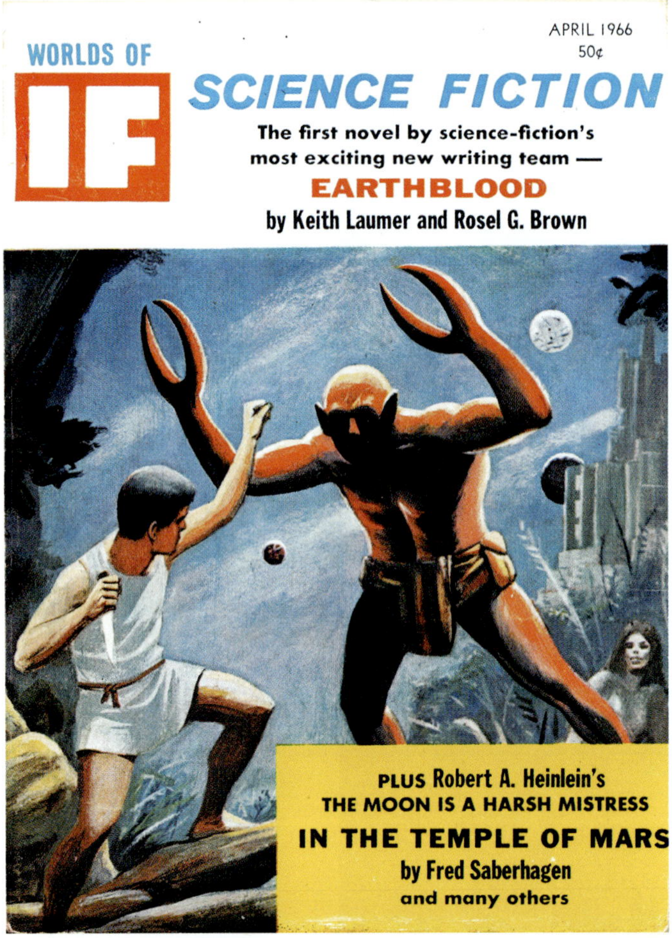
APRIL 1966
50¢
WORLDS OF
IF
SCIENCE FICTION
The first novel by science-fiction's most exciting new writing team —
EARTHBLOOD
by Keith Laumer and Rosel G. Brown
PLUS Robert A. Heinlein's
THE MOON IS A HARSH MISTRESS
IN THE TEMPLE OF MARS
by Fred Saberhagen
and many others

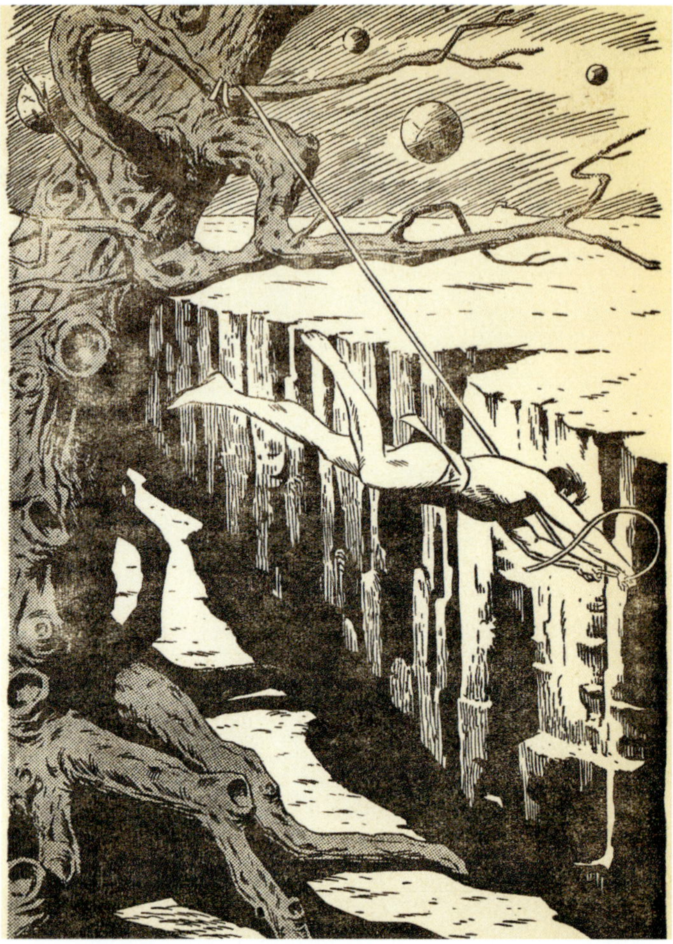

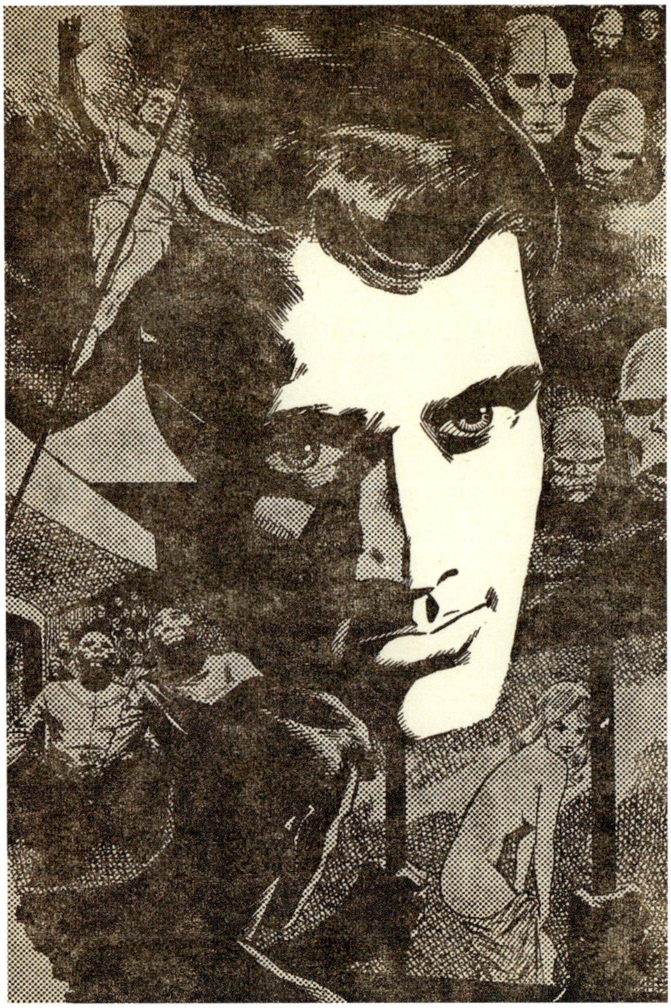

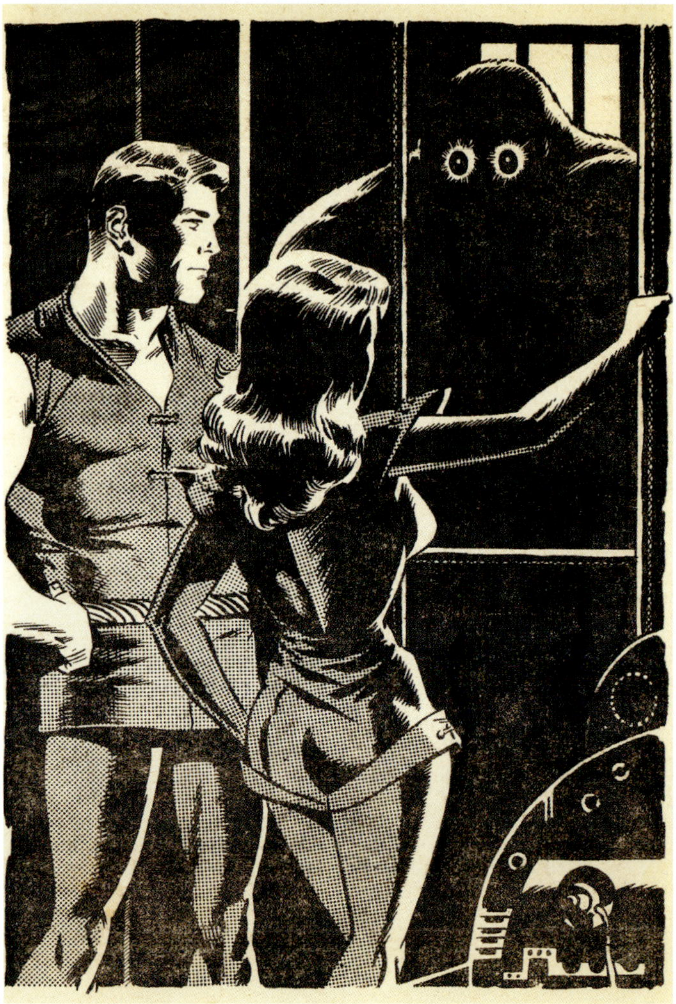

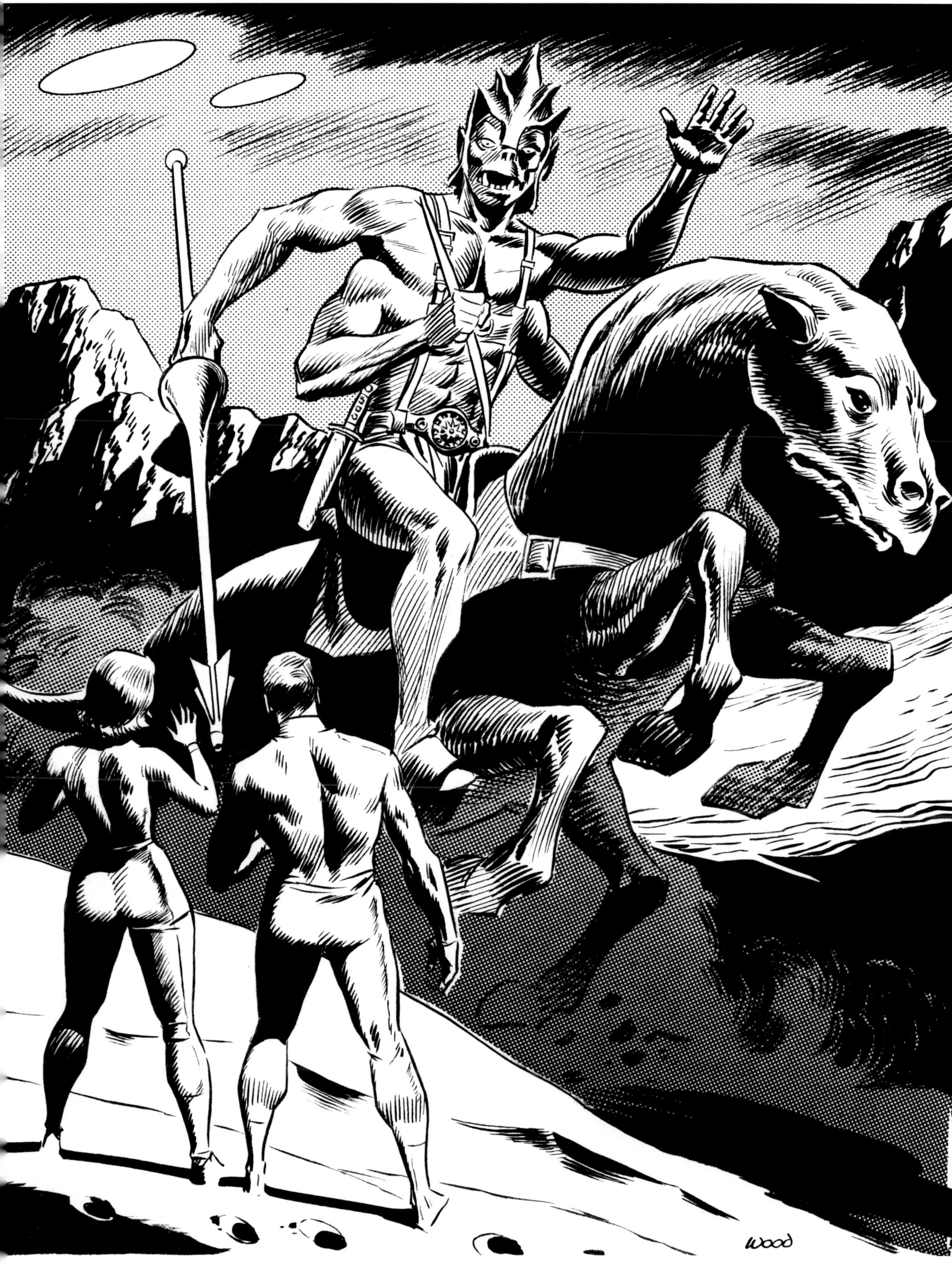

Wood

Previous page:
If Science Fiction September 1965
Illustration for "Under Two Moons"
by Frederik Pohl

This page, clockwise from upper right:
If Science Fiction May 1966
Illustration for "Earthblood" part 2
by Keith Laumer and Rosel George Brown

Both bottom illustrations:
If Science Fiction July 1966
Illustration for "Earthblood" part 4
by Keith Laumer and Rosel George Brown

If Science Fiction June 1966
Illustration for "Earthblood" part 3
by Keith Laumer and Rosel George Brown

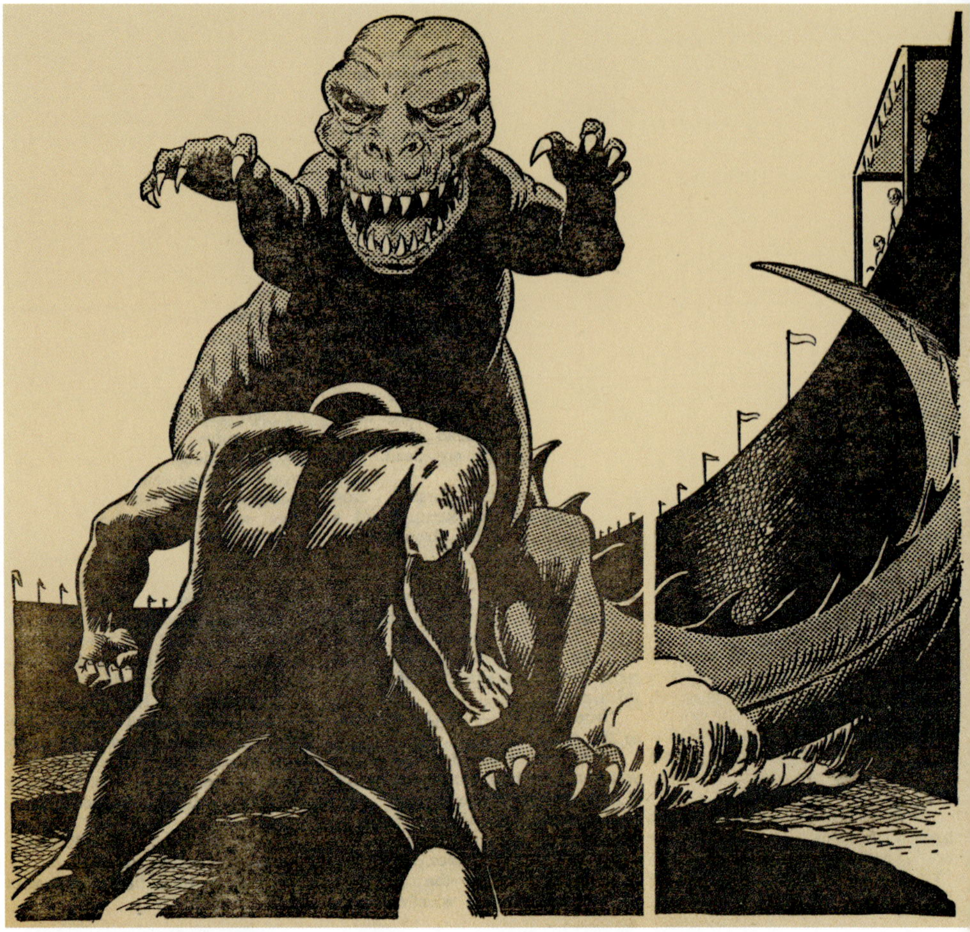

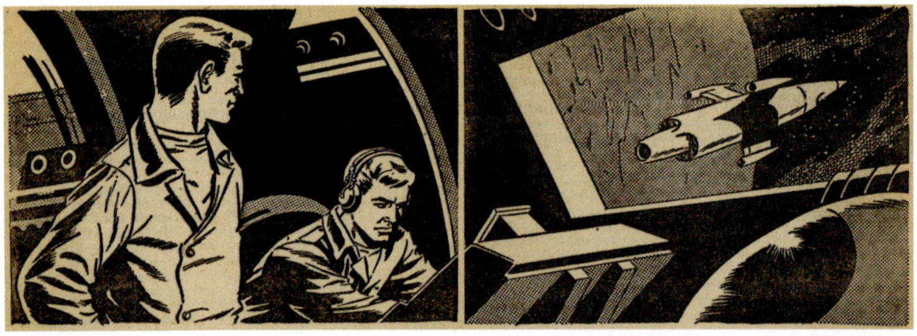

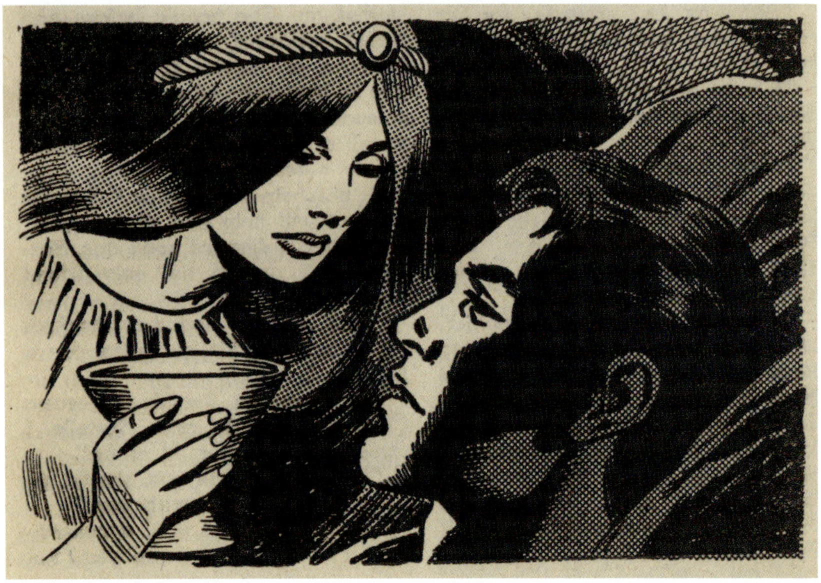

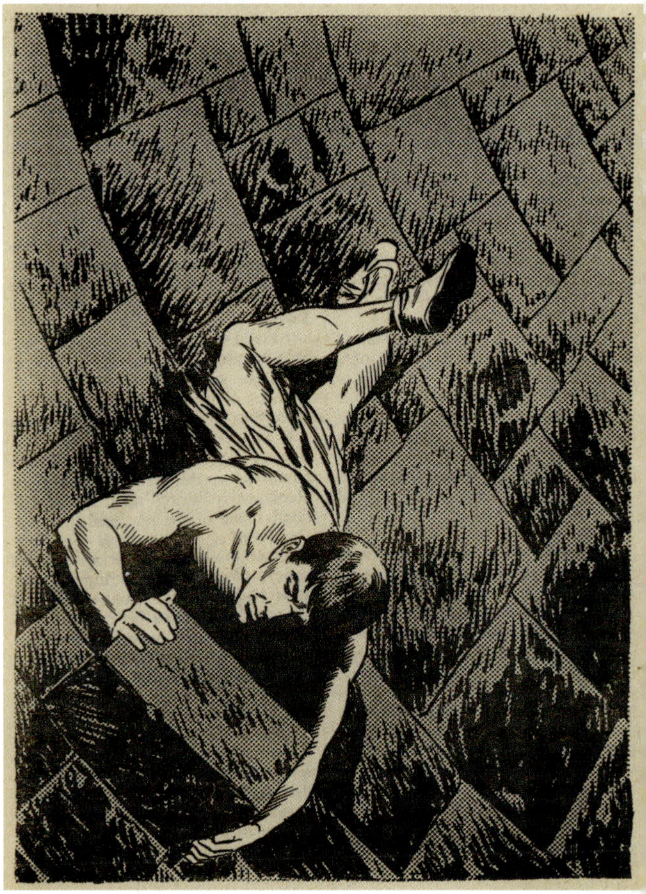

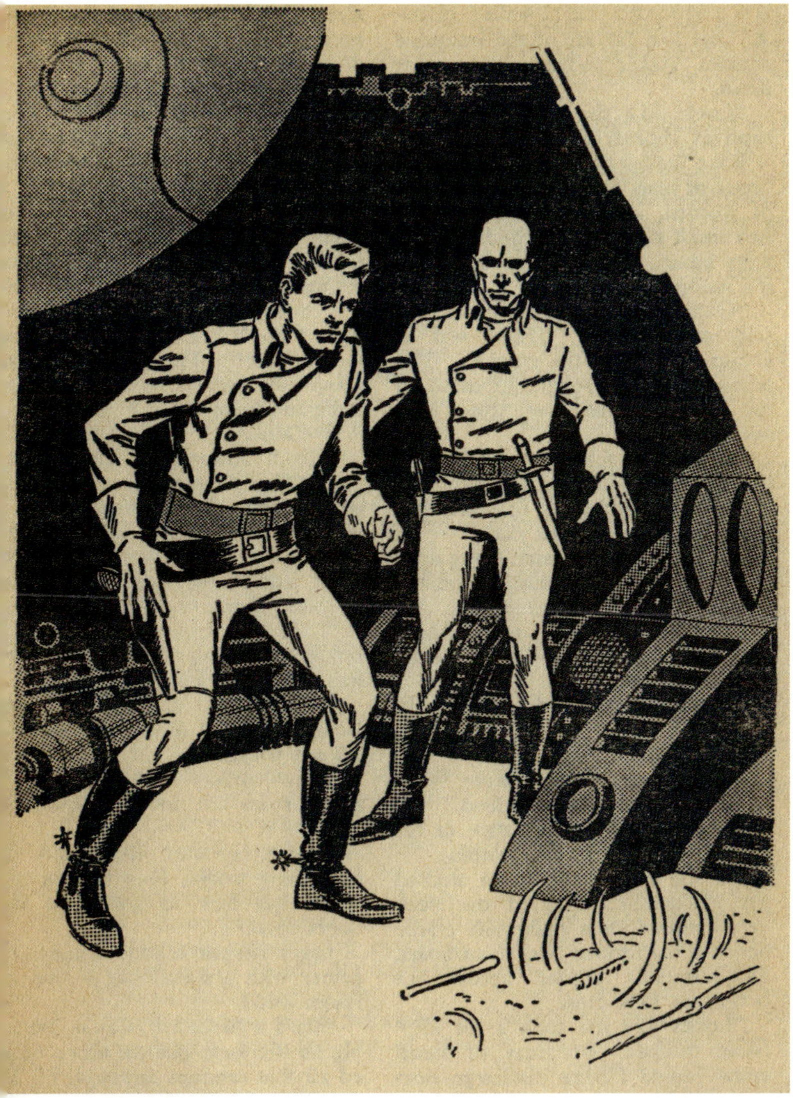

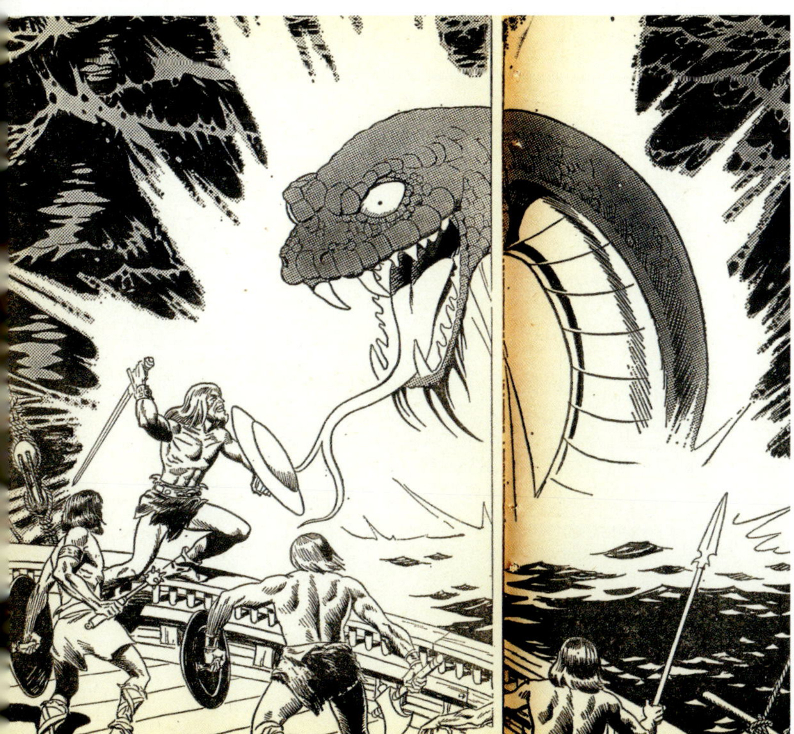

Top:
If Science Fiction June 1966
Illustration for "Earthblood" part 3
by Keith Laumer and Rosel George Brown

Bottom:
If Science Fiction August 1967
Illustration for "The Winged Helmet"
by Fred Saberhagen

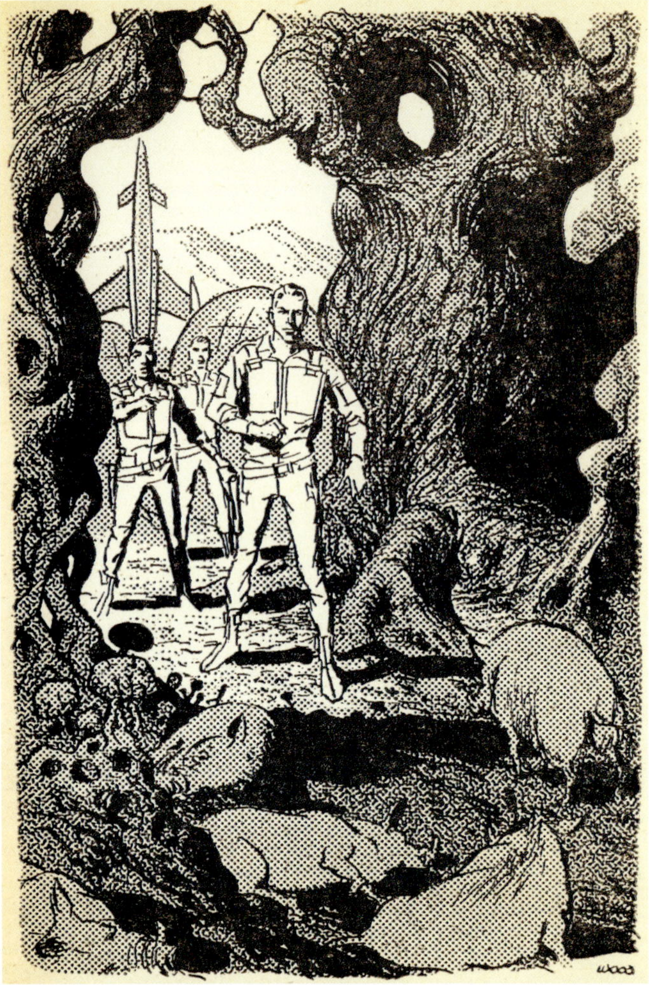
WOOD

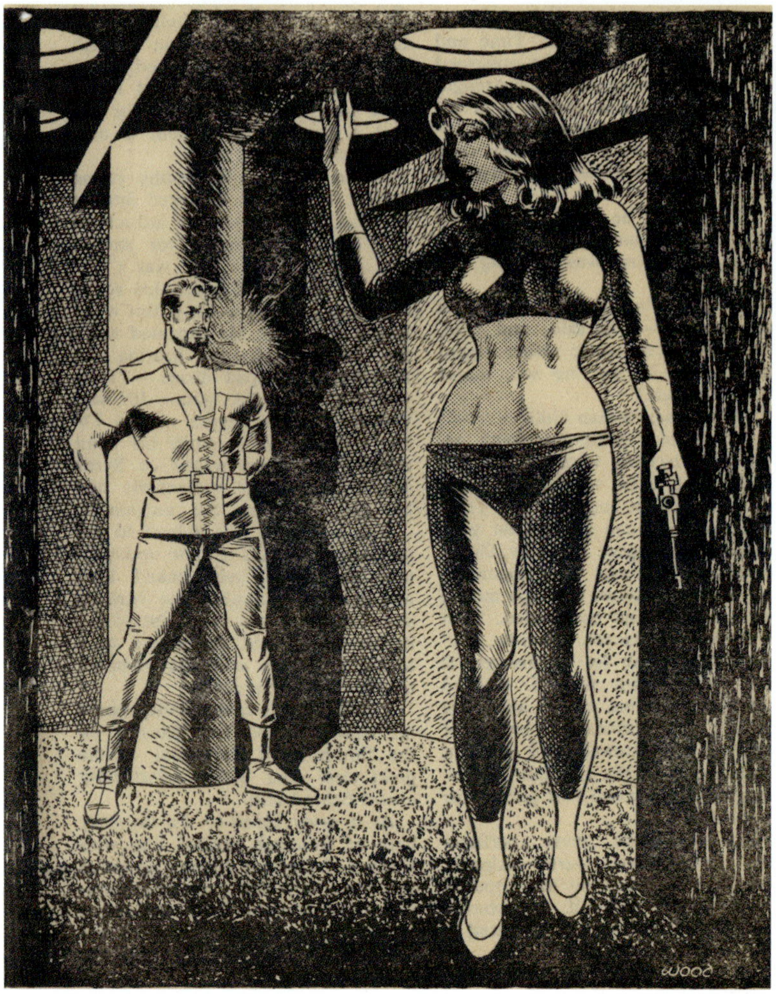
WOOD

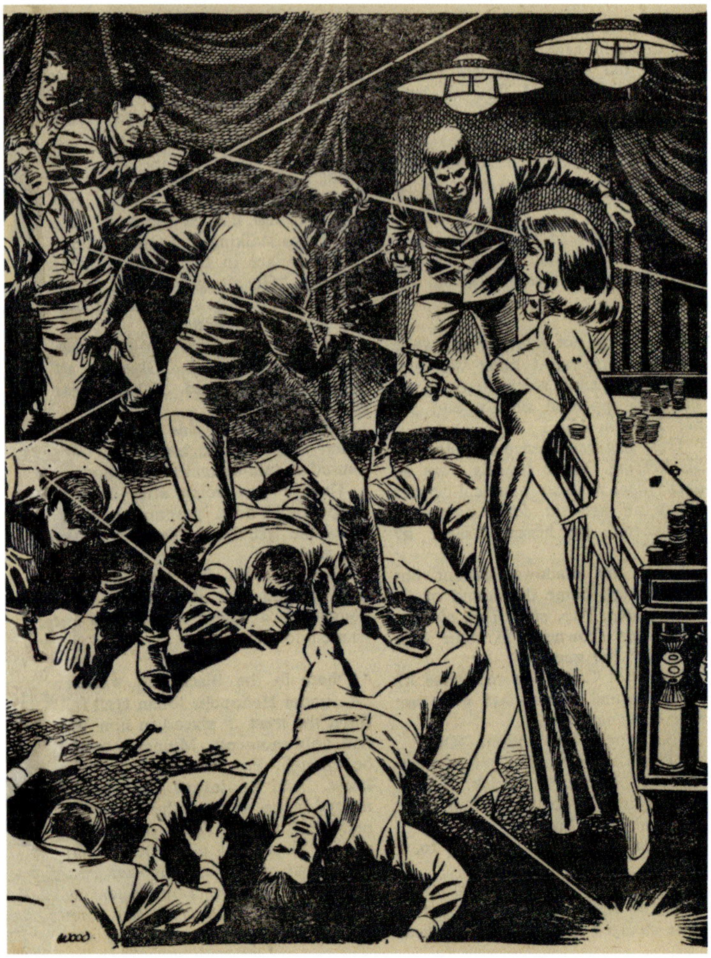
WOOD

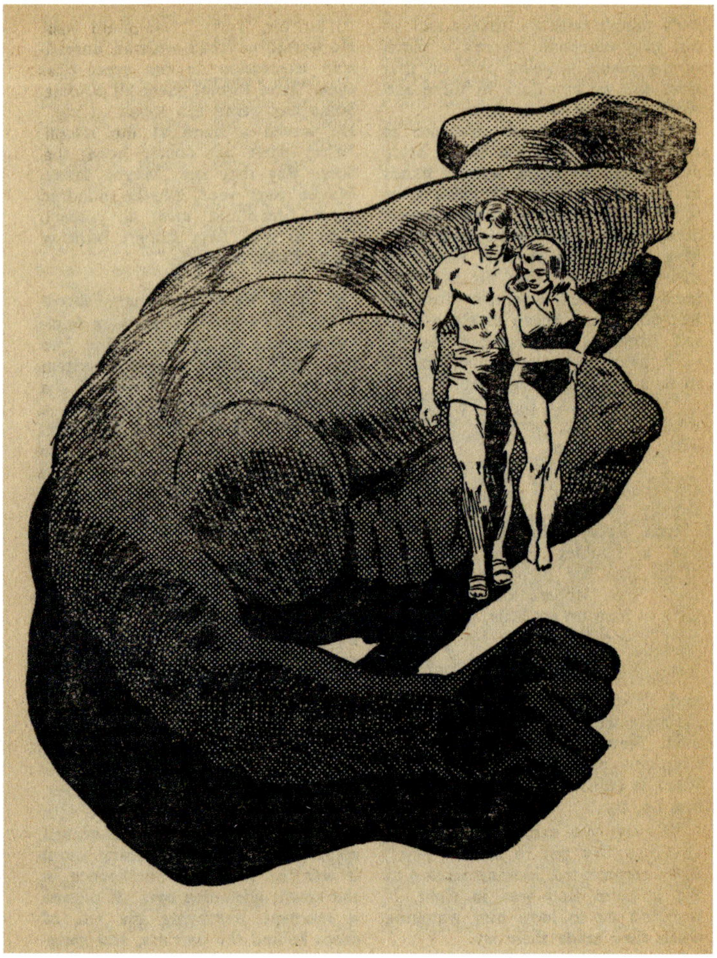

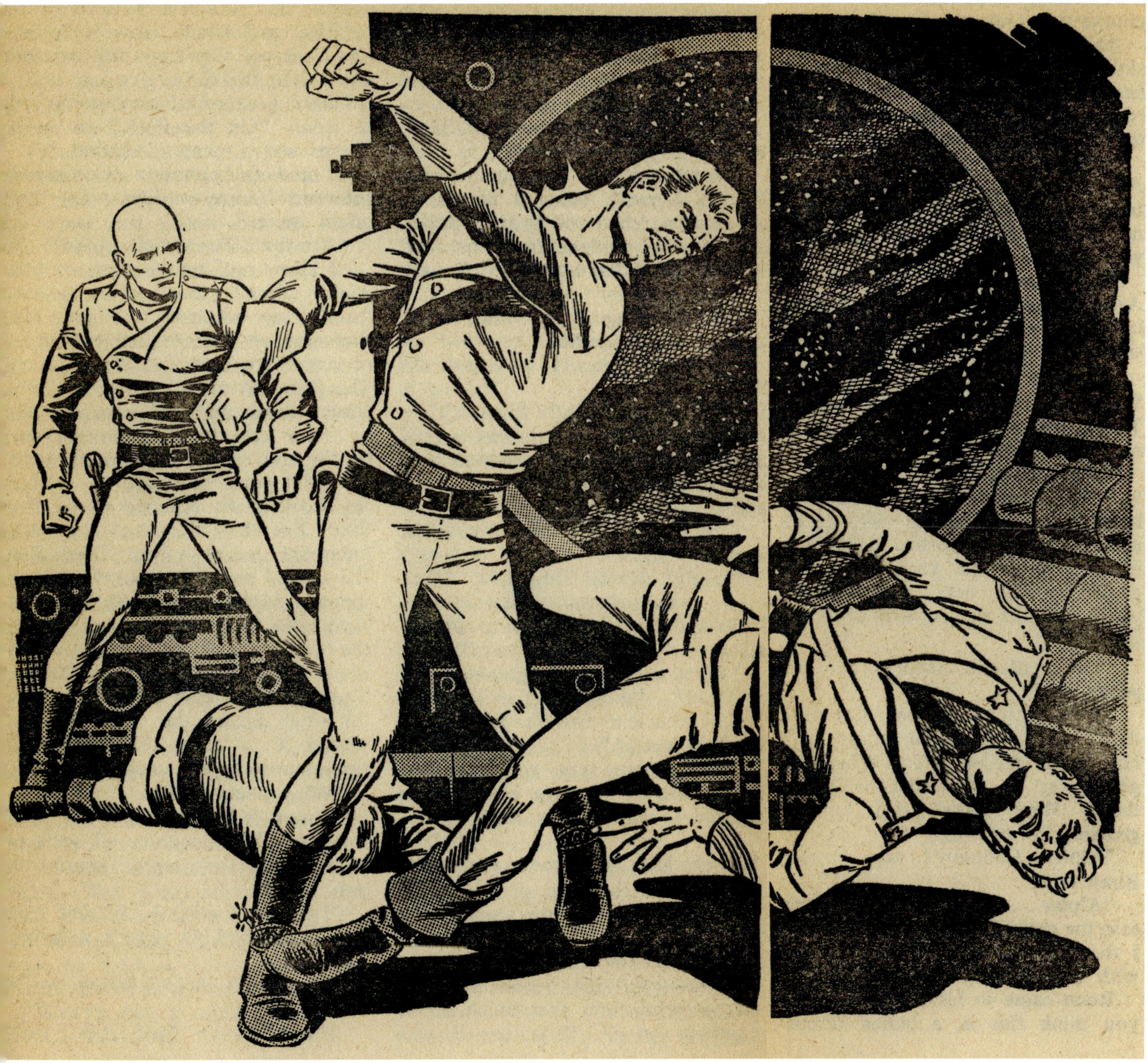

If Science Fiction June 1966
Illustration for "Earthblood" part 3 by Keith Laumer and Rosel George Brown

Opposite:
Uppper left:
If Science Fiction July 1961
Illustration for "The Planet with No Nightmare" by Jim Harmon

Upper right and lower left:
If Science Fiction September 1965
Illustration for "Under Two Moons" by Frederik Pohl

Lower right:
If Science Fiction May 1966
Illustration for "Earthblood" part 2 by Keith Laumer and Rosel George Brown

Next spread: *Galaxy Science Fiction* August 1963
Illustration for "Here Gather the Stars" part 2 by Clifford D. Simak

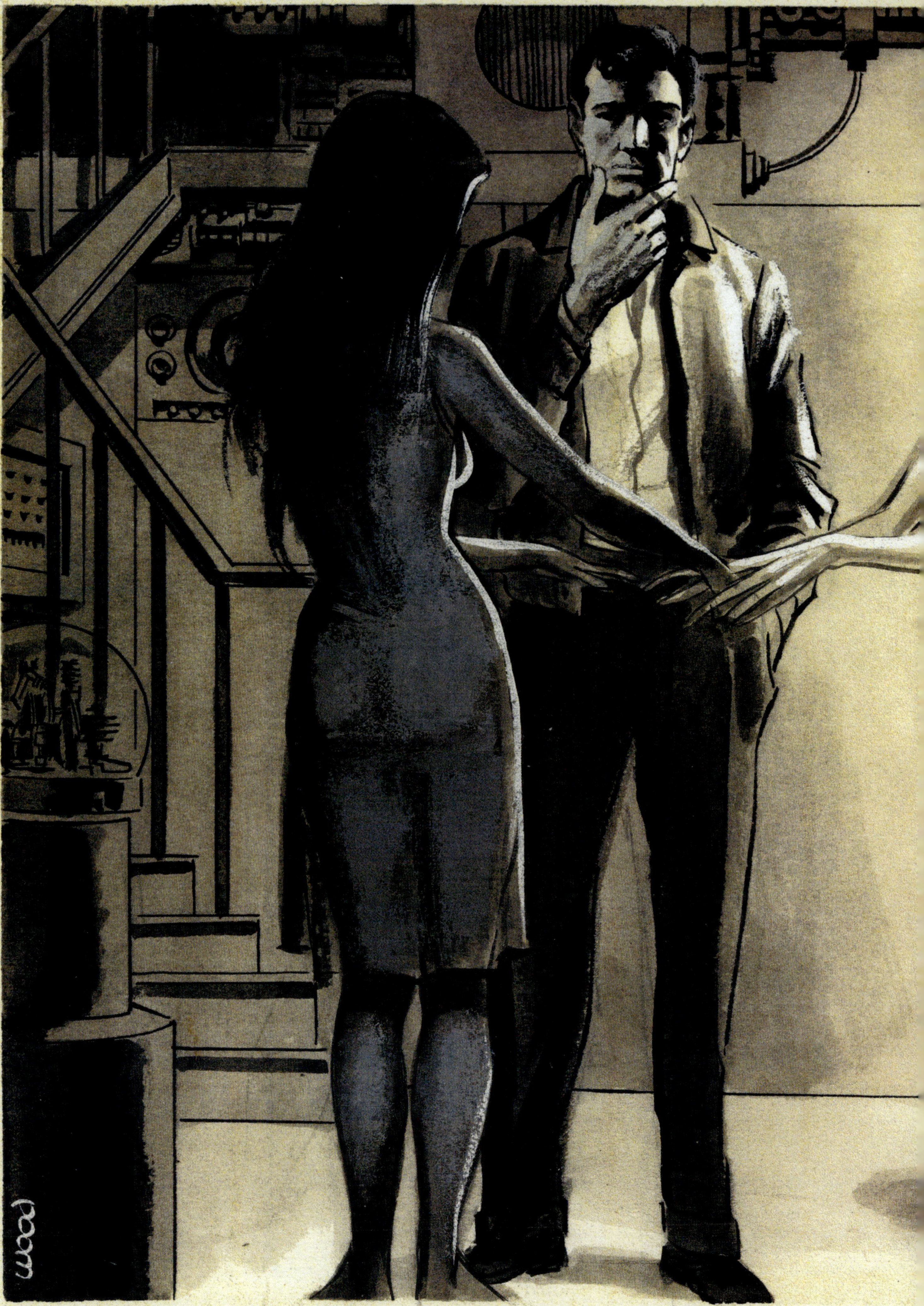
wood

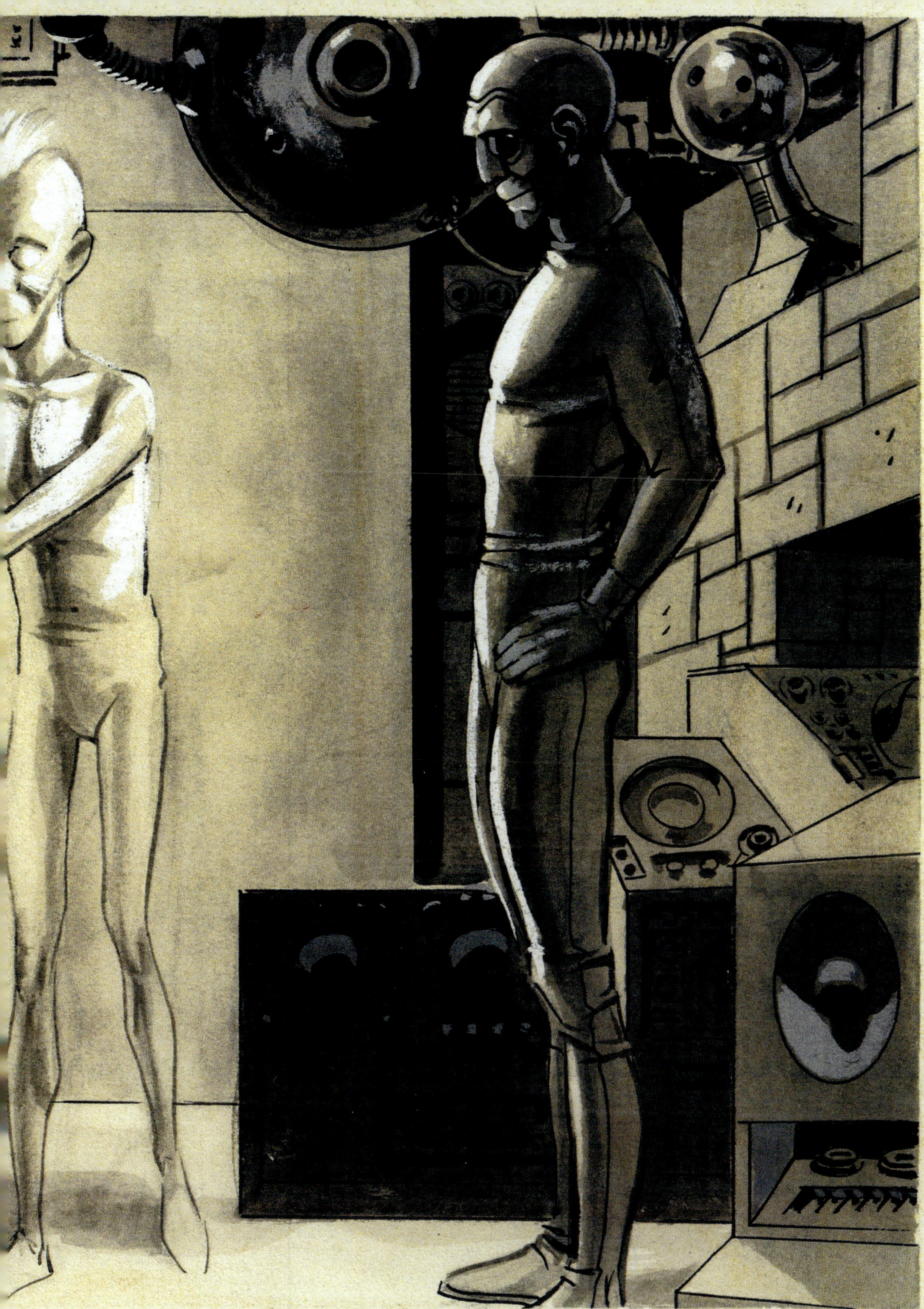

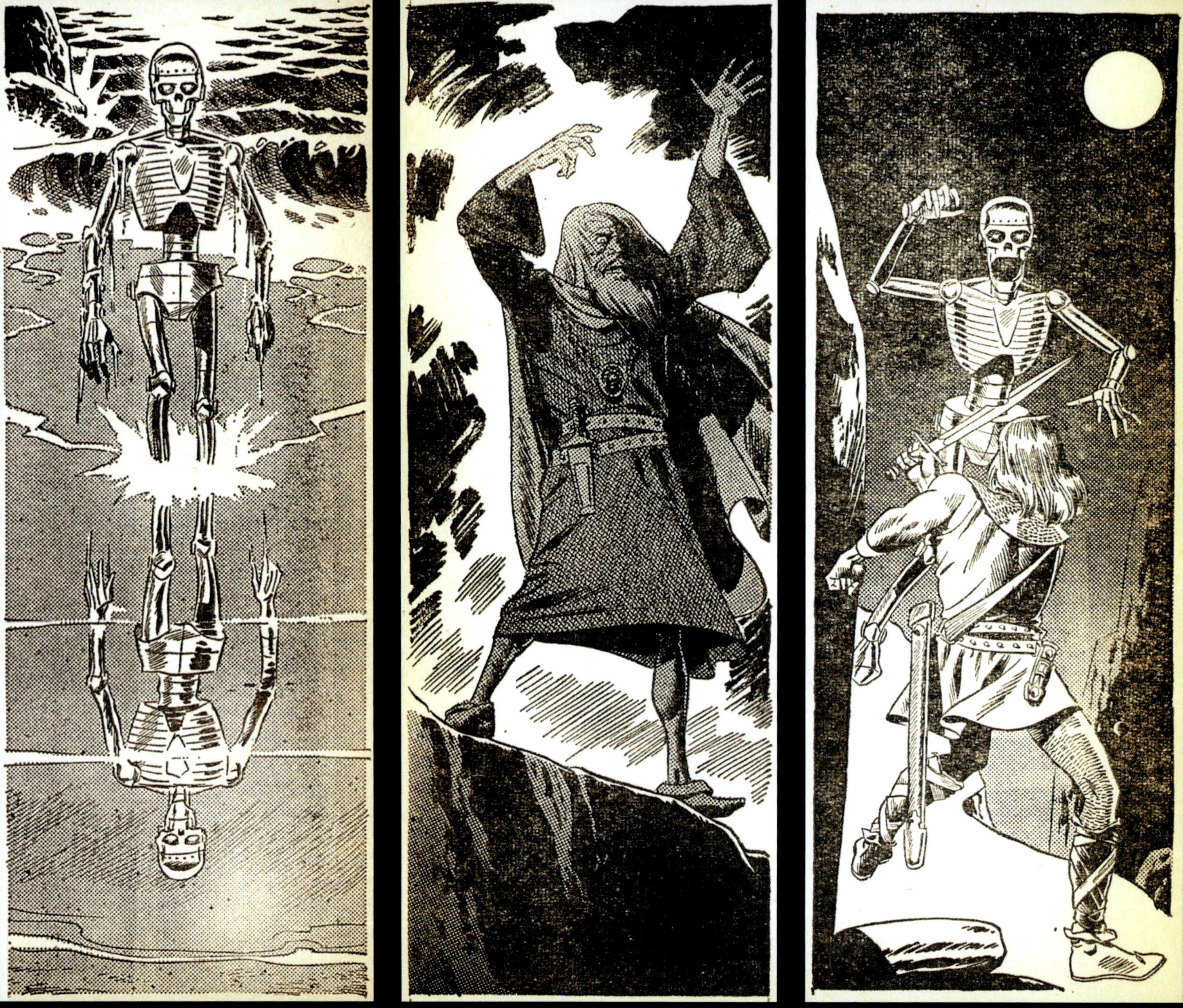

All: *If Science Fiction* August 1967
Illustration for "The Winged Helmet"
by Fred Saberhagen

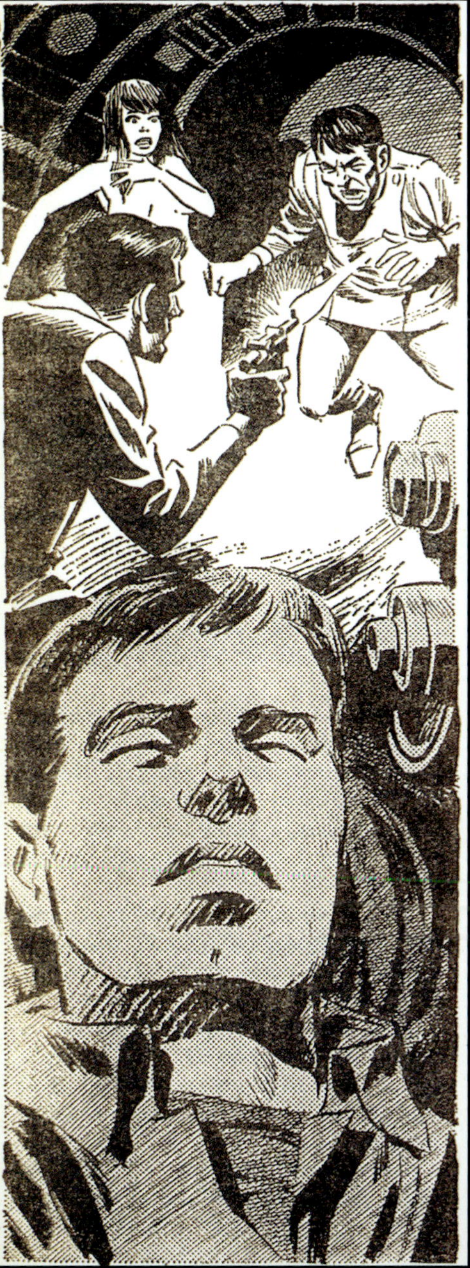
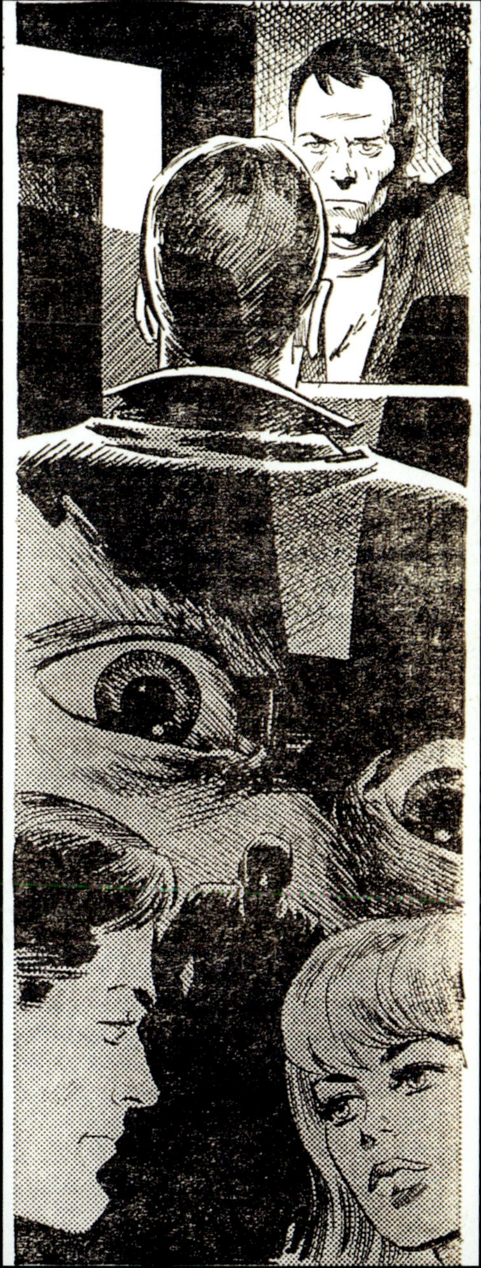
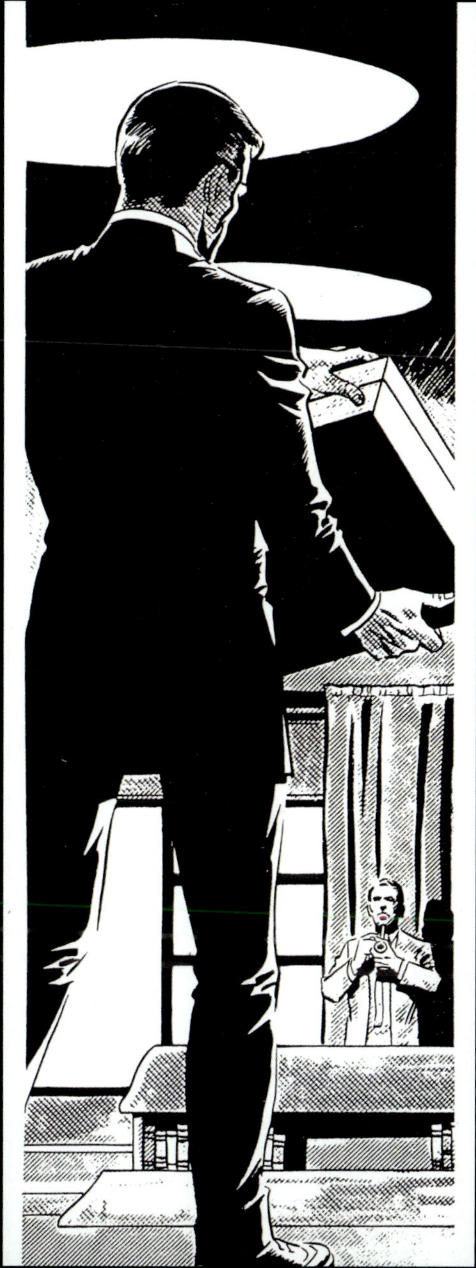

Left and center: *If Science Fiction* November 1967
Illustration for "Dreamhouse" by Gary Wright

Right: *If Science Fiction* November 1968
Illustration for "The Perfect Secretary" by Mike Kirsch

Clockwise from upper right:
Worlds of Tomorrow December 1963
Illustration for "The Creature Inside"
by Jack Sharkey

If Science Fiction December 1968
Illustration for "The Starman of
Pritchard's Creek" by Julian F. Grow

If Science Fiction November 1968
Illustration for "The Perfect Secretary"
by Mike Kirsch

Opposite:
If Science Fiction November 1968
Illustration for "The Perfect Secretary"
by Mike Kirsch

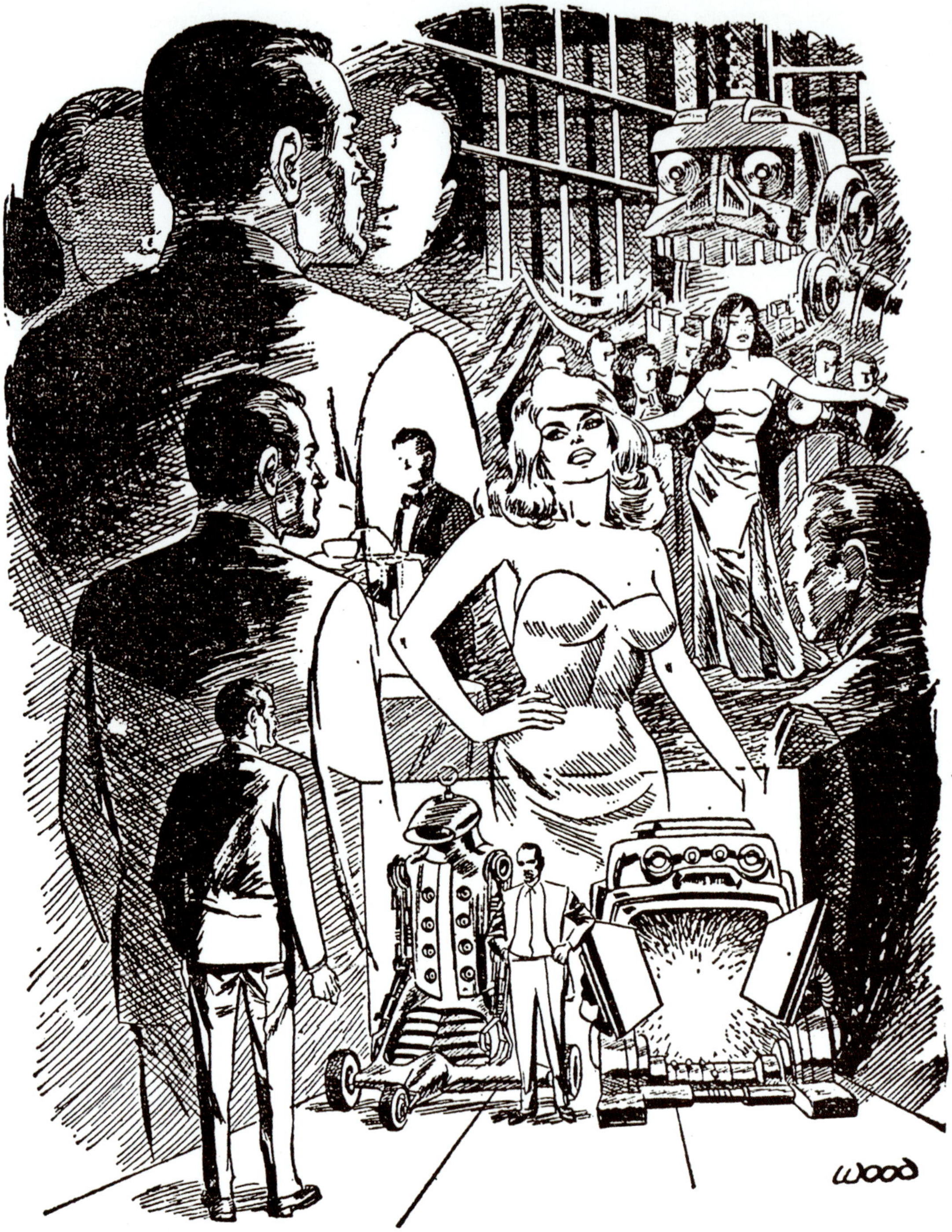

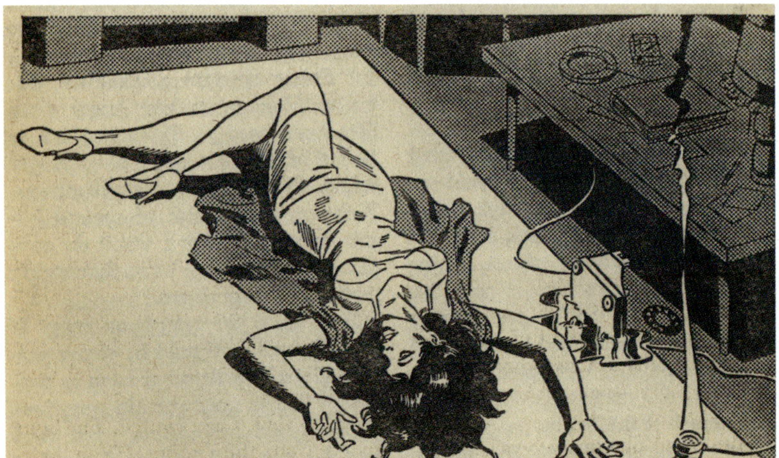

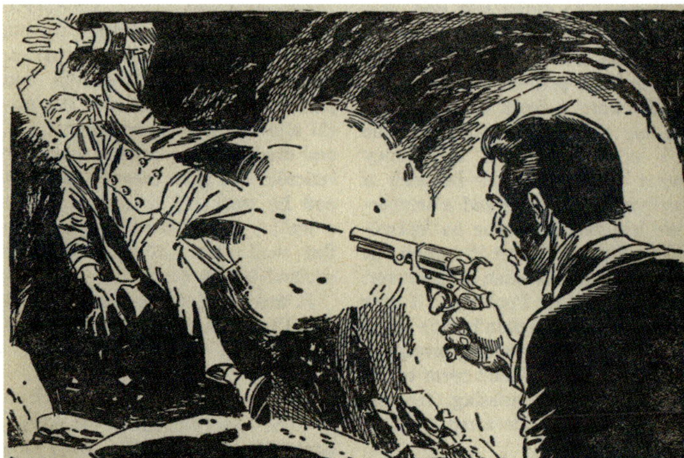

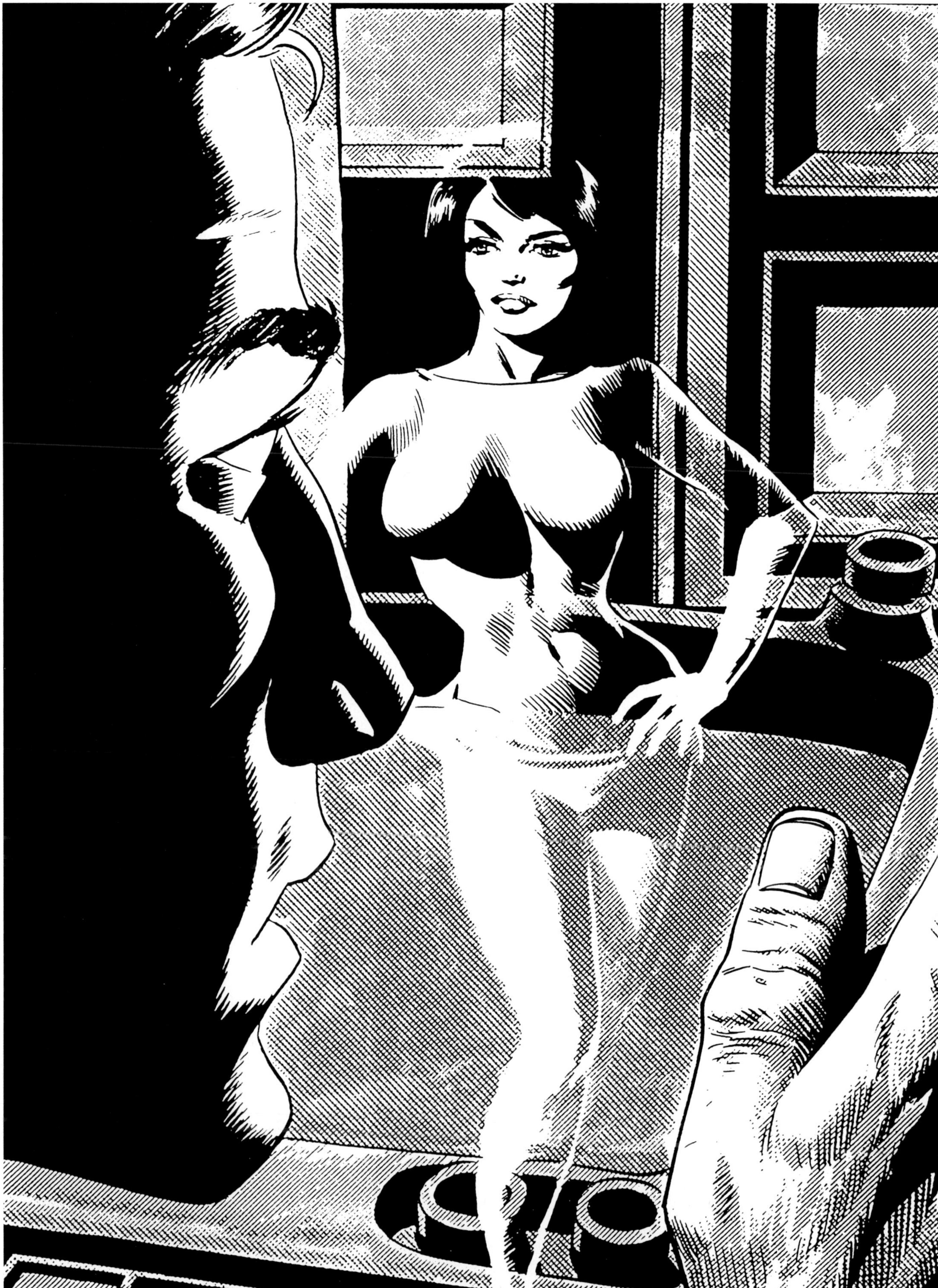

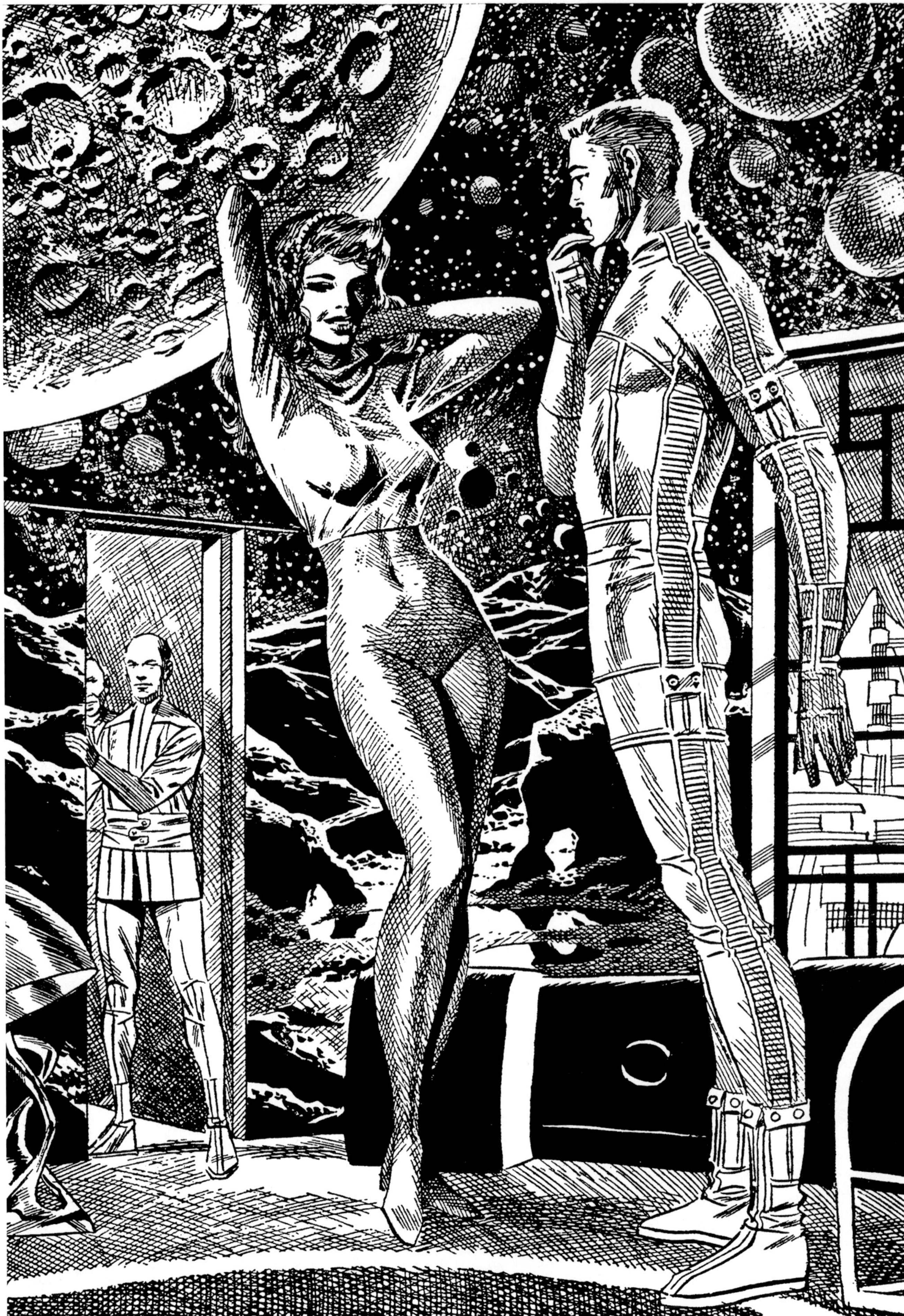

Original Science Fiction Stories May 1959 Illustration for "There's No Place Like Space" by Robert Silverberg

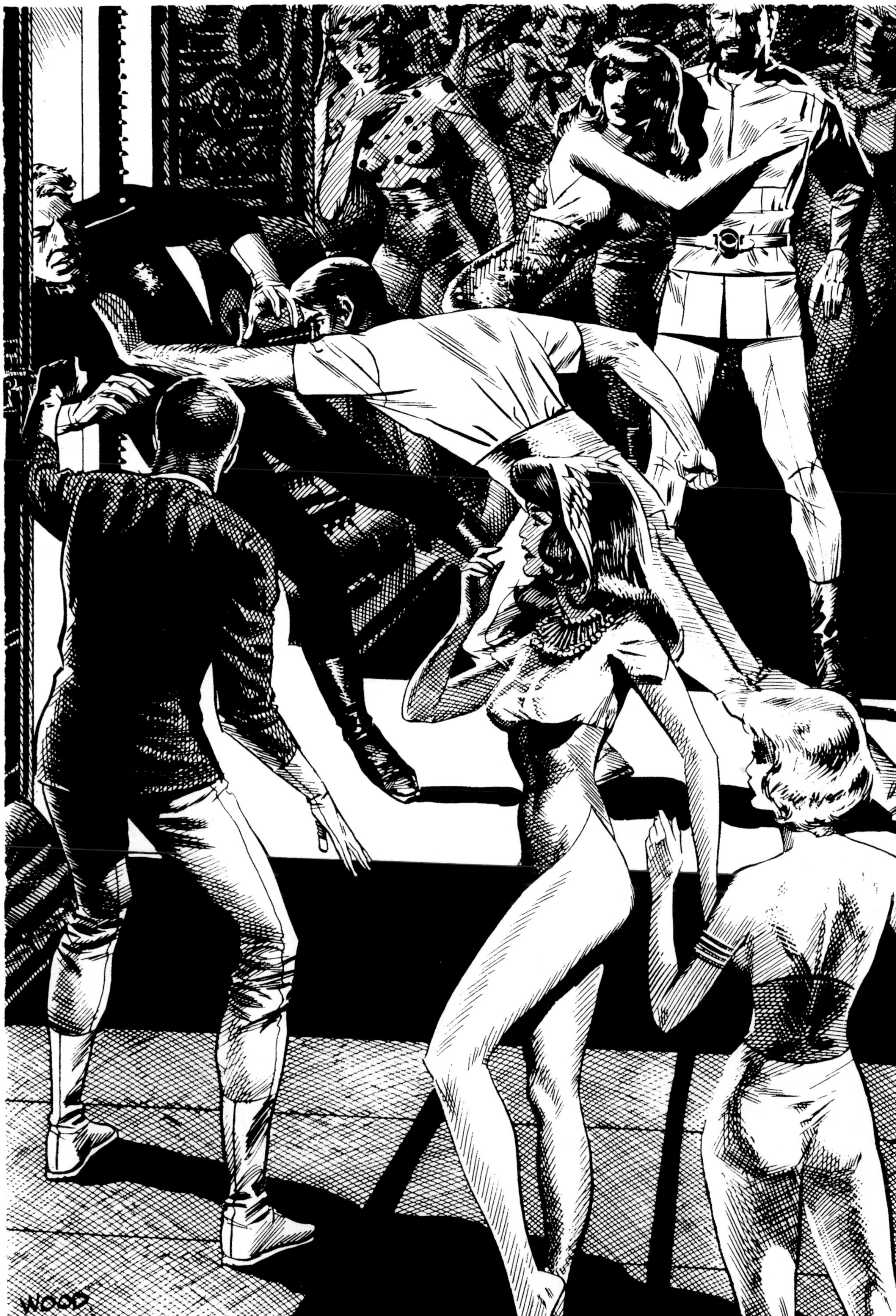

Original Science Fiction Stories May 1959 Illustration for "There's No Place Like Space" by Robert Silverberg

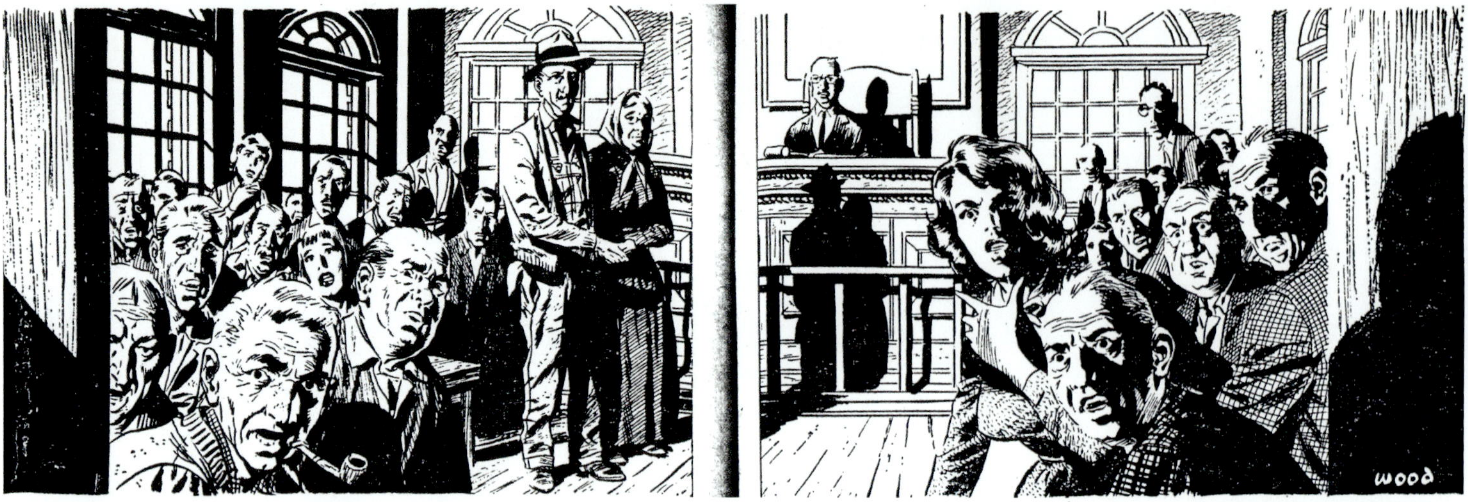

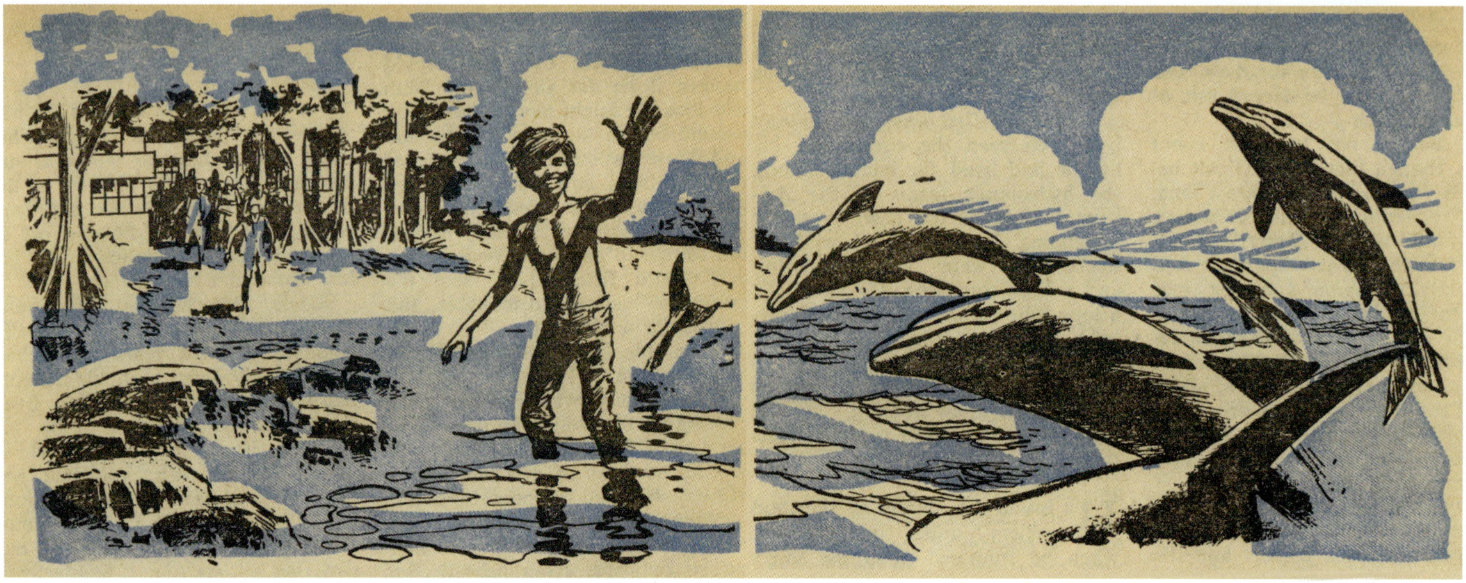

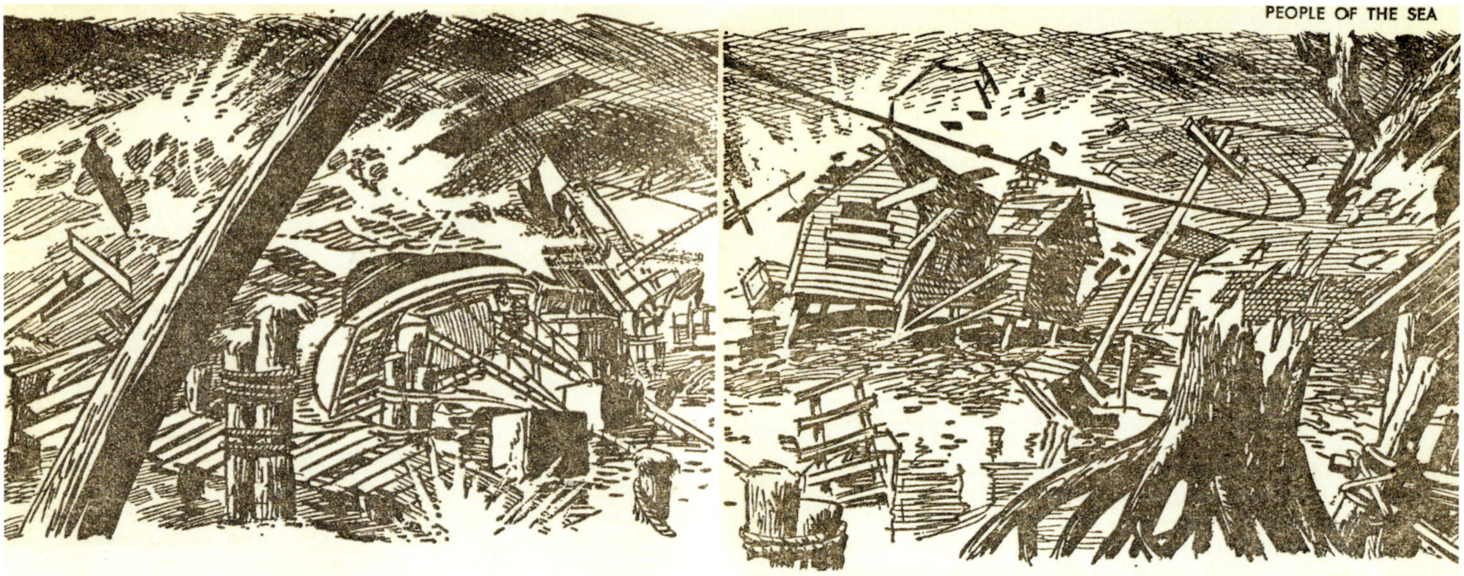

Top to bottom: *Amazing Stories* July 1958 Illustration for "Everything's Different Up There" by Genevieve Haugen

Worlds of Tomorrow April 1963 Illustration for "People of the Sea" part 1 by Arthur C. Clarke

Worlds of Tomorrow June 1963 Illustration for "People of the Sea" part 2 by Arthur C. Clarke

Opposite, both top illustrations and lower left: *Worlds of Tomorrow* April 1963 Illustration for "People of the Sea" part 1 by Arthur C. Clarke

Lower right: *Worlds of Tomorrow* August 1963 Illustration for "To the Stars" by J. T. McIntosh

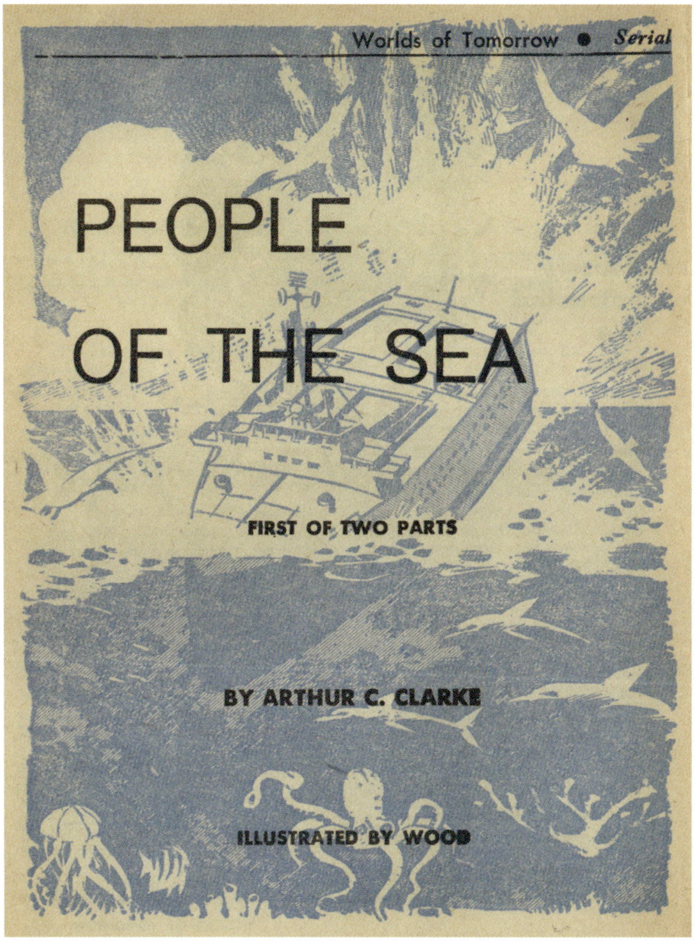
Worlds of Tomorrow ● Serial
PEOPLE OF THE SEA
FIRST OF TWO PARTS
BY ARTHUR C. CLARKE
ILLUSTRATED BY WOOD

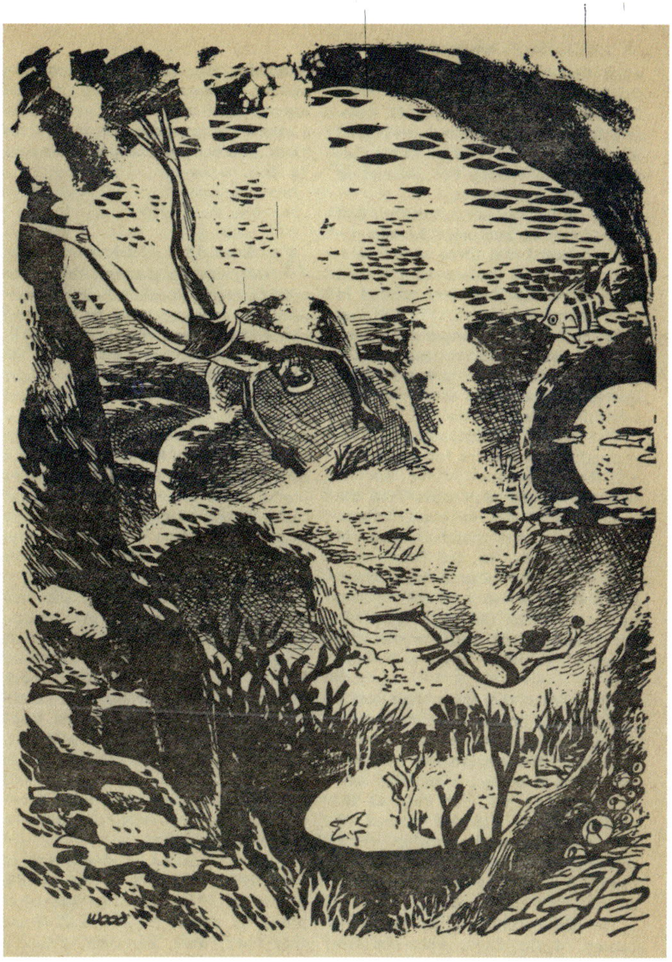
Wood

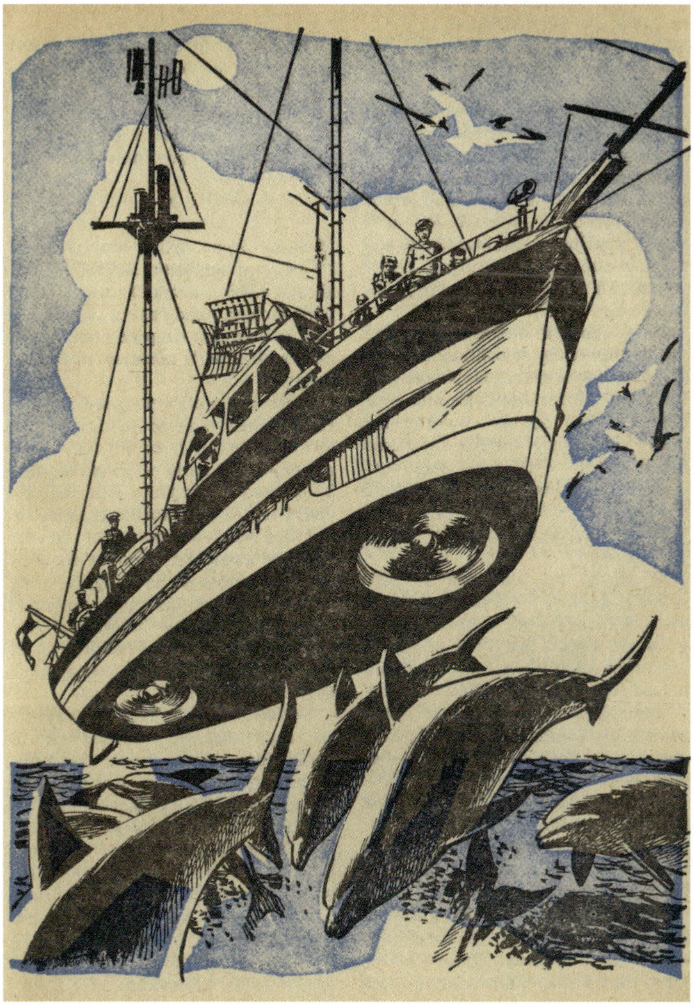

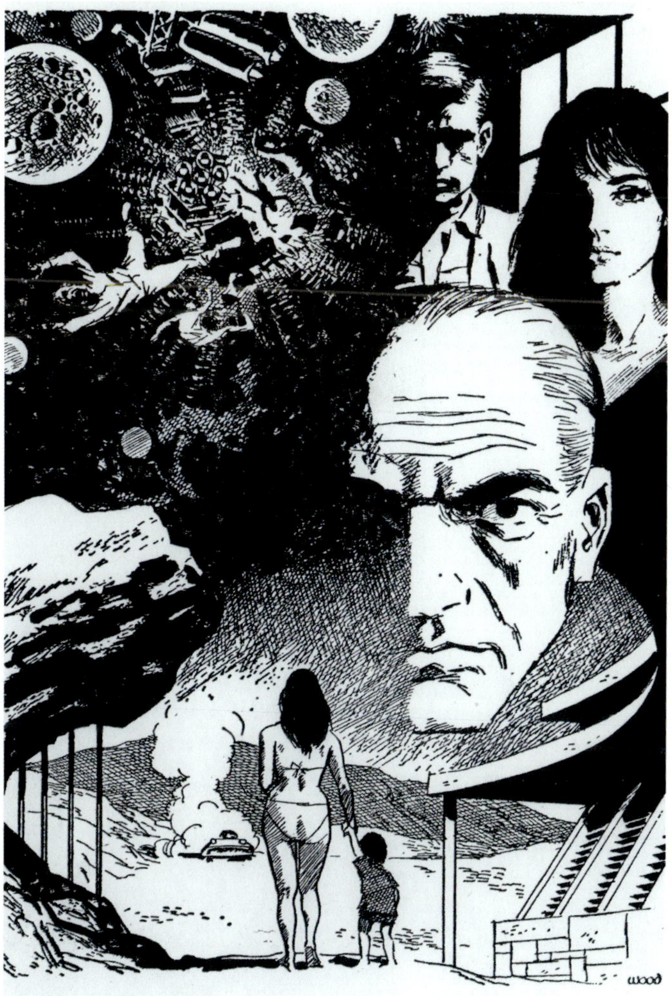
Wood

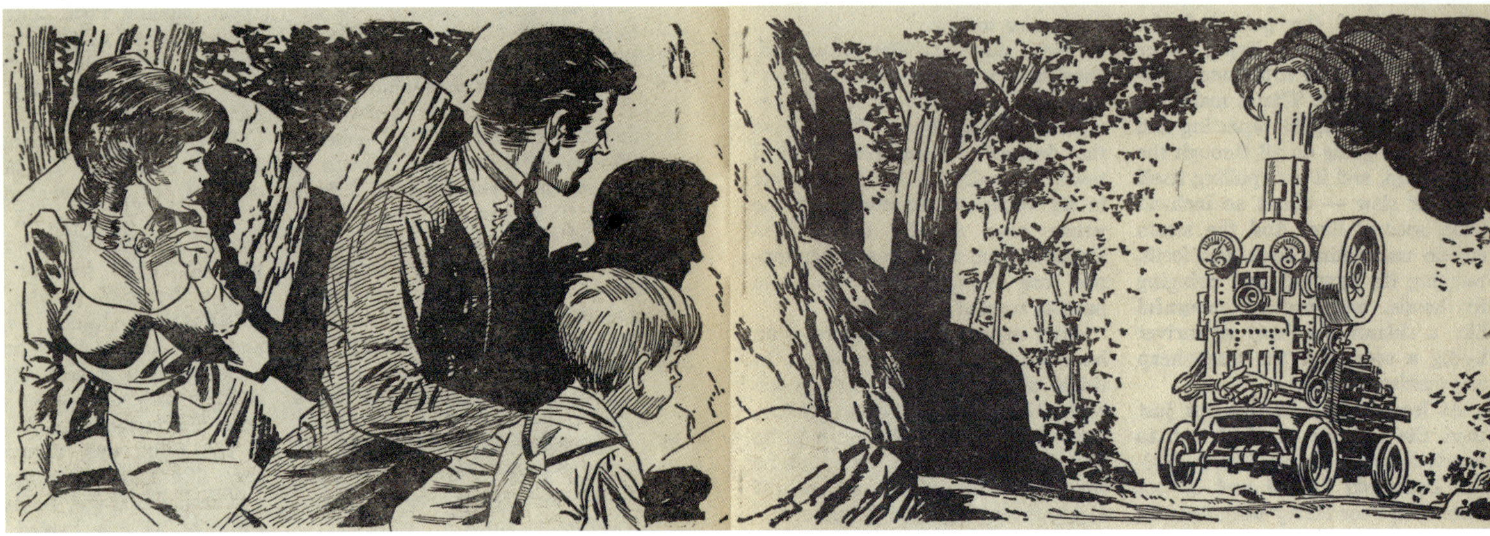

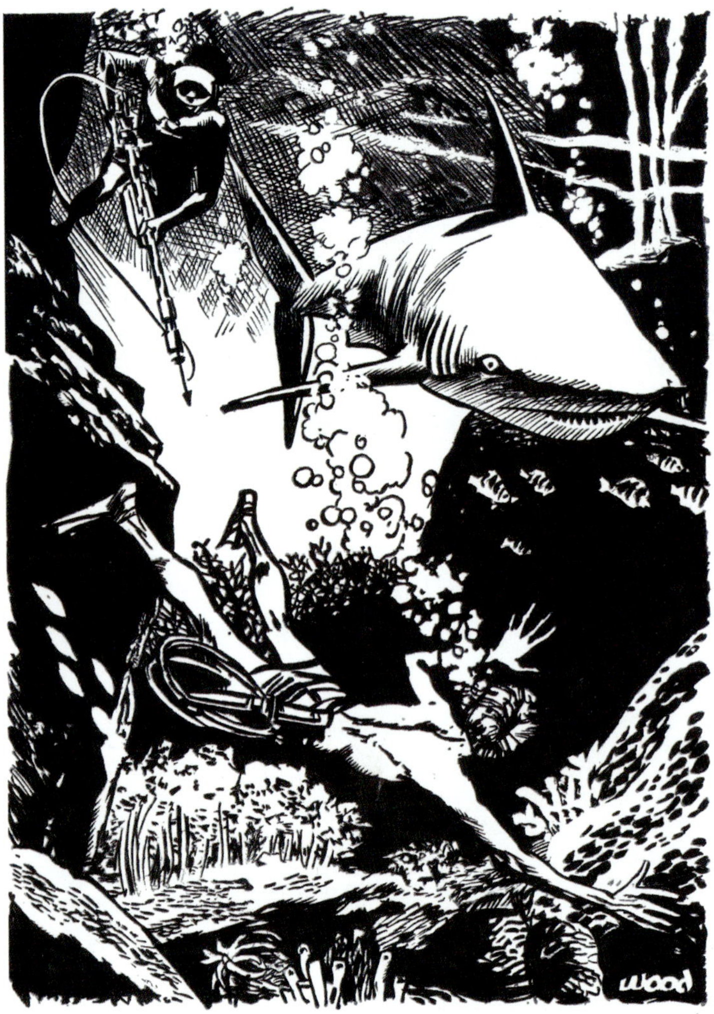

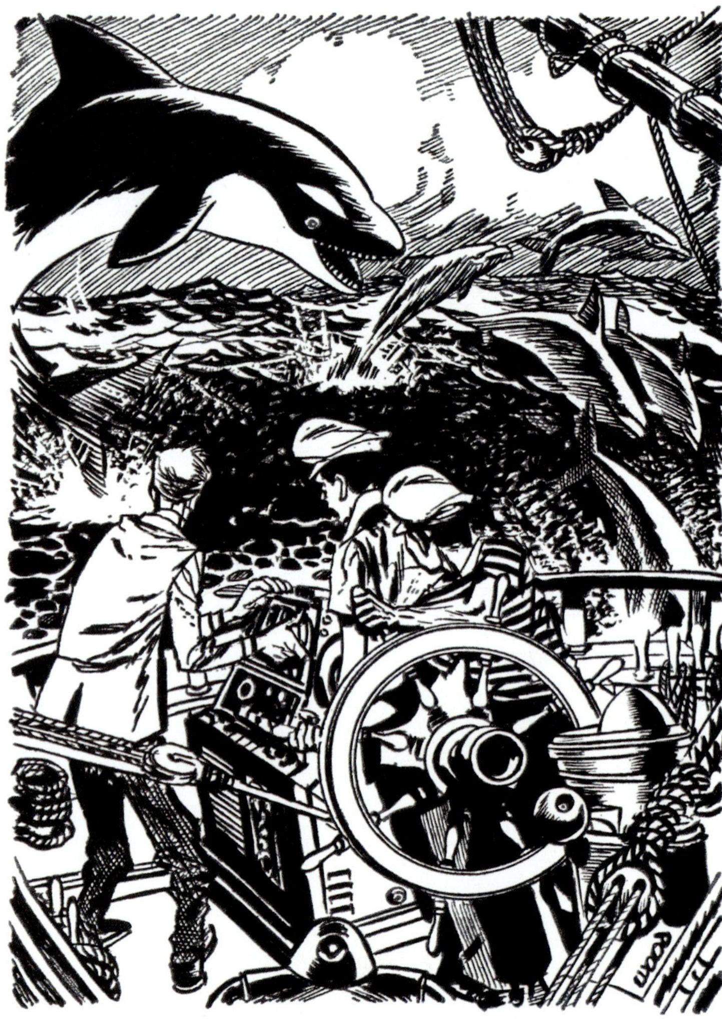

Top: *If Science Fiction* December 1968
Illustration for "The Starman of Pritchard's Creek" by Julian F. Grow

Both bottom illustrations:
Worlds of Tomorrow June 1963
Illustration for "People of the Sea" part 2 by Arthur C. Clarke

Opposite: *Worlds of Tomorrow* August 1963
Illustration for "To the Stars" by J. T. McIntosh

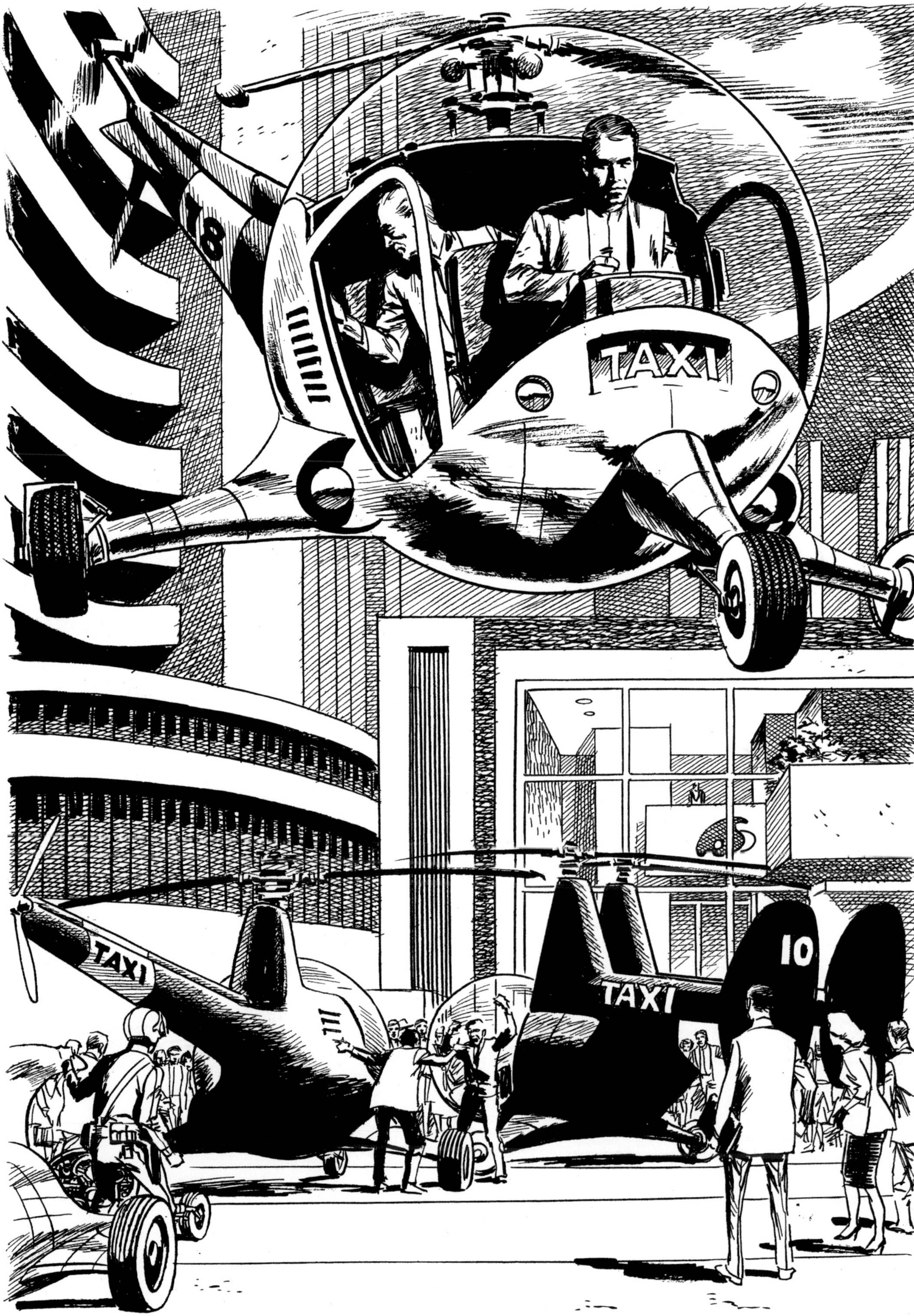
18
TAXI
TAXI
10
TAXI

WALLY WOOD CHECKLIST OF PULP AND MAGAZINE WORK

Compiled by Roger Hill and David Pardee

AMAZING STORIES

1958

July:
One interior illustration for "Everything's Different Up There" by Genevieve Haugen
pages 80-81 (double-page spread)

GALAXY SCIENCE FICTION

1957

September:
Three interior illustrations for "The Pod in the Barrier" by Theodore Sturgeon
pages 8-9 (double-page spread)
page 33 (full page)
page 46 (full page)

October:
Cover painting (color)
Three interior illustrations for "Wolfbane" part 1 by Frederik Pohl and C. M. Kornbluth
pages 8-9 (double-page spread)
page 33 (full page)
pages 40-41 (double-page spread)

November:
Three interior illustrations for "Wolfbane" part 2 by Frederik Pohl and C. M. Kornbluth
pages 54-55 (double-page spread)
pages 84-85 (double-page spread)
pages 94-95 (double-page spread)

December:
Three interior illustrations for "Payload" by Raymond F. Banks
pages 76-77 (double-page spread)
page 87 (full page)
pages 94-95 (double-page spread)

1958

February:
Three interior illustrations for "The Rule of the Door" by Lloyd Biggle, Jr.
pages 8-9 (double-page spread)
page 20 (full page)
pages 32-33 (double-page spread)
Three interior illustrations for "Bread Overhead" by Fritz Leiber
page 75 (full page)
page 79 (full page)
pages 86-87 (double-page spread)

March:
One interior illustration for "My Fair Planet" by Evelyn E. Smith (credited to Dillon but actually by Wood)
page 64 (full page)
Three interior illustrations for "Spare the Rod" by Lloyd Biggle, Jr.
pages 86-87 (double-page spread)
page 101 (full page)
page 103 (full page)

April:
Three interior illustrations for "The Sitters" by Clifford D. Simak
page 6 (full page)
page 21 (full page)
page 23 (full page)
Three interior illustrations for "First Man" by Clyde Brown
pages 56-57 (double-page spread)
page 63 (full page)
pages 68-69 (double-page spread)

May:
Three interior illustrations for "The Iron Chancellor" by Robert Silverberg
pages 120-121 (double-page spread)
page 131 (full page)
pages 138-139 (double-page spread)

June:
Three interior illustrations for "Mars by Moonlight" by Paul Flehr
pages 6-7 (double-page spread)
page 33 (full page)
pages 36-37 (double-page spread)
Three interior illustrations for "The Gentlest Unpeople" by Frederik Pohl
pages 68-69 (double-page spread)
page 78 (half page)
page 85 (full page)

July:
Three interior illustrations for "Innocent at Large" by Poul and Karen Anderson
page 125 (full page)
page 139 (full page)
pages 142-143 (double-page spread)

August:
Cover painting (color) depicting "Firemen of 2025 Fighting Industrial Blasts"
Three interior illustrations for "Seven Deadly Virtues" by Paul Flehr
pages 96-97 (double-page spread)
page 107 (full page)
pages 120-121 (double-page spread)

September:
Three interior illustrations for "Lastborn" by Isaac Asimov
pages 6-7 (double-page spread)
page 33 (full page)
pages 38-39 (double-page spread)
One interior illustration for "The Stroke of the Sun" by Arthur C. Clarke
page 75 (full page)
One interior illustration for "Personal Problem" by H. L. Gold
page 109 (full page)
Three interior illustrations for "Thing of Beauty" by Damon Knight
pages 116-117 (double-page spread)
page 133 (full page)
pages 140-141 (double-page spread)

October:
Three interior illustrations for "Time Killer" part 1 by Robert Sheckley
pages 4-5 (double-page spread)
page 21 (full page)
pages 32-33 (double-page spread)
Three interior illustrations for "The Wizards of Pung's Corners" by Frederik Pohl
pages 64-65 (double-page spread)
page 79 (full page)
pages 84-85 (double-page spread)
Three interior illustrations for "Lisbon Cubed" by William Tenn
pages 94-95 (double-page spread)
page 107 (full page)
pages 126-127 (double-page spread)

November:
Three interior illustrations for
"Birds of a Feather" by Robert Silverberg
pages 52-53 (double-page spread)
page 63 (full page)
pages 68-69 (double-page spread)
Three interior illustrations for
"Time Killer" part 2 by Robert Sheckley
pages 96-97 (double-page spread)
pages 112-113 (double-page spread)
pages 130-131 (double-page spread)

December:
Three interior illustrations for
"Ullward's Retreat" by Jack Vance
pages 66-67 (double-page spread)
page 76 (full page)
pages 84-85 (double-page spread)
Three interior illustrations for
"Time Killer" part 3 by Robert Sheckley
page 105 (full page)
pages 120-121 (double-page spread)
pages 132-133 (double-page spread)

1959

February:
Cover painting (color)
Three interior illustrations for
"Installment Plan" by Clifford D. Simak
pages 8-9 (double-page spread)
page 27 (full page)
pages 48-49 (double-page spread)
Three interior illustrations for
"I Plinglot—Who You?" by Frederik Pohl
pages 68-69 (double-page spread)
page 81 (full page)
pages 90-91 (double-page spread)
Three interior illustrations for
"Insidekick" by J. F. Bone
pages 106-107 (double-page spread)
page 121 (full page)
pages 132-133 (double-page spread)
Three interior illustrations for
"Time Killer" part 4 by Robert Sheckley
page 157 (full page)
pages 176-177 (double-page spread)
pages 190-191 (double-page spread)

April:
Cover painting (color) depicting
"Alien Poker Game"
One interior illustration for
"Security Plan" by Joseph Farrell
page 60 (full page)
Three interior illustrations for
"Kingslayer" by J. T. McIntosh
pages 96-97 (double-page spread)
page 107 (full page)
pages 112-113 (double-page spread)
Three interior illustrations for "The Man in
the Mailbag" by Gordon R. Dickson
pages 160-161 (double-page spread)
page 177 (full page)
pages 186-187 (double-page spread)

June:
Four interior illustrations for
"Whatever Counts" by Frederik Pohl
pages 10-11 (double-page spread)
page 29 (full page)
pages 46-47 (double-page spread)
pages 64-65 (double-page spread)

August:
Cover painting (color) depicting
"1 Down 1 Up"
Three interior illustrations for
"No Life of Their Own" by Clifford D. Simak
pages 10-11 (double-page spread)
page 25 (full page)
pages 48-49 (double-page spread)
One interior illustration for
"License to Steal" by Louis Newman
pages 114-115 (double-page spread)
Two interior illustrations for
"Lex" by W. T. Haggert
pages 118-119 (double-page spread)
page 131 (full page)
Three illustrations for "The Waging
of the Peace" by Frederik Pohl
pages 166-167 (double-page spread)
page 187 (full page)
pages 190-191 (double-page spread)

October:
Cover painting (color) for "Someone to
Watch Over Me" by Christopher Grimm

December:
One interior illustration for
"War Game" by Philip K. Dick
page 97 (full page)
One interior illustration for
"The Snowmen" by Frederik Pohl
page 143 (full page)
Three illustrations for "Blacksword"
by A. J. Offutt
page 164 (quarter page)
page 177 (full page)
pages 182-183 (double-page spread)

1960

February:
One interior illustration for
"The Nuse Man" by Margaret St. Clair
page 48 (full page)
Three interior illustrations for
"Crying Jag" by Clifford D. Simak
pages 128-129 (double-page spread)
page 137 (full page)
pages 144-145 (double-page spread)

April:
Three interior illustrations for
"Success Story" by Earl Goodale
pages 8-9 (double-page spread) Reduced
and printed again as half-page illustration
on page 32.
page 19 (full page)
pages 28-29 (double-page spread)
One interior illustration for
"The Airy Servitor" by Margaret St. Clair
page 48 (full page)
Three interior illustrations for
"Don't Look Now" by Leonard Rubin
pages 112-113 (double-page spread)
page 123 (full page)
pages 130-131 (double-page spread)

June:
One interior illustration for "The Good
Neighbors" by Edgar Pangborn
page 76 (full page)
One interior illustration for
"The Dope on Mars" by Jack Sharkey
page 98 (full page)

August:
Two interior illustrations for
"Sordman the Protector" by Tom Purdom
page 89 (full page)
pages 98-99 (double-page spread)

October:
One interior illustration for
"Beach Scene" by Marshall King
(Also reproduced on Italian cover)
page 57 (full page)
Three interior illustrations for "The
Stentorii Luggage" by Neal Barrett, Jr.
pages 104-105 (double-page spread)
page 113 (full page)
pages 120-121 (double-page spread)

December:
Two interior illustrations for
"Round-And-Round Trip" by H. B. Fyfe
pages 128-129 (double-page spread)
page 138 (full page)
Note: Two illustrations for the story
"Blueblood" by Jim Harmon on pages 153
and 158 are attributed to Wood but are
actually the work of Dick Francis.

1961

June:
Three interior illustrations for
"The Weirdest World" by R. F. Lafferty
pages 76-77 (double-page spread)
page 83 (full page)
page 91 (full page)

December:
One interior illustration for
"Oh Rats!" by Miriam Allen Deford
page 75 (full page)

1962

December:
Three interior illustrations for
"The Creature from the Cleveland Depths"
by Fritz Leiber
page 11 (full page)
pages 26-27 (double-page spread)
page 42 (full page)

1963

June:
Five interior illustrations for
"Here Gather the Stars" part 1 by
Clifford D. Simak
pages 20-21 (double-page spread)
page 29 (full page)
page 38 (full page)
page 48 (quarter page)
pages 58-59 (double-page spread)

August:
Four interior illustrations for
"The Great Nebraska Sea" by Allan Danzig
page 35 (half page)
pages 38-39 (double-page spread)
page 39 (quarter page)
page 42 (quarter page)
Four interior illustrations for
"Here Gather the Stars" part 2 by
Clifford D. Simak
page 99 (full page)
pages 102-103 (double-page spread)
pages 132-133 (double-page spread)
page 175 (half page)

October:
One interior illustration for
"On the Gem Planet" by Cordwainer Smith
page 105 (full page)

1965

October:
Three interior illustrations for
"The Age of the Pussyfoot" part 1
by Frederik Pohl
pages 8-9 (double-page spread)
page 25 (half page)
page 41 (half page)

December:
Two interior illustrations for
"The Age of the Pussyfoot" part 2
by Frederik Pohl
page 163 (full page)
page 177 (full page)

1966

February:
Two interior illustrations for
"The Age of the Pussyfoot" part 3
by Frederik Pohl
page 159 (full page)
page 183 (full page)

April:
Three interior illustrations for
"The Primitives" by Frank Herbert
page 143 (full page)
page 154 (full page)
page 167 (full page)

1967

August:
Three interior illustrations for
"Travelers Guide to Megahouston"
by H. H. Hollis
page 99 (full page)
pages 106-107 (double-page spread)
page 114 (half page)

GALAXY SCIENCE FICTION NOVELS

1957

Cover painting for #30 "Double Jeopardy"
by Fletcher Pratt

1958

Cover painting for #31 "Shambleau"
by C. L. Moore
Cover painting for #32 "Address Centauri"
by F. L. Wallace
Cover painting for #33 "Mission of Gravity"
by Hal Clement
Cover painting for #34 "Twice in Time"
by Manly Wade Wellman
Cover painting for #35 "The Forever
Machine" by Mark Clifton and Frank Riley

GNOME PRESS HARDCOVER BOOKS

1957

Dust-jacket design for "The Return of
Conan" by Björn Nyberg and
L. Sprague de Camp
Dust-jacket design for "The Shrouded
Planet" by Robert Randall
Dust-jacket design for "Colonial Survey"
by Murray Leinster

1958

Dust-jacket design for
"The Survivors" by Tom Godwin
Dust-jacket design for "Undersea City"
by Frederik Pohl and Jack Williamson

IF SCIENCE FICTION

1959

July:
One interior illustration for
"Growing Season" by F. L. Wallace
pages 4-5 (double-page spread)

September:
One interior illustration for
"Summer Guests" by James H. Schmitz
pages 5-6 (double-page spread)
One interior illustration for "Homecoming"
by Gordon R. Dickson
pages 66-67 (double-page spread)

1960

January:
One interior illustration for
"The Good Seed" by Mark Mallory
pages 5-6 (full page)
One interior illustration for
"The Divers" by James Stamers
pages 18-19 (double-page spread)
One interior illustration for
"The Last Leap" by Daniel F. Galouye
pages 64-65 (double-page spread)

March:
One interior illustration
for "Cleaners" by Clifford D. Simak
page 7 (full page)
One interior illustration
for "Gravy Train" by Daniel F. Galouye
pages 110-111 (double-page spread)

May:
One interior illustration for "A Tourist
Named Death" by Christopher Anvil
pages 6-7 (double-page spread)
One interior illustration for "A Pride of
Islands" by C. C. MacApp
pages 56-57 (double-page spread)
One interior illustration for
"Matchmaker" by Charles L. Fontenay
pages 106-107 (double-page spread)

September:
One interior illustration for
"Kangaroo Court" by Daniel F. Galouye
pages 6-7 (double-page spread)

One interior illustration for
Parallel Beans" by Margaret St. Clair
pages 42-43 (double-page spread)
One interior illustration for
To Choke an Ocean" by J. F. Bone
pages 56-57 (double-page spread)

November:
One interior illustration for
Superjoemulloy" by Scott F. Grenville
page 23 (half page)
One interior illustration for
The Quality of Mercy" by Daniel Keyes
pages 30-31 (double-page spread)

1961

July:
One interior illustration for "The Planet
with No Nightmare" by Jim Harmon
page 15 (full page)
One interior illustration for "The Bem
Called Windigo" by Theodore Sturgeon
page 88 (full page)

1965

September:
Three interior illustrations for
Under Two Moons" by Frederik Pohl
page 11 (full page)
page 22 (full page)
page 33 (full page)

1966

April:
Cover painting (color) by Wood and Adkins
for "Earthblood" by Keith Laumer and
Rosel George Brown (incorrectly credited
to Gray Morrow on contents page)
Three interior illustrations for "Earthblood"
part 1 by Keith Laumer and Rosel George
Brown (incorrectly credited to Norman Nodel)
page 7 (full page)
page 23 (full page)
page 43 (full page)

May:
Three interior illustrations for
Earthblood" part 2 by Keith Laumer and
Rosel George Brown (Wood and Adkins)
page 107 (full page)
pages 124-125 (double-page spread)
page 131 (half page)

June:
Four interior illustrations for
Earthblood" part 3 by Keith Laumer and
Rosel George Brown (Wood and Adkins)
page 103 (full page)
pages 116-117 (double-page spread)
pages 128-129 (double-page spread)
page 141 (half page)

July:
Three interior illustrations for "Earthblood" part
4 by Keith Laumer and Rosel George Brown
page 112 (full page)
page 129 (half page)
pages 136-137 (double-page spread)

1967

August:
Four interior illustrations for
"The Winged Helmet" by Fred Saberhagen
pages 58-59 (double-page spread)
page 67 (half page)
page 86 (half page)
page 95 (half page)

November:
Three interior illustrations for
"Dreamhouse" by Gary Wright
pages 98-99 (double-page spread)
page 102 (full page)
page 103 (full page)

1968

November:
Three interior illustrations for
"The Perfect Secretary" by Mike Kirsch
page 134 (full page)
page 146 (full page)
page 149 (full page)

December:
Four interior illustrations for
"The Starman of Pritchard's Creek" by
Julian F. Grow (Wood and Ralph Reese)
page 16 (full page)
page 17 (full page)
pages 20-21 (double-page spread)
page 25 (full page)

JUNGLE STORIES

1953

Spring:
One interior illustration for
"Gorilla! Gorilla!" by Anson Slaughter
page 52 (full page)

ORIGINAL SCIENCE FICTION STORIES

1959

May:
Two interior illustrations for "There's No
Place Like Space" by Robert Silverberg
page 7 (full page)
page 11 (full page)

PLANET STORIES

1953

January:
One interior illustration for
"War Drums of Mercury Lost"
by John Jakes
page 63 (full page)

September:
One interior illustration for
"The Un-Reconstructed Woman" by
Hayden Howard
page 37 (full page)

1954

Summer:
One interior illustration for "Color Blind"
by Charles A. Stearns
page 35 (full page)

WORLDS OF TOMORROW

1963

April:
Four interior illustrations for
"People of the Sea" part 1 by
Arthur C. Clarke
page 6 (full page, printed in blue tone)
pages 10-11 (double-page spread, printed
in blue tone)
page 17 (full page)
page 41 (full page, printed in blue tone)

June:
Three interior illustrations for
"People of the Sea" part 2 by
Arthur C. Clarke
page 117 (full page, printed in green tone)
page 133 (full page, printed in green tone)
pages 142-143 (double-page spread)

August:
Two interior illustrations for
"To the Stars" by J. T. McIntosh
page 89 (full page)
page 99 (full page)

December:
One interior illustration for
"The Creature Inside" by Jack Sharkey
page 27 (full page)

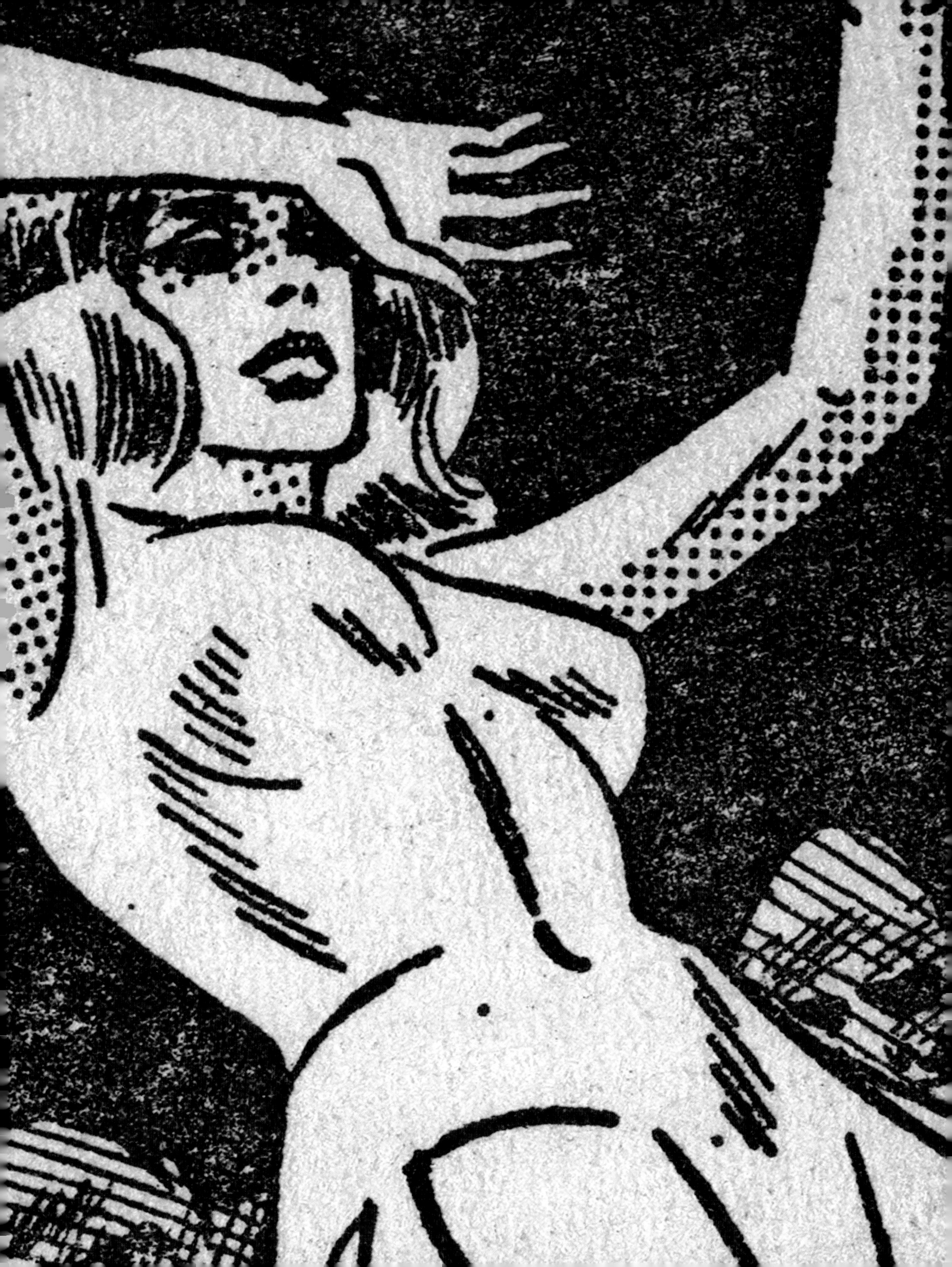

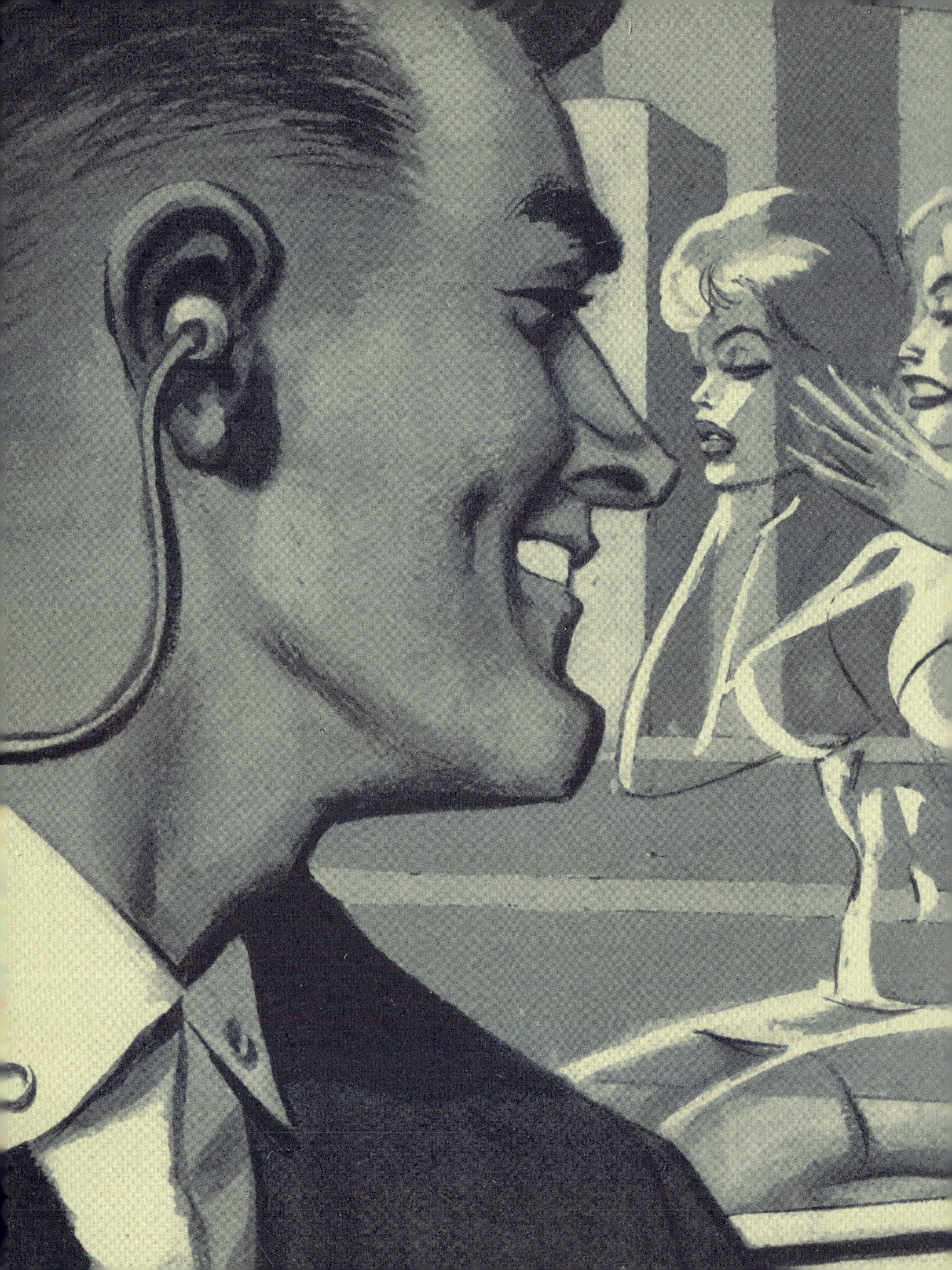